THE **MICROBIOLOGY** COLORING BOOK

Also available in the Coloring Concepts series:

THE ANATOMY COLORING BOOK, by W. Kapit and L.M. Elson
THE HUMAN BRAIN COLORING BOOK, by M.C. Diamond,
    A.B. Scheibel, and L.M. Elson
THE BIOLOGY COLORING BOOK, by R.D. Griffin
THE HUMAN EVOLUTION COLORING BOOK, by A.L. Zihlman
THE MARINE BIOLOGY COLORING BOOK, by T. Niesen
THE ZOOLOGY COLORING BOOK, by L.M. Elson
THE BOTANY COLORING BOOK, by P. Young

Contact: Coloring Concepts, Inc. 1-800-257-1516
        1732 Jefferson Street, Suite 7, Napa, CA 94559
        or
        HarperCollins College Publishers
        10 E. 53rd Street
        New York, NY 10022
        Customer Communication Number:
        1-800-8HEART1

$17.00

I. Edward Alcamo / Lawrence M. Elson

The MICROBIOLOGY COLORING BOOK

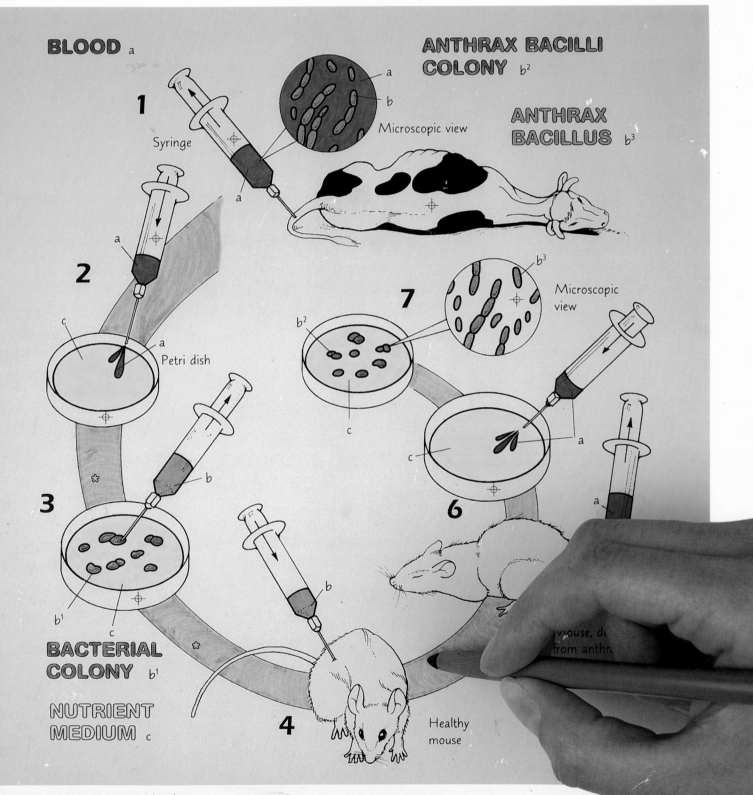

BLOOD a

ANTHRAX BACILLI COLONY b²

ANTHRAX BACILLUS b³

1

Syringe

a

b

Microscopic view

a

2

c

a

Petri dish

7

b²

b³

Microscopic view

c

3

b

a

b¹

c

6

c

a

BACTERIAL COLONY b¹

b

NUTRIENT MEDIUM c

4

Healthy mouse

THE
MICROBIOLOGY
COLORING
BOOK

by
I. Edward Alcamo
Lawrence M. Elson

HarperCollins*CollegePublishers*

I. Edward Alcamo, Ph.D., is one of two authors of this text. Professor of Microbiology at State University of New York at Farmingdale, he teaches medical microbiology to medical, premedical, and allied health science students. Dr. Alcamo is the author of *Fundamentals of Microbiology*, *Aids: The Biological Basis*, *DNA Technology - The Awesome Skill*, *Blond's Medical Guide to Microbiology*, and *Cliff's Reviews of Biology*. Dr. Alcamo provided the original manuscript for this work, and directed its scope and technical content. Dr. Alcamo resides in Long Island, New York.

Lawrence M. Elson, Ph.D. is the co-author of this text. He is a clinical and forensic human anatomist who taught at the Baylor College of Medicine in Houston, the University of California Medical School at San Francisco, and the City College of San Francisco. He is the author and co-author of several books, including *The Anatomy Coloring Book*. Dr. Elson is the founder and president of Coloring Concepts, Inc., and the director of graphic and textual content of its several publications. He created the layouts for the illustrations in this book, and contributed textual content, research, and editorial direction to it. Dr. Elson lives in the Napa Valley with his wife and daughter.

Charles W. Hoffman III is the illustrator for the book. He holds a baccalaureate degree in microbiology, and is a graduate of the medical illustration curriculum at the University of California at San Francisco. He has contributed numerous medical and scientific illustrations to several publications. Mr. Hoffman undertook considerable research into developing accurate renditions of the many, some unusual, microorganisms presented in this book. He lives in San Francisco with his wife and two daughters.

Production Notes

This book was produced by Coloring Concepts, Inc.
1732 Jefferson Street, Suite 7, Napa, CA 94559.
Pre-press digitial layout was by Mark Jones.
The book development editor was Joan W. Elson.
The executive editor was Bonnie Roesch.

The typefaces used in this book come from the library of Monotype Typography Inc.
Headlines are set in ITC Eras demi.
Small page titles are set in ITC Eras medium.
Coloring titles and subheads are set in AG Book Rounded bold outline.
Text is set in Times roman.
Callouts are set in Monotype's Blueprint, designed by Steve Matteson.

The art was prepared in ink on acetate, then scanned on a HP ScanJet IIc scanner into Adobe Photoshop. Adobe Pagemaker 5.0 was used for page layout, and Microsoft Word 5.1 was used for the manuscript.

Hardware included a Macintosh IIci, Quadra 610, Quadra 950 as well as assorted storage devices by MicroNet Technology Inc.

Final output was on a Linotronic 330 RIP 50, provided by California Graphics in San Rafael, CA.

ISBN: 0-06-500941-X

I am pleased to dedicate this book to you, the student of microbiology. May you become all you are capable of being.

— I.E.A.

To Joan. Her attention to detail during the development of this book was simply incredible. Not only that, but our egos survived the inevitable clashes. And so did our love. I am very grateful for her remarkable contribution to the goodness of this book. I am grateful for her in my life.

— L.M.E.

TABLE OF CONTENTS

PREFACE

This book is about microorganisms and the roles they play in our lives. It deals with bacteria, viruses, fungi, protozoa, and multicellular parasites. It describes their structure, physiology, biochemistry, cultivation, and replication patterns. The book pays particular attention to the pathology of microbial disease and the mechanisms of nonspecific and immune resistance exhibited by the body. The book finishes with very interesting plates on industrial microbiology and the nitrogen cycle.

In developing the plates of *The Microbiology Coloring Book*, we made no assumptions that users had prior knowledge of microbiology. Further, the art and text were created with the awareness that users may have had limited exposure to both physical and biological science subjects. The graphic and textual presentation of the principles of microbiology in this book have teaching value to a broad range of users, from high school to professional and graduate school.

This directed coloring book-text includes the basic information and sequence of presentation found in most microbiology texts used in the United States. We selected those microbiology topics that are particularly amenable to understanding from and learning by graphic work (coloring). The result is 105 discrete, especially-designed plates of colorable, meaningful microbiological structure and related characteristics, laboratory techniques, tests and procedures, life cycles of certain human pathogens, introductory immunology, and mechanisms of disease and associated clinical signs and symptoms. The text associated with each plate of art is designed to facilitate verbal understanding of what is colored.

The important principles of microbiology can be learned from this book without another textbook or outside instruction. Students taking their first microbiology course will find this coloring text with its low-key, graphics approach to learning extremely helpful. Advanced students will find the book valuable for reviewing information. Instructors may wish to use the various plates as adjuncts to their lectures.

If this is your first Coloring Concepts book, you are in for a pleasant surprise. The coloring activity is an integral part of what has proven to be a highly-effective method for learning reinforcement and retention enhancement. You are asked to study actively with coloring pencil/pen in hand rather than passively reading a textbook or lecture notes. With your hand involved, there is a greater chance that your mind will have a more intense focus while it engages single microbiology topics at a time. Simultaneously, the mind forms associations with other concepts, and transfers from short-term memory to long-term memory are rapidly facilitated.

Your physical involvement with each plate begins with the overview assessment: starting at the upper part of the plate, observe the orientation of the illustrations to be colored, and the color pattern to be employed. Having read the Coloring Instructions and Getting the Most Out of Color, plan your use of colors and their shades and tints. This is quite important, as it will be the colors that you will instantly recall in a subsequent examination, and the recall of these colors is associated with the structures or steps in the microbiological organism or process that you are being tested on. Soon you will become totally immersed in the project of coloring the titles and related structures, one after another, and experiencing the rush of positive feelings (ah hah!) as insight and understanding are rapidly acquired, and facts integrate into a meaningful whole. On finishing, you will have an aesthetically pleasing picture that you completed, color-coded for rapid and effective review at a later time. Your sense of accomplishment and pride at completing a well-designed, colored instructional set will uplift you and motivate you to do more. Just as uplifting will be the newly acquired knowledge base that will permit even greater, in-depth focus on the subject. A References section provides some guidance in this respect. The more time you spend coloring and reading this text, the more your understanding of microbiology will prosper, and your high grades will reflect your creative work.

Best wishes for a successful study of the microorganisms.

I. Edward Alcamo, Ph.D.
Long Island, NY

Lawrence M. Elson, Ph.D.
Napa Valley, CA

ACKNOWLEDGMENT

We had a lot of help in making this book.

Reviewers of the first part of the text included Beverly Dixon, Ph.D. from the Department of Microbiology, California State University at Hayward, Kathy Smith, Ph.D., from the Department of Microbiology, University of California at Davis, and Sarah McHatton, a graduate student in microbiology at the University of California at Davis. We are very grateful to them for their suggestions, corrections, and criticisms.

Clarence Wolfe, Ph.D., professor of microbiology at Northern Virginia Community College, reviewed the entire manuscript and offered excellent suggestions. His professionalism, attention to detail, and his responsiveness to our deadlines are very much appreciated.

Susan Conroy, clinical microbiologist at St. Helena Hospital, St. Helena, CA, reviewed the laboratory tests and procedures plates and kept us relevant. She was especially helpful with the fungi plates, providing corrections, and offering suggestions for both art and text. It turns out she has a particular penchant for these tiny but complex organisms, and led us to excellent literature sources on fungi.

Our thanks to Celia E. Ramsay who undertook the copy editing of the first twenty plates of the book, and helped us develop a style sheet.

To Christine and Jay Golik, our appreciation for their fine contribution to understanding the principles of color and for creating the Getting the Most Out of Color section.

We are very grateful to our development editor Joan W. Elson for making sure the book "works." In developing a book like this, it is critical that all of the subscripts, titles, callouts, and leaders are correctly placed, and match with the art. She checked and corrected these things again and again. Just as importantly, she edited the work for clarity and logical progression.

The fine illustrations, and the extensive research that went into them, by Charles W. Hoffman, III, are very much appreciated.

Our layout and digital prepress designer, Mark B. Jones, responsible for computerized layout and typesetting, was especially helpful in researching typefaces and fonts, and helping us find the best collection of these for this book. We are grateful for his patience and sharp eyes.

The assistance of Christopher L. Elson in researching the industrial microbiology plates is very much appreciated. He found the key people (noted below), set up personal tours with the microbiologists, researched material, and took great notes that enabled these plates to be created.

Thanks to Jennifer J. Elson for her research efforts and continuing support during the development of this book.

The administrative support of Carolyn Scott, office manager for Coloring Concepts, is particularly appreciated.

She made schedules, arranged events, coordinated activities, and made it possible for one of the authors (LME) to work on the book in many three day or week long periods with minimal distraction.

Our thanks to Cecile Maclean for providing her hand for the cover and to Emanuel Moutsos for the photography of the cover.

For the plate on milk pasteurization, we were fortunate enough to be assisted by Marcia McGlochlin, microbiologist and quality assurance manager at Clover Dairy in Santa Rosa, CA. She took us through the entire milk production sequence, gave us Federal and State Standards and data to review, as well as dairy workers' educational and testing materials, took us through the microbiology lab, and even brought out sections of an HTST pasteurization machine to ensure our understanding of what goes on there. She corrected our text and our illustrations, and offered suggestions. We are very grateful for her direction and assistance.

For the plate on water treatment, we consulted Turan Ramadan, water operations supervisor at the Napa Water Department. He took us through the water treatment operation explaining every step of the operation, gave us diagrams, drawings, and data to pour over, and critiqued the illustrations until we had it right. We are very grateful for the time, effort, and expertise he extended us.

William Gaffney, chemist at the Napa Sanitation District wastewater management plant, and technician Tracy Talbott, ably assisted us in the development of the wastewater treatment plate. They took us through the treatment plant, discussed proper terminology with us, reviewed and corrected our illustrations and text, and made themselves available to us on quick notice. We are grateful to them and the time they gave us.

Arthur Lim, microbiologist, is responsible for quality assurance at the Fairfield (CA) brewery of Anheuser-Busch, Inc. He took us on a full, one-on-one tour of the brewery from malt storage to distribution. He gave us materials to study, went over our drawings and text and cleaned them up until we came up with a respectable plate of visual and textual information. We are very appreciative of the expertise and time he devoted to us.

We went to two wineries of renown in the Napa Valley for insight into red wine production. Ken Shrivers, microbiologist at the Robert Mondavi Winery in Oakville, took us through the operation there, kindly explaining various aspects of microbiological and enological significance. Ken Bernards, winemaker at Truchard Vineyard in Napa, explained in detail each aspect of red wine production, led us through the operation there, and evaluated and corrected our art and text on the subject of red wine. We are very grateful to him for his work in our behalf.

COLORING INSTRUCTIONS

1. This is a book of illustrations (plates) and related text pages. Each of the illustrations, or part of an illustration to be colored is labeled with a lower case letter called a **subscript.** Each of those illustrations or parts of an illustration has a name called a **title.** The title is printed in outline (colorable) letters. The title and its related structure are linked by identical subscripts. You color each title and its identical subscripted structure the same color. When each plate is completed, the relationship between titles and their structures can be seen at a glance. The arrangement of color among the structural parts or processes speeds visual orientation. The color-coded parts help immensely during attempts to recall facts, parts, or processes in an examination or discussion setting.

2. You will need coloring pencils or felt tip markers. Good quality coloring pencils are preferred if you wish to extend your basic set by adding varying degrees of white or black to a color to create a series of tints and shades (be sure to see Getting the Most Out of Color before you start coloring this book). Laundry markers and crayons are not recommended: the former because they stain through the paper and obscure the text on the following page, the latter because they are coarse, messy, and produce unnatural colors.

3. To achieve maximum benefit of instruction, you should color the plates in the order presented, or at least within each group or section. Some plates may seem intimidating at first glance, even after reviewing the coloring notes. However, once you begin to color the plate in order of presentation of titles or stations (1,2, and so on), and you begin to read the text, the illustrations will begin to have meaning, and the relationships of different parts will become clear.

4. As you come to each plate, look over the entire illustration; note the arrangement of titles and station numbers, if applicable. Scan the coloring notes (printed in boldface type) for further guidance. Be sure to follow the coloring instructions with each plate. The authors and editors have colored these plates many times during development of this book, and they are giving you the benefit of that experience in those notes. Suggestions for use of color or for sequence of coloring are based on the experience of having colored these plates.

5. Most of the time, you will start coloring at the top of the plate with a subheading (*) or with a title subscripted with (a), followed by its related structure or part. Then simply color in alphabetical order of subscripts. Any outline-lettered word followed by a small lower case letter subscript should be colored unless it is followed by a "do not color" symbol ($\diamondsuit$). In all cases, there is a structure with the identical subscript that requires the same color as its title. Contemplate a number of color arrangements before starting. In some cases, you may want to color related forms with different tints/shades of the same color, or with analogous colors; in other cases, contrast is desired, and complementary colors may be selected.

6. In some plates, more colors will be required than you have in your possession. If you have pencils, you can extend the colors available significantly by adding white or black (see Getting the Most out of Color). You can also use different patterns of the same color (diagonal marks, dots, and so on). Whatever you choose to do, take care to avoid confusion in identification and review. On occasion, you may be asked to use colors on a plate that were used for the same structure on a previous, related plate. In this case, set aside those colors until you reach the appropriate title/structure.

7. Now turn to any plate and note:
    a. Areas to be colored are separated from one another or from "do not color" areas by heavy lined borders. Lighter lines represent background, suggest texture, or define form, and they can be colored over. Some boundaries between coloring zones may be represented by dotted lines. These represent a division of names or titles, and indicate that a physical boundary may not exist or, at least, is not clearly visible in this view.

b. As a general rule, large areas to be colored should receive a light color; dark colors should be reserved for small areas. Take care with very dark colors: they may obscure illustration detail, and identification of the part by its structural character will be lost.

c. In the event a structure is repeated on the plate several times, only one or a few of them will be identified by a subscript. Without boundary restrictions, or instructions to the contrary, these like structures should all be given the same color.

d. The following symbols are used throughout the book:

✿ = color gray.

⊕ = do not color.

• = color black.

(a^1), (a^2) = identical letter with different superscripts denotes that the parts so labeled are sufficiently related to receive the same color or shades/tints of the same color, unless otherwise specified in the bold face coloring instructions.

(f) = a subscript enclosed by parentheses following a title signals that the titled part is made up of parts that follow as (f^1), $(f.^2)$, and so on. The color used for (f) can be used for any of its parts or none of its parts.

1
IMPORTANCE OF MICROORGANISMS TO HUMANS

We live at the center of a microbial universe. Every breath we take brings thousands of microorganisms into our body. A single pinch of rich soil may contain a billion microorganisms. The great majority of microorganisms benefit humans by recycling the elements of life, producing many foods and industrial products, and serving as research tools. Other microorganisms are the agents of disease. This plate briefly surveys some examples of how microorganisms influence the quality of our lives.

Look over the plate and plan the use of colors for each of the six groups. Color the titles (a) through (b²), and related foods. Use light colors for larger areas. Try to select colors that do not obscure illustration details.

Over the centuries, social customs and traditions have led people to accept and enjoy certain foods produced by microorganisms. Through fermentation, microorganisms transform the organic constituents of food. For example, bacteria naturally present in cucumbers grow in the cucumbers and partially digest the plant tissue into a flavorful pickle (a¹). Sausages (a²) are the product of microbial growth occurring within meat and spices. Different sausages can be made depending on the type of meat and spices used and the microorganisms that alter the chemistry of the meat. To bake bread and rolls (a³), the baker includes yeast in the dough to metabolize carbohydrates and form carbon dioxide, and the dough rises. Many dairy products are produced through the activity of microorganisms. For instance, cheese is manufactured by heating milk, adding enzymes to curdle the milk protein, and combining bacteria or molds with the milk curds. During the ripening process, microorganisms deposit the chemical compounds that give different cheeses their distinctive tastes. For example, cheddar cheese (b¹) takes its flavor from the acids produced by lactobacilli and streptococci growing within the ripening curd. Yogurt (b²) is a form of milk in which bacteria have produced acid from the milk sugar, and made the milk sour.

Now color titles (c) and (c¹) and the potato.

Microorganisms are nature's great recyclers of elements. Some bacteria, for example, release nitrogen from animal waste. Other microorganisms grow on the roots of pod-bearing plants and bring nitrogen back into the cycle of life by using nitrogen to make organic compounds. These compounds are released into the soil where they are utilized as nutrients by plants, such as potatoes (c¹).

Continue by coloring the titles (d) through (d³) and related objects.

The range of industrial products of microorganisms is broad and includes such diverse products as perfumes and the antibiotic penicillin. Also, the bacterial enzyme protease removes organic debris from animal hides in the manufacture of leather (d¹). Certain bacteria produce enzymes that accelerate the transformation of organic compounds. For example, the enzyme pectinase decomposes the pectin fibers that bind cellulose fibers in plants. Once the pectin is dissolved, the cellulose can be spun to form linen (d²). Other bacteria are employed to produce enzymes to form a starch used in the sizing of porous material during paper processing (d³).

Next color the titles (e) and (e¹) and the insulin in the enlargement of the dispensing device (also shown hanging on the spectator's belt). Use a color light enough not to obscure the illustration detail.

Many microorganisms are used as models and tools in research laboratories to study life processes. This is because the chemistry of microorganisms is similar to that of all other living organisms. In one technique, called genetic engineering, genes from other organisms are inserted into bacteria. This technique programs the bacteria with genetic instructions for the manufacture of a number of chemicals, such as insulin (e¹). Insulin is a hormone that maintains the metabolic machinery of the body and is essential to life. In juvenile onset diabetes mellitus, the hormone is in short supply. Insulin produced by genetically engineered bacteria can be taken by the sophisticated dispensing device shown.

Complete the plate by coloring the titles (f) and (f¹) and the infected areas of the foot.

Throughout history, microorganisms have posed formidable challenges to humans as infectious agents of disease. When transmitted from one person to another, certain microorganisms have been responsible for diseases such as plague, smallpox, typhoid fever, syphilis, and AIDS. Modern populations continue to be confronted with life-threatening, as well as irritating and inconvenient infections. One example of the latter is the fungal foot disease (f¹; athlete's foot).

IMPORTANCE OF MICROORGANISMS TO HUMANS

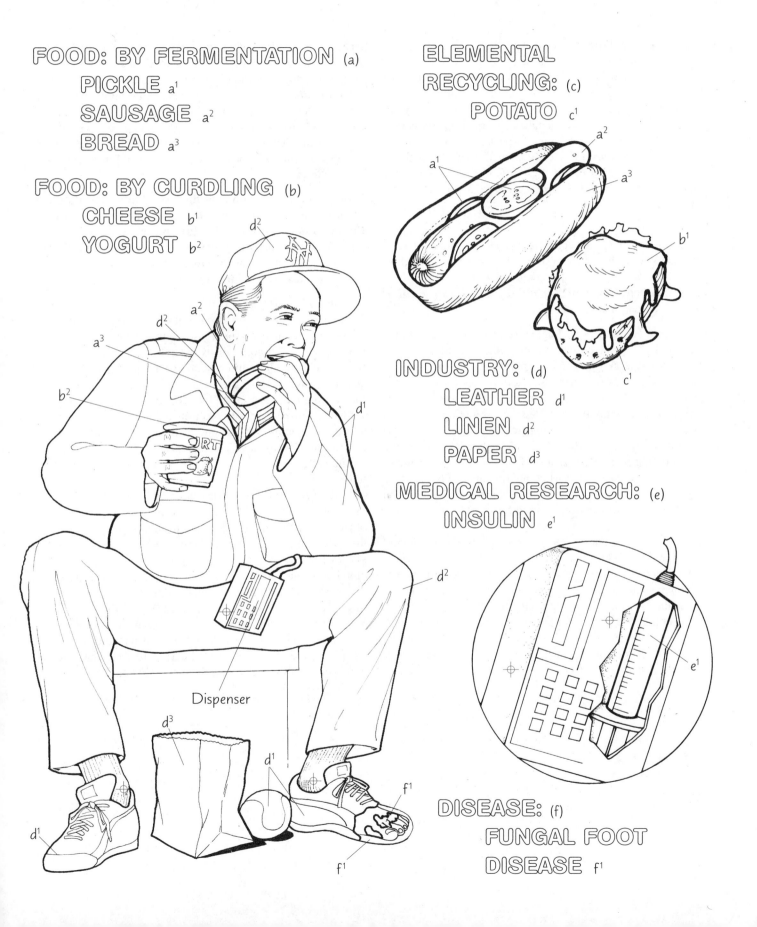

FOOD: BY FERMENTATION (a)
PICKLE a¹
SAUSAGE a²
BREAD a³

FOOD: BY CURDLING (b)
CHEESE b¹
YOGURT b²

ELEMENTAL
RECYCLING: (c)
POTATO c¹

INDUSTRY: (d)
LEATHER d¹
LINEN d²
PAPER d³

MEDICAL RESEARCH: (e)
INSULIN e¹

DISEASE: (f)
FUNGAL FOOT
DISEASE f¹

Dispenser

2
EARLY MICROSCOPES

Microbiology is the branch of biology concerned with microscopic forms of life. Because these forms of life cannot be seen with the naked eye, the development of microbiology depended in large measure on the development of the microscope and the science of microscopy. No one is sure who invented the microscope (although that distinction is generally attributed to Zaccharias Janssen, a spectacle maker from the Netherlands who lived in the early 1600s). However, scientists are certain that two of the first investigators to use the microscope for scientific purposes were Robert Hooke and Anton van Leeuwenhoek.

Note the layout before coloring, selecting your colors for contrast. Begin by coloring the subheading in the upper half of the plate. A very light color (e.g., yellow) for (b) is recommended. Color title (a) and the flame, and continue through title and structure (f).

Robert Hooke was an English scientist who lived in the mid-1600s. Hooke lectured widely on the scientific value of the microscope and in 1665, he published *Micrographie*, a book in which he described a method of constructing a microscope. His book also contained 57 illustrations of microscopic structure that revealed the fascinating details of a hitherto unseen world of biological objects. Among his illustrations were a feather of a bird, an eye of a fly, and a stinger of a bee. Hooke also drew the honeycomb structure of a slice of cork and named its compartments "cells" because they resembled the small rooms of a monastery. His recorded descriptions of cells in cork secured for Hooke a place in the history of cell biology.

By today's standards, Hooke's microscope is considered primitive. The light source (a) consisted of a flame, which was reflected off a mirror. The reflected light (b) passed through a focusing lens (c), which directed a beam of light onto a specimen (d) stuck with a pin attached to the base of the microscope. The microscope consisted of two magnifying lenses (e), one at either end of a six-inch tube. The eye (f) visualized the magnified image (cells) of the specimen (cork). Although Hooke's microscope suffered from surface irregularities that often caused distorted images, his descriptions were among the first to awaken the scientific community to the world of microscopic specimens. As far as scientists know, Hooke even observed microorganisms. His writings include descriptions of two types of fungi and at least one type of protozoan seen among the sand particles he chose for viewing. However, the first detailed description of microorganisms was furnished by Anton van Leeuwenhoek.

Now color the subheading at the lower half of the plate. As before, start with title and structure (a). Consider using the same colors as above for (a), (b), (e), and (f). Note (c) and (d) are not used in this lower part of the plate. Use a light color for (h); color the entire magnified view and then overcolor (h^1) with a darker shade of (h).

In the late 1600s, a Dutch merchant named Anton van Leeuwenhoek, without university training or scientific experience, learned to grind lenses with flawless accuracy. He developed a microscope consisting of a single lens, in contrast to Hooke's two-lens microscope. A flame (a) or beam of sunlight provided the light. On a vertical plate was mounted a magnifying lens (e) secured by two silver or brass rivets (not shown). Adjustable screws positioned the specimen holder (g) so that the specimen (h) fit into the light path (b) and the focal length could be adjusted. The magnified view of the specimen (animalcules; h^1) seen by the eye (f) is shown encircled at lower right.

Van Leeuwenhoek's lenses were no larger than the head of a hat pin but, by most accounts, they could magnify an object over 200 times (sophisticated student microscopes can magnify objects 1000 times). During the course of 40 years, van Leeuwenhoek drew careful illustrations of hair fibers, blood cells, and sperm cells. In 1674, he made the startling discovery that a droplet of greenish marsh water contained tiny forms of life. Van Leeuwenhoek named them "animalcules." Today we know them as microorganisms. Some of the different microorganisms he found are shown in the illustration (h^1). Van Leeuwenhoek was the first to describe microorganisms so thoroughly and in such vivid detail.

EARLY MICROSCOPES

ROBERT HOOKE'S MICROSCOPE ✿

LIGHT SOURCE a

LIGHT PATH b

FOCUSING LENS c

SPECIMEN: CORK d

MAGNIFYING LENS e

EYE f

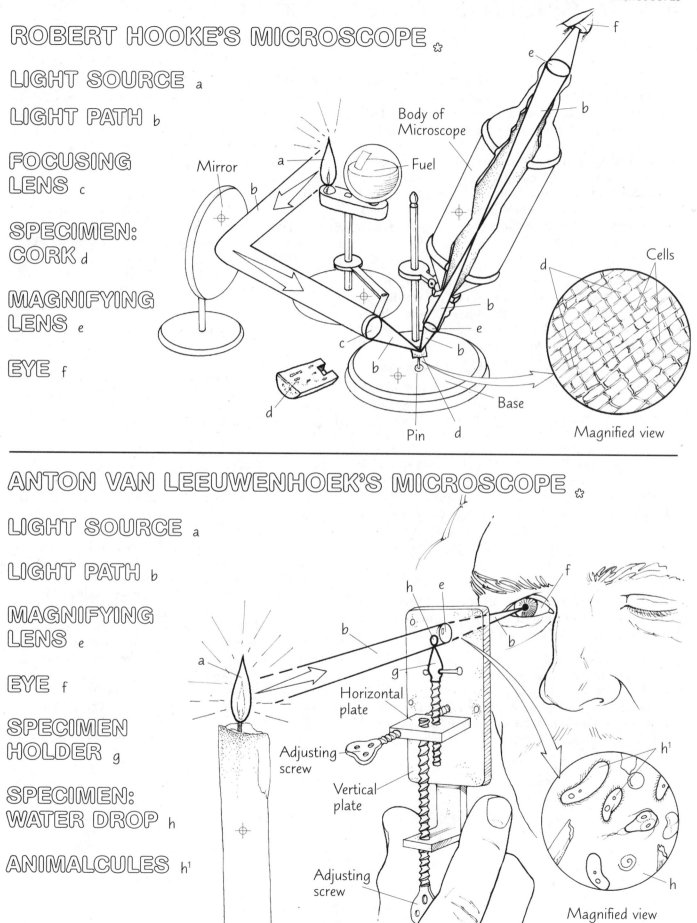

Mirror
Body of Microscope
Fuel
a
b
d
Cells
c
e
b
b
Base
Pin
d
Magnified view

ANTON VAN LEEUWENHOEK'S MICROSCOPE ✿

LIGHT SOURCE a

LIGHT PATH b

MAGNIFYING LENS e

EYE f

SPECIMEN HOLDER g

SPECIMEN: WATER DROP h

ANIMALCULES h¹

h
e
f
b
a
g
Horizontal plate
Adjusting screw
Vertical plate
Adjusting screw
b
h¹
h
Magnified view

3
SPONTANEOUS GENERATION

In the fourth century, B.C., the Greek philosopher Aristotle wrote that small animals could arise from substances that were not living, e.g., wood, decaying matter, or vegetation. Over time, his observations formed the basis for the doctrine of spontaneous generation, a belief that lifeless substances could give rise to living organisms. The concept of sexual or asexual reproduction was not appreciated in organisms too small for reproductive behavior to be observed. Thus, the idea of spontaneous generation took hold in scientific communities, and it governed their thinking for centuries, especially with respect to microorganisms.

Begin by coloring the subheading, Needham's Experiment: Pro. Select contrasting colors for (b) and (c), making (b) the darker. Color titles (a) through (c) and related structures in the upper two flasks and then the lower two flasks. Color the stopper and its title (d).

John Needham, a British clergyman of the mid-1700s, undertook experiments to show that microorganisms arise by spontaneous generation. Needham poured clear, sterilized animal broth (a) into flasks and left them open to air. After some days, the broth became cloudy, and microscopic observations showed that the broth was contaminated (c) with microorganisms (b). Needham then repeated the experiment but this time sealed the broth-containing flasks with stoppers (d). Once again, the broth became cloudy after several days, and microscopic observations of the broth showed populations of microorganisms. Needham hypothesized that microorganisms came into existence "spontaneously" from the clear, lifeless broth.

Color the subheading, Pasteur's Heated Air Experiment: Con, and related titles and structures in the middle set of illustrations. Use light shades for (e) through (e²).

The French scientist Louis Pasteur was one of those who did not accept the spontaneous generation theory. In the late 1850s, he conducted a series of experiments to disprove the spontaneous generation of microorganisms. In one of these experiments, he drew out the neck of some flasks and formed long tubes with them, leaving the end of the tubes open. He placed an insulation coil around the tube to enhance sterilization of the air when heated by a burner (f). He partially filled each flask with broth, and boiled it for a long time, effectively sterilizing the broth (a) and the air (e) in the flask and the first part of the tube. With the flame from a gas burner, he heated and sterilized the air (e^1) in the tube. The open end of the tube was left exposed to unheated room air (e^2). After several days, the broth remained clear and sterile, and no microorganisms appeared in the flask. Pasteur explained this result by pointing out that heating the air in the extended tube killed incoming microorganisms (b $\oplus$). With the air in the long tubes sterilized, there was no means for microorganisms to fall into the broth. Had the theory of spontaneous generation been operative, microorganisms would have "spontaneously" appeared in the broth within a few days whether or not the tube had been sterilized. In defense, Pasteur's critics questioned whether the heat had killed the "life force."

Color the subheading Pasteur's Swan-Neck Flask Experiment: Con, and related titles and structures. First color the single flask and flame on the left, then go to the flasks at right, coloring the broken flask last.

Seeking to silence his critics, Pasteur devised another experiment to disprove the notion of spontaneous generation. He lengthened the neck of a flask to resemble a swan's neck. He left the neck open to the room air. The broth (a) and air (e) in the flask were sterilized by boiling from the heat source (f).

In time, only the dusty bottom part (g) of the neck trapped microorganisms (b). Pasteur had proposed that no microorganisms would appear in the broth because they could not move up the long neck beyond the trapped dust. He was sure microorganisms were not capable of spontaneous generation, and given their inability to move up the long neck, he postulated that none would enter the flask. Several days after the flask had been left undisturbed, no microorganisms appeared. In later experiments, he cut the necks off the flasks and exposed the broth to the air; microorganisms soon appeared in the flasks. It was clear that the room air contained microorganisms and that the air could transmit microorganisms as possible agents of disease. Pasteur's classic experiment with the swan-neck flasks put to rest the notion of spontaneous generation.

SPONTANEOUS GENERATION

NEEDHAM'S EXPERIMENT: PRO ✿

STERILE BROTH a

MICROORGANISM b

CONTAMINATED BROTH c

STOPPER d

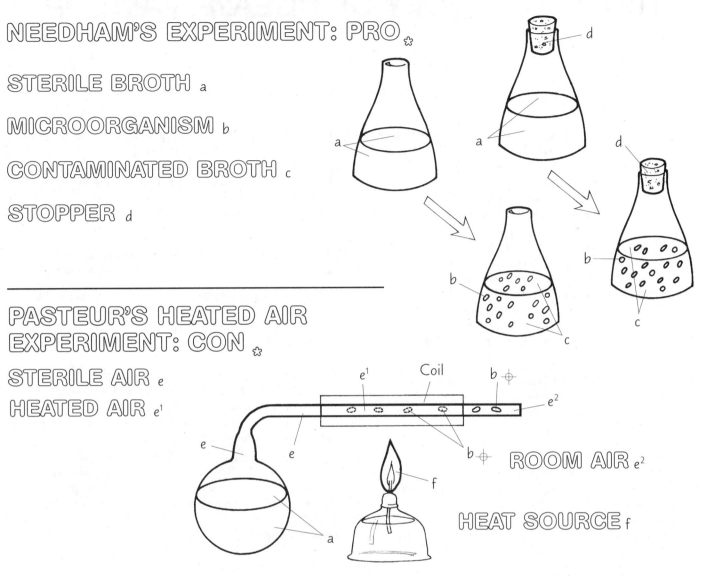

PASTEUR'S HEATED AIR EXPERIMENT: CON ✿

STERILE AIR e

HEATED AIR e¹

ROOM AIR e²

HEAT SOURCE f

PASTEUR'S SWAN-NECK FLASK EXPERIMENT: CON ✿

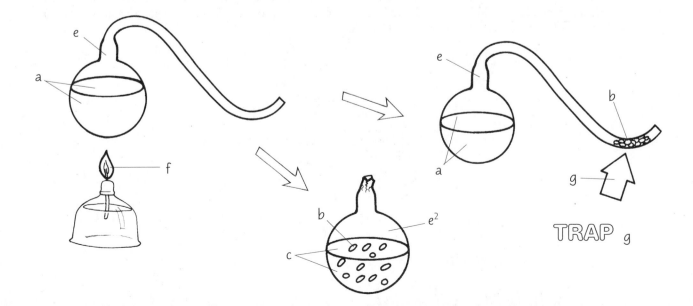

TRAP g

4
THE GERM THEORY OF DISEASE: PASTEUR

The germ theory of disease was set down in a short paper by Louis Pasteur in 1857. The theory holds that microorganisms (germs) are capable of living in the body and causing disease. The germ theory is one of the fundamental principles of biology. Before the germ theory was established and proven, it was believed that disease was due to a mysterious, indefinable chemical quality of the atmosphere called "miasma." It was believed that miasma originated in the dead and spread to infect living people. When Louis Pasteur proved that microorganisms were present in the air (see Plate 3), he not only raised the possibility that the spread of miasma could be interrupted, he also implied that microorganisms could cause disease. In this plate, two of the experiments that led Pasteur to propose the germ theory of disease will be examined.

Color the subheading Yeast Experiment. Color the titles and structures associated with stations 1 through 5 on the left side of the page. Use light colors for the fluids.

Pasteur believed that discoveries of science should have practical application and responded to a friend's request to try and unravel the mystery of why local wines were turning sour. Having worked extensively with a microscope in his study of chemistry, he used this instrument to observe samples of normal and sour wines. Pasteur's microscope consistently revealed large numbers of microscopic oval bodies, called yeasts, in the wine. Yeast cells are slightly larger than bacteria, but little was known about their function. Yeast cells are regarded as a type of fungus by microbiologists, and are now known as key elements in the production of bread, beer, wine, and spirits. In pursuit of the role of yeast in wine production, Pasteur performed the following experiment: he poured grape juice (a), with its natural population of yeast cells (b), into a flask (station 1). He heated (c) the juice, killing the yeast cells (b¹; station 2). He plugged the flask with cotton (d), set it aside

to ferment, and found that the grape juice failed to ferment (station 3). After some time, he added yeast back to the grape juice (station 4), and again set the juice aside to ferment. In this case, fermentation did occur and wine was formed (station 5). Pasteur concluded that the tiny oval yeast cells were necessary for the fermentation process.

Color the subheading Bacteria Experiment, and the titles and structures (a) through (g¹) on the right side of the plate.

While conducting his experiments with yeast, Pasteur found that sour wines were consistently laden with microscopic rods, commonly known as bacteria. He reasoned that if the bacteria were removed from the grape juice, then the wine might not become sour.

Pasteur devised an experiment in which he heated a flask of grape juice (a) known to contain bacteria (g) (station 2). By this process he hoped to kill the bacteria (g¹) as well as the yeast cells in the grape juice. He then added yeast (b) to the grape juice, carefully avoiding bacterial contamination (station 3). He then plugged the flask with cotton (d). Following fermentation (e) (station 4), the resulting wine (f) had its distinctive taste, aroma, and bouquet without any trace of sourness. It was apparent that yeast alone fermented the grape juice to wine. It was also clear that bacterial contamination of the yeast soured the wine.

Pasteur's work indicated that bacteria could be agents of chemical change. Physicians of the time had seen bacteria in the blood of the dead but they assumed that the presence of bacteria was an effect of disease not a cause of it. Now their thinking began to change.

Pasteur was unable to isolate the bacteria basic to his theory. A young physician, Robert Koch, would furnish the definitive proof that microorganisms caused disease.

THE GERM THEORY OF DISEASE: PASTEUR

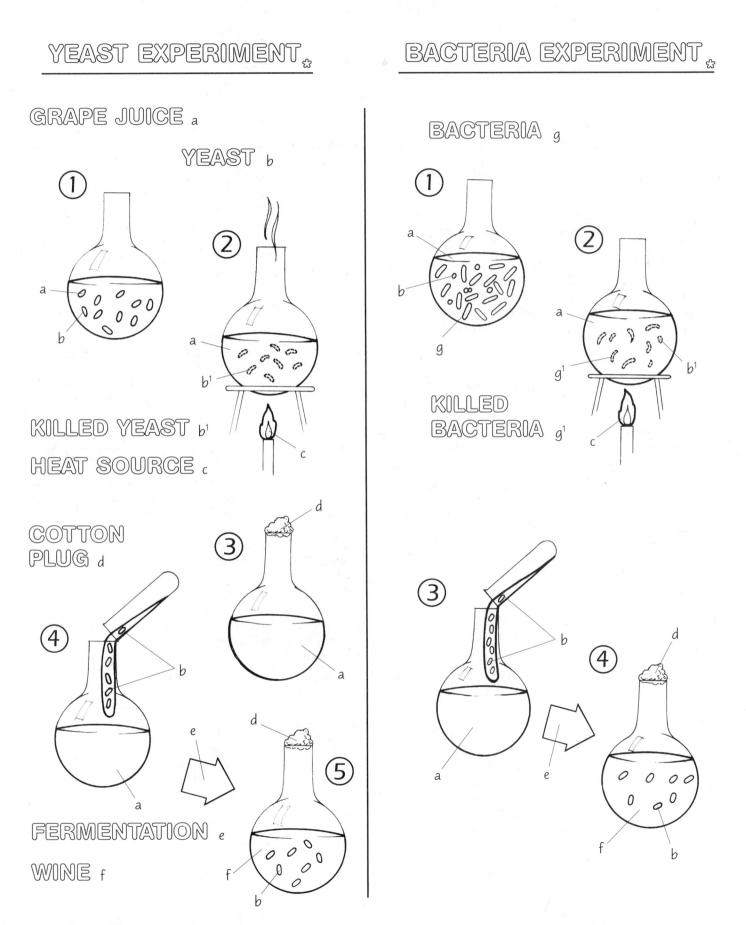

YEAST EXPERIMENT ✿

GRAPE JUICE a

YEAST b

KILLED YEAST b¹

HEAT SOURCE c

COTTON PLUG d

FERMENTATION e

WINE f

BACTERIA EXPERIMENT ✿

BACTERIA g

KILLED BACTERIA g¹

5
GERM THEORY OF DISEASE: KOCH'S POSTULATES

Louis Pasteur proposed the germ theory of disease but was unable to prove it because he could not isolate bacteria in pure culture. Robert Koch, a young country doctor from East Prussia (now Germany), offered the necessary proof. The experiments of Koch proved the germ theory of disease and related a single microorganism to a single disease. The procedures in his experiments became known as Koch's Postulates.

Orient yourself with the plate by following the numbers representing stations in order from 1 to 7. Select a red color for (a), and a darker color for (b). Use a light color for (c). Start with the syringe and related title at station 1, and color counter-clockwise to station 7 as you read the relevant text. Coloring the pathway (labeled *) a light gray from station 1 through 7 will help you appreciate the sequence, especially on review. Titles and structures (b) and (b³) should receive the same color.

As a young doctor living and working in an agricultural community in the 1870s, Robert Koch was interested in a deadly disease characterized by excessive coagulation and discoloration of blood. The disease was called anthrax (*anthrac*, coal) because of the black color of the blood in the affected cattle, sheep, and goats. Whole herds were destroyed by the disease. Determined to learn more about anthrax, Koch undertook the following experiment.

He filled a syringe with blood (a) from a cow that had died of anthrax (station 1) and examined the sample under the microscope. He noted sticks and rods commonly known to physicians as bacteria (b). He then prepared a plate of solid nutrient medium (c) consisting of sterile beef broth solidified with gelatin. He sprayed the blood sample taken from the dead cow onto the surface of the medium (station 2), and set the medium aside to incubate at body temperature (37° C). Two days later, bacteria from the blood sample appeared as visible masses, or colonies (b^1) on the surface of the nutrient medium (station 3). All the colonies consisted of only one kind of bacterium, presumably the bacterium that caused anthrax. Koch now had isolated a single type of bacterium, a step that had eluded Pasteur earlier.

Koch's next experiment in the series was to prove that the isolated bacterium growing on the nutrient medium caused anthrax. He secured pure bacteria from one of the colonies and injected the sample into a healthy mouse (station 4). Within seven days, the mouse died. Koch performed an autopsy on the dead mouse (not shown) and observed the typical signs of anthrax. He then drew a sample of the black blood from the dead mouse (station 5) and sprayed it on a plate of fresh nutrient medium (station 6). He set the plate aside to incubate at 37° C. Two days later, the anthrax bacilli present in the mouse's blood sample had grown into visible colonies (b^2) on the surface of the medium (station 7). Koch took a sample of the bacteria from the colony and examined it under the microscope (b^3), discovering that the bacteria he now observed (microscopic view, station 7) were the same kind he discovered in the blood of the dead cow (b; microscopic view, station 1). The experiment was completed.

Koch's experiments showed several things: bacteria from the blood of the original dead animal could be isolated and cultivated in a pure culture; a sample of the culture could be used to reproduce the disease in laboratory animals; and blood could be taken from the laboratory animals and the bacteria cultured to produce a pure culture of the same bacteria found in the dead, diseased cow. Koch's postulates were derived from these steps.

In 1876, Koch was invited to present his findings to fellow scientists at the University of Breslau in Germany. The audience was astonished. Here was the proof for the germ theory of disease that they had been searching for. Scientists returned to their laboratories and, using Koch's postulates as a guide, began isolating the bacteria of such diseases as plague, tetanus, diphtheria, typhoid fever, and cholera. With the knowledge provided by Koch, the means of bacterial transmission could be examined and epidemics of bacterial origin could be interrupted. The science of microbiology matured significantly as an outgrowth of Pasteur's concept of the germ origin of disease and Koch's confirming experiments.

GERM THEORY OF DISEASE: KOCH'S POSTULATES

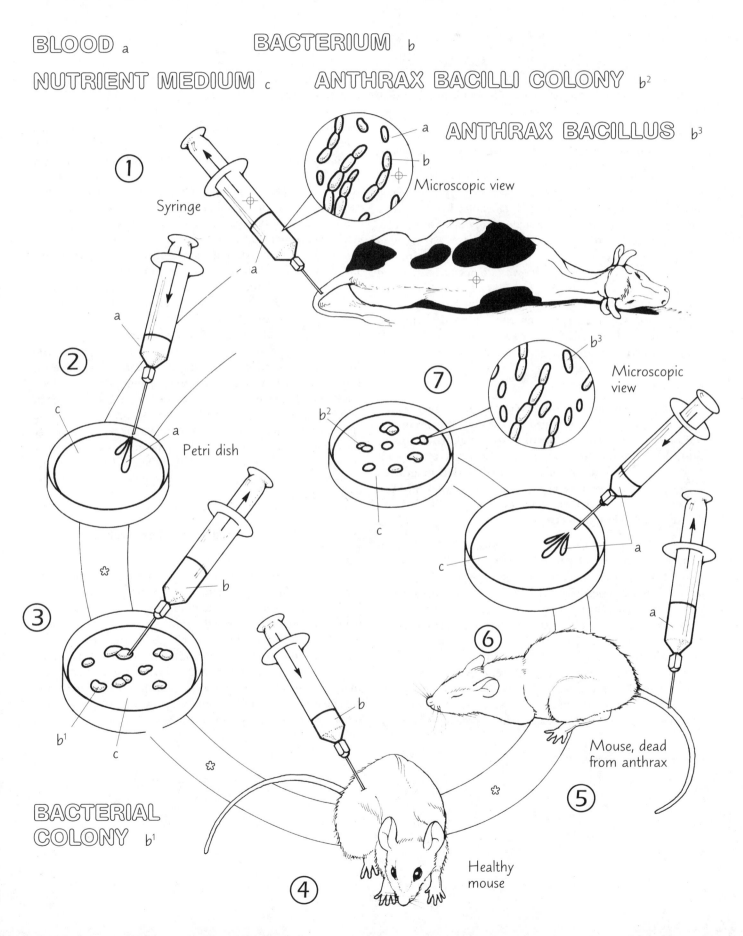

BLOOD a

BACTERIUM b

NUTRIENT MEDIUM c

ANTHRAX BACILLI COLONY b²

ANTHRAX BACILLUS b³

Microscopic view

Syringe

Petri dish

Microscopic view

BACTERIAL COLONY b¹

Healthy mouse

Mouse, dead from anthrax

6
LIGHT MICROSCOPE

Modern technologists have made available to scientists several types of instruments for observing microorganisms. One such instrument is the light microscope.

Look over the titles and related structure as you read below.

The light microscope operates on the principle that light energy will pass through and around a suitably thin object, such as a microorganism and, with the aid of lenses, form a magnified impression on the visual sensory layer of the eye. The light microscope uses two sets of lenses (ocular and objective) and is therefore referred to as a compound microscope. The microscope shown is binocular because it has paired ocular lenses. The two sets of lenses and their housing are supported by a base and an arm. The stage is the platform on which the specimen or a slide supporting the specimen is placed. Clips hold the slide in place. The slide may be moved by fingers or by a mechanical apparatus associated with the stage.

Now color the titles (a) through (f) and related parts of the microscope. Use a light color for (a^2); do not overcolor it with colors for adjacent structure. Use a light color for (d) and (d^1). As you color each of the objectives (e) through (e^3), color the outer rim of the related circled magnified image of the specimen (as visualized through the microscope) at the lower part of the plate. Color the calculations below each circle.

A light microscope is usually equipped with a built-in light source (a) controlled by an on/off switch (a^1) on the base. The light (a^2) is projected upward through a condenser (b) which houses a series of lenses under the stage. These lenses concentrate the passing light into a strong beam which is projected onto the specimen above. The condenser lenses are focused by a controlling knob (b^1). Mounted on the condenser is the iris diaphragm (c), a series of thin brass sheets (not shown) adjusting the size of the aperture through which the light passes. The diaphragm is operated by a lever (c^1) which spreads the sheets apart (opening the aperture) or overlaps them (closing the aperture). This action controls the angular width of the light beam, increasing the contrast between the light and dark portions of the specimen.

The concentrated beam of light passes through and around the thin or thinly-cut specimen (d) placed on a glass slide and usually covered with a glass or plastic coverslip (d^1). Specimens for microscopic observation must often be specially prepared, e.g., fixing the material to make it chemically stable, removing the water, putting it into a medium for cutting, such as paraffin wax, cutting thin slices with a microtome, placing the thin layer of the specimen on the glass slide, staining it, and attaching the coverslip.

Light passing up through and around the specimen enters the objective placed in the light path. Four objectives, housed on a movable turret, are shown in this illustration. Each objective has a certain magnifying power; for example, a scanning objective (e) magnifies 4 times, a low power objective (e^1) magnifies 10 times, a high power objective (e^2) magnifies 40 times, and an oil immersion objective (e^3) magnifies 100 times. It is wise to start out with the scanning objective for orientation, then switch to progressively higher powers. Upon passing through the objective, the light then enters the tube of the microscope where an image forms. An ocular or eyepiece (f) is located at the top of the microscope. The oculars magnify the image projected 10 times. Thus, the images visualized by the eye can be magnified 40, 100, 400, and 1000 times, the product of the magnifying powers of the ocular and the objective in use.

Now color the titles (g) and (h) and the focusing adjustment knobs.

Light microscopes have two focusing controls. The coarse adjustment control (g) is generally the larger of two grooved knobs at the lower side of the microscope's arm. It is used to bring the specimen roughly into focus. This is usually accomplished by moving the stage up/down a track on the arm of the microscope. A more refined, precise focus can be achieved by rotation of the smaller fine adjustment knob (h). In this manner, the microscope permits the observer to see the best possible image of the form, shape, size, arrangement and other features of the specimen.

LIGHT MICROSCOPE

LIGHT SOURCE a / SWITCH a¹

LIGHT PATH a²

CONDENSER LENS b / KNOB b¹

IRIS DIAPHRAGM c / LEVER c¹

SPECIMEN d / COVER SLIP d¹

OBJECTIVE LENSES:
    SCANNING O. (4X) e
    LOW POWER O. (10X) e¹
    HIGH POWER O. (40X) e²
    OIL IMMERS. O. (100X) e³

OCULAR LENS (10X) f

COARSE ADJUSTMENT g

FINE ADJUSTMENT h

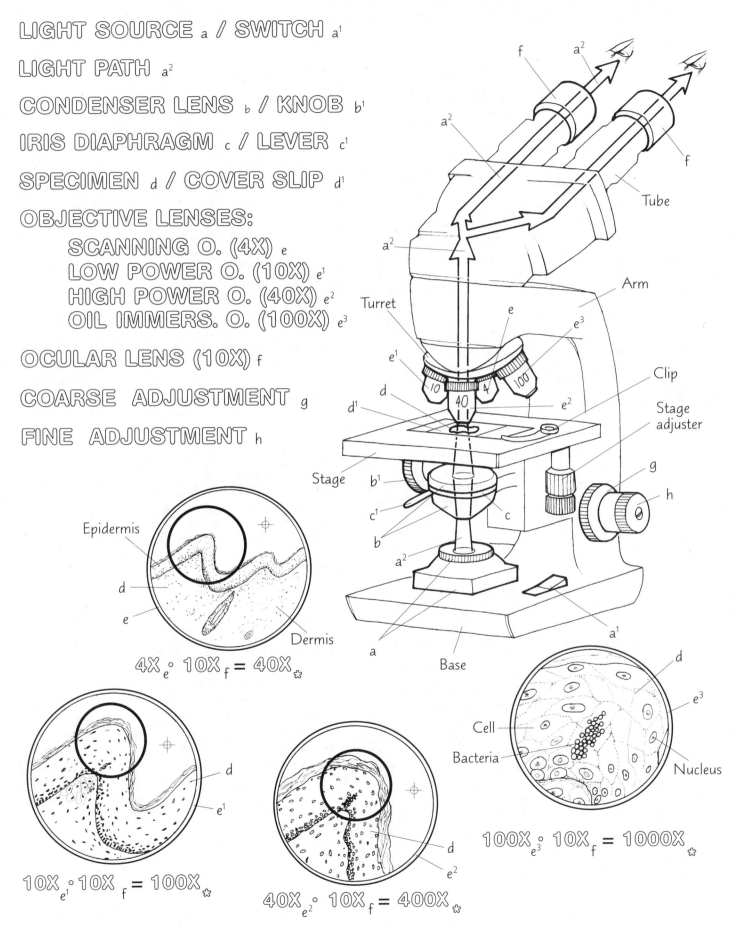

Tube

Arm

Turret

Clip

Stage adjuster

Stage

Base

$4X_{e} \circ 10X_{f} = 40X_{✿}$

Epidermis

Dermis

$10X_{e^1} \circ 10X_{f} = 100X_{✿}$

$40X_{e^2} \circ 10X_{f} = 400X_{✿}$

$100X_{e^3} \circ 10X_{f} = 1000X_{✿}$

Cell

Bacteria

Nucleus

7
ELECTRON MICROSCOPE

The electron microscope is one of the great technological advances of this century. Based on electron motion research in the 1920s, it was first developed and made commercially available in the 1930s. By the 1960s, improved versions of the electron microscope were in widespread use as research instruments in universities throughout the world. The electron microscope uses accelerated electrons that have a wavelength a fraction of the wavelength of light. This short wavelength greatly increases the discrimination of small parts (resolution) over that of the light microscope (LM). The result is an image magnified tens and hundreds of thousands of times. Microbiologists have used this remarkable instrument to upgrade their knowledge of the fine structure of micro-organisms and substantially increase their understanding of microbial function in relation to structure, structural relationships, and disease processes.

There are several types of electron microscopes available today for use. In microbiology, the transmission electron microscope (TEM) and the scanning electron microscope (SEM) are the most widely used imaging instruments.

Color the title and instrument (a), titles (b) through (h^1) and the related parts in both external and schematic views. Use a light color for (a) and (b^1). The rim of the circled light microscope image is identified by subscript (e^3) from the previous plate and should be colored similarly. Color the rim of the TEM image (a^1).

The transmission electron microscope or TEM (a) employs an electron gun as a source of electrons (b). Electrons discharged by a cathode accelerate toward a circular anode. The electrons are emitted through an aperture in the anode at high speed, forming an electron beam (b^1). The beam, traveling within an evacuated space (vacuum), is directed toward the specially prepared, ultra-thin specimen. An electromagnetic condensing lens (c) concentrates and focuses the beam, much like the condenser lens focuses light in the light microscopes. The electron beam passes through the specimen chamber (d) where the electrons are absorbed or deflected by the specimen. The residual beam of electrons is focused and magnified by electromagnetic lenses (e; usually an intermediate and a projector lens), and projected onto a fluorescent screen (f) or photographic plate. Here an image (a^1) is formed. The eyepiece (h) permits rapid visualization of the image on the fluorescent screen through the viewing window (g). The image may be magnified further by binoculars (h^1). Photographic enlargement can provide even greater magnification. Refer to the circled light microscope image (e^3). Note the group of bacteria. Compare that image with the TEM image (a^1) which shows a portion of the cell membrane of one bacterium.

Color the titles and parts associated with the scanning electron microscope (i) through (k). Use a light color for (i). Refer to the title list above for coloring (b) through (e). Lastly, color the rim of the circled SEM image (i^1) at the center of the plate.

The scanning electron microscope or SEM (i) was developed in the 1960s. The SEM is easier to operate than the TEM, and permits the scanning of whole and sectioned objects without extensive preparation. It also permits the visible surface of the specimen to be studied in vivid detail. This is unlike both the LM and TEM where special thin-section preparation of the specimen is required and the entire structure (not just the surface) is visualized. SEM methodology usually results in relatively low magnification (under 100,000 times).

To use the scanning electron microscope, the specimen is placed into the specimen chamber (d) and covered with a thin film of gold to increase electrical conductivity and decrease blurring. The electron source (b) then emits the electron beam (b^1) toward the specimen. An electromagnetic condensing lens (c) and other electromagnetic lenses (e) concentrate and focus the beam. Consistent with its name, the SEM employs a scanner to permit the electron beam to sweep back and forth across the rough "hills and valleys" surface of the specimen, inducing secondary emissions of electrons from the surface. A signal detector (j) captures the electrons and, line by line, builds an image on a monitor (k) much like a television receiver does. Electrons that strike a sloping surface are deflected such that not all of the deflected electrons are captured by the signal detector. This generates a greater degree of contrast and produces a three-dimensional effect.

Now refer to the center of the plate. A photograph prepared from a TEM image can be enlarged to achieve a total magnification of over 20 million times, although as a practical matter, magnifications of 5000 times to 500,000 times are most commonly employed. The TEM image here (a^1) is a micrograph of a portion of a single Gram-positive bacterium magnified 75,000 times. Compare this with the same specimen magnified 1000 times with the light microscope (e^3). Whereas the density of the light microscope (LM) image is determined by the amount of light coming through the specimen, the density of the TEM image is determined by electron absorption. And whereas the specimen studied with the LM is usually stained with dyes for color differentiation of structure, the specimen destined for TEM and SEM study is not stained. Chemical fixation is necessary, however. Note that the SEM image (i^1) with respect to the TEM image, is at a lower level of magnification and reveals a greater area of coverage. It shows only the surface features of the specimen, and it appears as a three-dimensional image.

ELECTRON MICROSCOPE

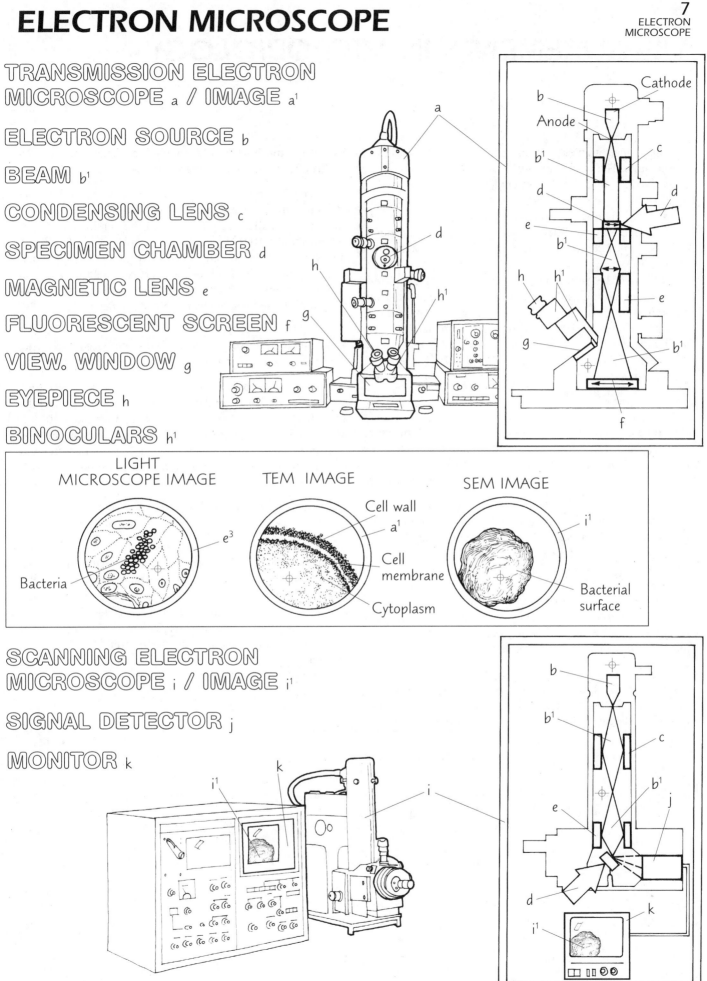

TRANSMISSION ELECTRON
MICROSCOPE a / IMAGE a¹

ELECTRON SOURCE b

BEAM b¹

CONDENSING LENS c

SPECIMEN CHAMBER d

MAGNETIC LENS e

FLUORESCENT SCREEN f

VIEW. WINDOW g

EYEPIECE h

BINOCULARS h¹

Cathode
Anode
b
b¹
d
e
b¹
h
h¹
g
e
b¹
f
c
d

LIGHT
MICROSCOPE IMAGE

TEM IMAGE

SEM IMAGE

Bacteria
e³

Cell wall
a¹
Cell
membrane
Cytoplasm

i¹
Bacterial
surface

SCANNING ELECTRON
MICROSCOPE i / IMAGE i¹

SIGNAL DETECTOR j

MONITOR k

i¹
k
i

b
b¹
c
b¹
e
j
d
k
i¹

8
MEASUREMENTS IN MICROBIOLOGY

All linear measurements in microbiology are expressed in metric units. The basic unit of the metric system is the meter (expressed as "m"). The equivalent length in the U.S. system is about three feet. Multiples and fractions of meters are expressed in factors of 10, and the related term is attached to the front of the word "meter": hence, kilometer (1000 meters), centimeter (1/100th of a meter), millimeter (1/1000th of a meter), and so on.

Illustrated here is a series of rulers of progressively shorter lengths. In each ruler, a fraction of the length is extracted and expanded for analysis and coloring. For each level of measurement, examples of structure from humans (left side) and microorganisms (right side) are shown. Coloring this plate will give you a feeling for the relationship of one level of measurement to another and introduce you to common units of measurement in microbiology.

Plan on using sharply contrasting colors for the four levels of measurement. Color title (a) and the small 1 centimeter long part of the lower half of the inch/centimeter rule (a). Color the corresponding 10 cm measure bars on either side of the uppermost ruler.

The centimeter (a; abbreviated cm) is commonly the largest unit of length used for measuring microorganisms. The length of an index finger taken from the web between the thumb and index finger is about 4 inches long. One inch = 2.5 cm; therefore, a 4 inch long finger is equivalent to 10 cm. A parasitic roundworm common to the intestines of those who have eaten contaminated food or food prepared with unsanitary practices is *Ascaris lumbricoides*. It is not unusual to see a 10 cm or longer *Ascaris* in the intestines of infected persons or animals.

Color over the thin double line (b) extending down from the center of the centimeter rule, and the diverging lines leading to the arrow heads (b). Color title (b) and the upper half of the millimeter ruler labeled (b). Then color the 10 millimeter bars at either side of the millimeter ruler.

Referring to the lower part of the centimeter ruler, note that a millimeter (b; abbreviated mm) is the distance between two of the smallest vertical lines in the centimeter rule. A millimeter is 1/10th of a centimeter. Look at your thumb and note the three or four skin creases over the joint on the nail side of your thumb. Each of those creases is about 1 mm in width. A colony of bacteria growing in a Petri dish of nutrient medium is quite visible when 1 mm in diameter.

Continue with the micrometer ruler and related lines, title, and measure bars labeled (c) as you did above.

Structures measured in micrometers (c; abbreviated μm) are visible only with a high powered microscope. An average epithelial cell of human skin has a length of about 10 μm. Most bacteria measure about 1 to 5 μm in length. The micrometer is the unit of measurement most frequently used in microbiology. It has replaced the term "micron."

Continue with the nanometer ruler, and related lines, title, and measure bars labeled (d) as you did above.

A nanometer (d; abbreviated nm) is the unit of length commonly used by microbiologists to measure the dimensions of viruses. The smallest viruses are about 20 nm across; the largest are about 300 nm across. The picometer (abbreviated pm; not shown) is equivalent to a tenth of a nanometer (or one ten-thousandth of a micrometer). The thickness of the viral envelope is about 1pm. The picometer is only rarely encountered in microbiology; it has replaced the equivalent unit of measurement known as the Ångstrom unit.

MEASUREMENTS IN MICROBIOLOGY

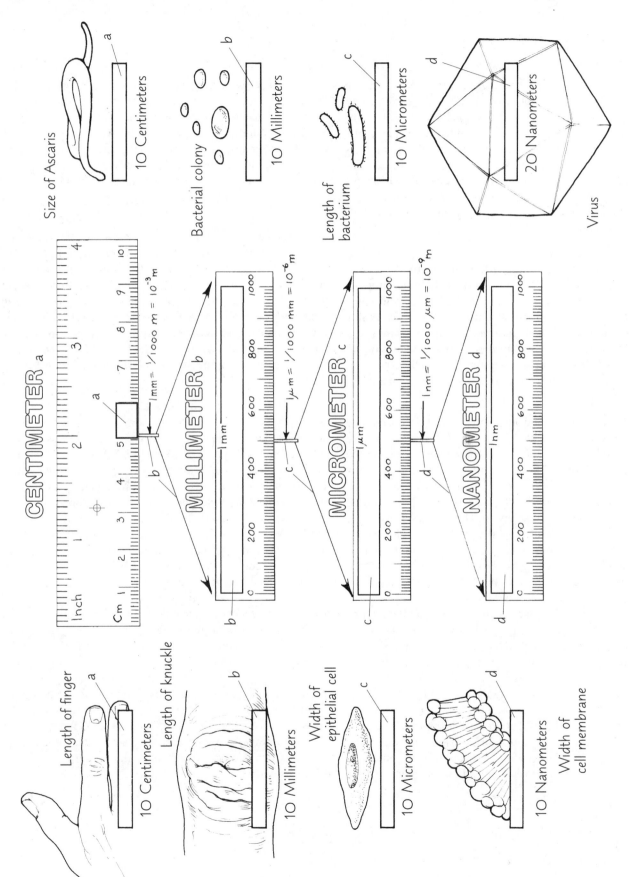

Size of Ascaris
a 10 Centimeters

Bacterial colony
b 10 Millimeters

Length of bacterium
c 10 Micrometers

d 20 Nanometers
Virus

CENTIMETER a
$1\,mm = \frac{1}{1000}\,m = 10^{-3}\,m$

MILLIMETER b
$1\,\mu m = \frac{1}{1000}\,mm = 10^{-6}\,m$

MICROMETER c
$1\,nm = \frac{1}{1000}\,\mu m = 10^{-9}\,m$

NANOMETER d

Length of finger
a 10 Centimeters

Length of knuckle
b 10 Millimeters

Width of epithelial cell
c 10 Micrometers

Width of cell membrane
d 10 Nanometers

9
EUKARYOTES AND PROKARYOTES

One of the most important generalizations to emerge in recent decades is that organisms on Earth may be categorized either as eukaryotes or prokaryotes, viruses excepted. Both terms are derived from the Greek word karyon, which means "nut." The nut, in this case, refers to the nucleus of a cell. Prokaryotes are organisms whose cells lack a nucleus, while eukaryotes are organisms whose cells have a well-defined nucleus. In this plate, we shall examine further differences between the eukaryotes and prokaryotes and see how microorganisms fit into the world of living things.

Color the subheading Eukaryotic Cell and the titles and related structures (a) through (j). Consider leaving the cytoplasm uncolored, or use a very light color and color over the entire area outside the nucleus, then overcolor with darker colors for structures within the cytoplasm. Generally, use light colors for large structures so as not to obscure cell detail. After coloring the cell, color the subheading Eukaryotes and the titles and organisms (k) through (n).

The cells of eukaryotic organisms possess a well-defined nucleus (arrow a), separated from the cell cytoplasm by a nuclear membrane/envelope (a^1). The nucleus contains the genetic information of the cell in multiple strands of DNA and protein called chromosomes. Here the DNA and proteins are shown as chromatin (b), the dispersed nuclear material seen between periods of cell division. The nucleus is the largest body in the eukaryotic cell. A smaller body, the nucleolus (c), exists within the nucleus and plays a role in the production of ribosomal RNA and ribosomes.

Eukaryotic cells also have membrane-bound structures, called organelles, suspended in the cytoplasm (d). Important cellular functions take place within the organelles. Mitochondria (e) are sites of energy production for cellular work. Ribosomes (f) are masses of RNA and proteins that function as the site of protein synthesis. The endoplasmic reticulum (g) consists of flattened tubules involved in the transport of newly-synthesized protein. The Golgi apparatus (h) is a set of tubules and vesicles in which protein is packaged for export. Lysosomes (i) are vesicles that contain enzymes for cellular digestive processes. In those organisms that carry on photosynthesis, the chloroplast (not shown) is an important organelle since it contains the pigment chlorophyll which traps the sun's energy and converts it into the energy in carbohydrates. The cell membrane (j) is the organelle that regulates the flow of materials in and out of the cytoplasm.

Many familiar organisms are eukaryotes. Animals (k), including humans, and all plants (l) are eukaryotic. Among the microorganisms there are several eukaryotes. The protozoa (m), fungi (n), and one-celled algae, for example, are composed of cells that are eukaryotic.

Color the subheading Prokaryotic Cell and the titles and structures listed on the left side of the lower half of the plate. These should receive the same color as their respective structures in the eukaryotic cell above except for (o). Note that the chromosome (b) and chromatin in the eukaryotic cell (b) receive the same color. Finally, color the subheading Prokaryotes and the titles and related structures on the right side of the lower half of the plate. Use the same color or shades of the same color for (p) through (p^2).

The cells of prokaryotes lack a nucleus or nuclear membrane, in contrast to the cells of eukaryotes where the nucleus is present. Instead, prokaryotic cells possess a single chromosome (b) composed solely of DNA. The chromosome exists as a closed loop. With the exception of ribosomes, there are no organelles in the cytoplasm of prokaryotic cells. In these cells, chemical activities take place in the cytoplasm instead of organelles. In prokaryotes that carry on photosynthesis, the necessary pigments are not found in chloroplasts, but are suspended in the cytoplasm (not shown). Prokaryotic cells contain ribosomes (f). Although they are relatively small compared to the larger eukaryotic ribosomes, they have the same function. Prokaryotic cells possess cell membranes (j) and these are generally surrounded by complex cell walls (o). Animal eukaryotes do not have cell walls. Plant eukaryotes do have cell walls but of a much simpler construction than the prokaryotes.

Several types of microorganisms are prokaryotes. The most familiar of these types is the bacteria (p), including the subgroups of rickettsiae (p^1) and cyanobacteria (p^2). An important group of microorganisms, the viruses, are not included in this survey because they are neither prokaryotes or eukaryotes. Viruses do not have the properties of living things, as discussed in plates ahead.

EUKARYOTES AND PROKARYOTES

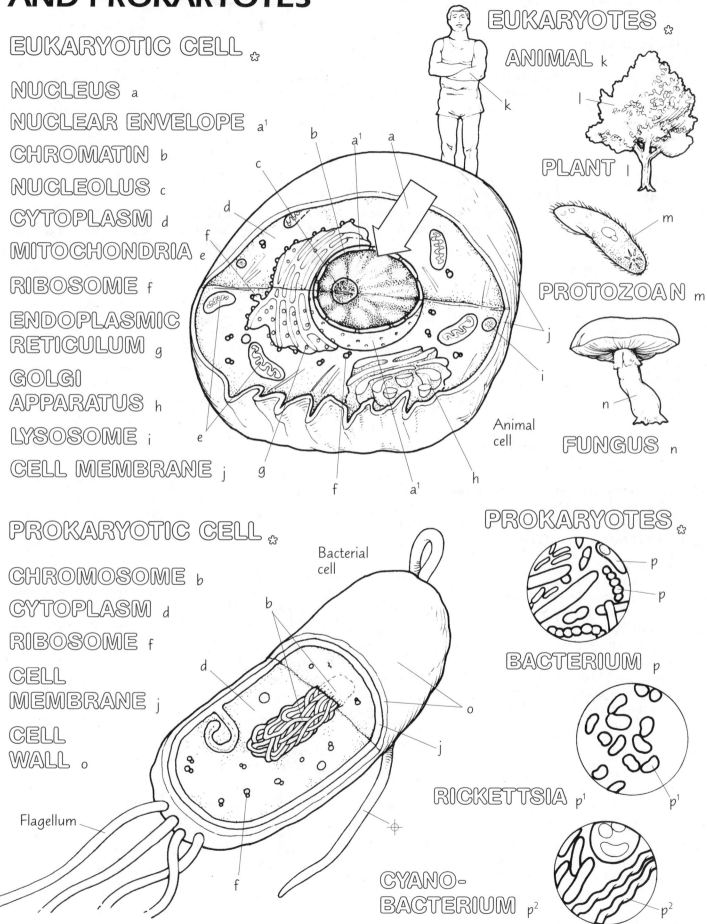

EUKARYOTIC CELL ✿

NUCLEUS a
NUCLEAR ENVELOPE a¹
CHROMATIN b
NUCLEOLUS c
CYTOPLASM d
MITOCHONDRIA e
RIBOSOME f
ENDOPLASMIC
RETICULUM g
GOLGI
APPARATUS h
LYSOSOME i
CELL MEMBRANE j

Animal cell

EUKARYOTES ✿
ANIMAL k
PLANT l
PROTOZOAN m
FUNGUS n

PROKARYOTIC CELL ✿

CHROMOSOME b
CYTOPLASM d
RIBOSOME f
CELL
MEMBRANE j
CELL
WALL o

Bacterial cell

Flagellum

PROKARYOTES ✿
BACTERIUM p
RICKETTSIA p¹
CYANO-
BACTERIUM p²

10
THE SPECTRUM OF MICROORGANISMS

The spectrum of microorganisms is very broad, ranging from the ultra-small viruses to the relatively large protozoa. A feature common to most microorganisms is that one may use a microscope to observe their structural details. Their sizes can be compared with the measuring bars as illustrated. Microorganisms are grouped by structural, functional, and biochemical qualities.

Color the title (a) and the three related viruses at upper left. Also color the measuring bar at upper left, noting the length of the bar (in nm) in relation to the viruses shown. Light pastel colors are recommended for the viruses.

Viruses (a) are among the smallest microorganisms, and require electron microscopes for visualization. Composed of fragments of nucleic acid surrounded by protein coats, viruses occur in three major shapes: the icosahedron (a 20-sided figure), the helix (or coil), and the complex form. Lacking synthetic "machinery," viruses do not grow or show any observable activity except replication, which can be accomplished only within living cells. Despite their structural simplicity, viruses are responsible for several important human diseases including influenza, hepatitis, chicken pox, and acquired immune deficiency syndrome (AIDS). In each of these cases, disease virulence is related to the viral capacity for replication, which alters cellular function and structure in the host.

Color the title (b) and the four different bacteria. Color the measuring bar at upper right, noting the difference in size between bacteria and viruses. Consult Plate 8 for review of comparison of linear dimensions.

Bacteria (b) are perhaps the most abundant organisms on Earth. Most species of bacteria are hundreds of times larger than viruses, and the great majority of species can be seen with an ordinary light microscope at a magnification of 1000 times (1000 X). Bacteria occur in three major shapes: the rod-shaped bacillus, the spherical coccus, and the spiral-shaped spirochete and spirillum. The vast majority of bacteria play positive, non-pathogenic roles in nature, such as digesting the remains of dead animals and plants. They also recycle the elements, extracting nitrogen from the air for protein production in plants. Moreover, they manufacture foods for human consumption. Bacteria that cause disease do so by growing in the body tissues, digesting healthy cells, and producing toxins that interfere with cellular functions. Such pathogenic activity can produce serious sickness or death in the host. Biologists estimate that the mass of bacteria on Earth outweighs the mass of all plants and animals.

Color the title (b¹) and the two members of this bacterial subgroup. Color the measuring bar and compare.

Cyanobacteria (b¹; formerly called blue-green algae but now called cyanobacteria to reflect their place as bacteria) are microorganisms that possess pigments which function in the solar energy-trapping process of photosynthesis. Their microscopic size and chemical features are typical of microorganisms. Cyanobacteria may occur in one-celled (unicellular) or multicellular filamentous forms. They inhabit fresh water and marine environments, sometimes giving a "pea soup" color to the water.

Color the title (c) and the three members of the group, and the related 50 μm measuring bar. Note these organisms are significantly larger than those microorganisms just colored.

Microscopic algae (c), like cyanobacteria, have pigments within their cells and are considered microorganisms. The pigments of these algae are more like those of plants, however, and have a complexity typical of plant cells. Two types of microscopic algae are important in microbiology: the diatoms, which serve as major sources of food in the oceans; and the dinoflagellates, which cause the red tides that occur periodically in ocean waters and coastal regions.

Color the title (d) and the four members of the group, and the related 50 μm measuring bar.

Protozoa (d) are microorganisms with complex cellular features. Means of locomotion is an important criterion in the classification of protozoa. Amoebas move by thrusting out portions of their cell membranes (pseudopods) then moving into the projections. Flagellates move as one or more whiplike flagella push or pull the organism along. Ciliates, covered with rows of hairlike cilia, move by synchronized beating of the cilia. Sporozoans have no method of locomotion in the adult form, but move with the flow of their external environment. Most species of protozoa are harmless, but certain ones cause malaria and sleeping sickness, among other diseases.

Color the title (e) and the three members of the group, and the three measuring bars (10 μm, 100 μm, and 1 cm).

Fungi (e) are complex microorganisms subdivided into two groups: the molds and yeasts. Molds, which include mushrooms, are long, branching chains of cells called hyphae. With vigorous growth, hyphae may result in a visible mass, called a mycelium. Fungi commonly employ spores as a means of reproduction. Many molds prefer acidic environments such as citrus fruits, cheeses, and bread. Yeasts are single-celled fungi, about the size of large bacteria. They are important in bread production and fermentation of juices to produce wine. Together with bacteria, the fungi are the prime decomposers of the world's organic matter.

THE SPECTRUM OF MICROORGANISMS

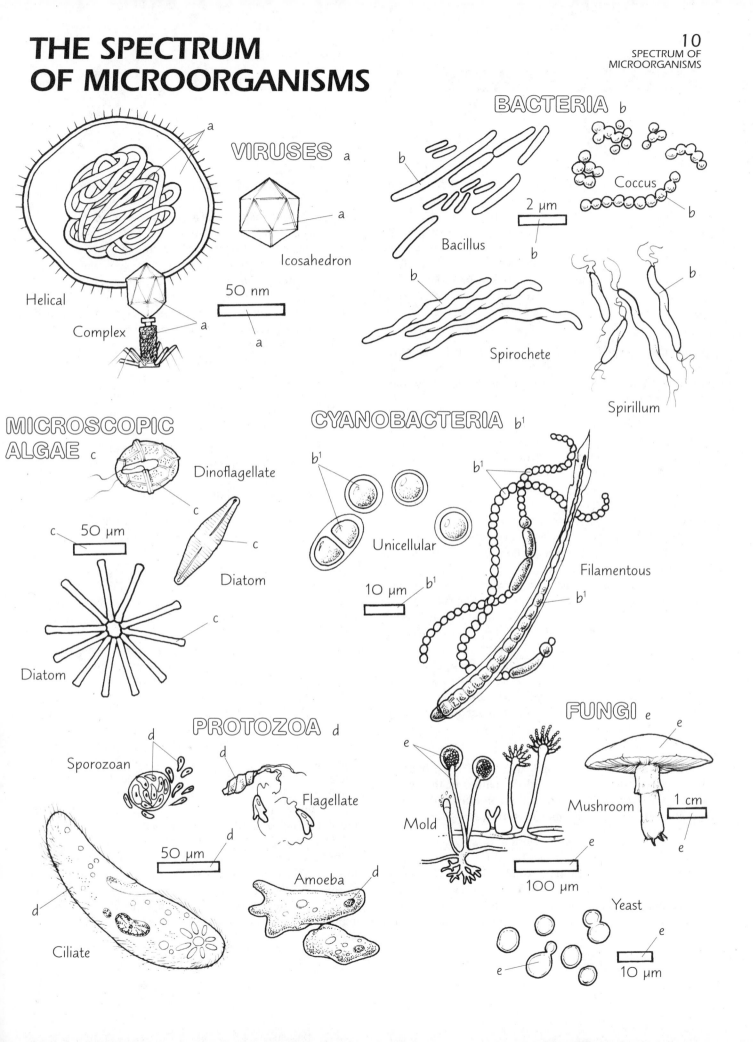

VIRUSES a

Helical

Icosahedron

Complex

50 nm

BACTERIA b

Bacillus

2 μm

Coccus

Spirochete

Spirillum

MICROSCOPIC ALGAE c

Dinoflagellate

Diatom

50 μm

Diatom

CYANOBACTERIA b¹

Unicellular

10 μm

Filamentous

PROTOZOA d

Sporozoan

Flagellate

50 μm

Amoeba

Ciliate

FUNGI e

Mold

Mushroom

100 μm

1 cm

Yeast

10 μm

11
FORMS OF BACTERIA

The word "bacterium" may have been used for the first time in the 1850s when the French investigator Casimir Davaine used the term to mean "rod" or "staff." As the years unfolded it became apparent that many bacteria are not rodlike, but the name remained and soon it was applied to all microscopic organisms of that general size and with properties similar to the rods. In this plate three basic forms of bacteria and their arrangements are examined and related to their role as disease agents.

Color the subheading (a), the four different forms of the bacillus (a^1) through (a^4) and their titles, and the names of the related diseases. Use four different shades of the same color for these forms.

The rod form of a bacterium is called a bacillus (a; pl. bacilli). Bacilli vary in size, and may be as long as 20 μm or as short as 0.5 μm. Certain bacilli (*Bacillus anthracis*) are rectangular with sharply rounded ends (a^1); these bacilli cause anthrax (a^1), a disease of such animals as cows, goats, sheep, and deer. The disease is communicable to humans by air, contaminated meat, and contact with animals. Certain rod-shaped bacilli are wide at one end and tapered at the other end (club-shaped, a^2; e.g., *Cornybacterium diphtheriae*). They are known to cause diphtheria (a^2). In this disease of the respiratory tract, bacterial toxins damage the nerves and the heart. One type of bacillus (*Clostridium tetani*) is rod-like but swollen at one end (a^3). These swollen ends contain endospores, a very resistant form of the bacterium. Tetanus (a^3), a disease caused by these bacteria, is characterized by muscle spasms, seizures, and paralysis of respiratory muscles. There are several species of bacilli that occur in chains (a^4). A streptobacillus is shown here; *strepto*, refers to bacteria linked end-to-end in chains. Certain streptobacilli cause rat bite fever (a^4), a disease characterized by chills, vomiting, and fever.

Color the subheading (b), and the five different forms of the coccus and their titles, and the names of the related diseases. Use five shades of the same color or similar colors which contrast with the color used for (a).

The spherical form of a bacterium is known as a coccus (b; pl. cocci). A coccus is about 0.5 μm in diameter. Some cocci called diplococci are paired (diplo-, double). One species of diplococcus (*Streptococcus pneumoniae*) has tapered sides (b^1) and causes pneumonia (b^1), an inflammation of the air spaces of the lungs accompanied by fluid formation. Another type of diplococcus (*Neisseria gonorrhoeae*) (b^2) resembles two tiny beans lying face to face. *N. gonorrhoeae* causes gonorrhea (b^2), a disease transmitted by sexual contact. The streptococcus is a well-known group of cocci characterized by individuals in a chain (b^3). "Strep throat" (b^3), a serious infection of the pharynx, is caused by a species of streptococcus. In contrast, a harmless species of streptococcus is one of the "active cultures" in a cup of yogurt. A cubelike packet of four or eight cocci (b^4) is called a sarcina. One sarcina called *Micrococcus luteus* is a common nonpathogenic inhabitant of the human skin (b^4). Another type of coccus, called staphylococcus, occurs in clusters (b^5) and produces toxins in food resulting in staphylococcal food poisoning (b^5). Other staphylococci enter hair follicles and inflame the skin causing boils or "staph infection" (not shown).

Color the subheading (c), the three different forms of the spiral and their titles, and the names of the related diseases. Use three shades of the same color which contrasts with the colors used for (a) and (b).

A third form of a bacterium is the spiral (c). These bacteria are about 15 μm in length. In the spiral form called the vibrio (c^1), the bacterium has only a single turn, appearing curved, like a comma. One vibrio causes cholera (c^1), a serious disorder characterized by vomiting, diarrhea, and cramps. Severe dehydration caused by *Vibrio cholerae* is induced by toxins that interfere with sodium absorption in the intestines. Another form of spiral bacteria is the spirillum (pl. spirilla). The spirillum resembles a corkscrew, with the spiral making several turns (c^2). The spirillum possesses a rigid cell wall with flagella for movement. This bacterium causes rat bite fever (c^2), which is similar in symptoms to the rat bite fever caused by streptobacilli. A spirochete is a spiral bacterium that has the corkscrew form but a flexible cell wall and no flagella (c^3). It uses axial filaments to move in a snakelike manner. A spirochete (*Treponema pallidum*) is responsible for syphilis (c^3), a disease in which the bacteria enter the tissues through breaks in the skin, such as the skin of the genital organs.

The anatomical pattern of a bacterium can be of great practical value. In the diagnostic laboratory, for example, a technician may note the characteristic diplococci of gonorrhea in a patient's urine sample and report this observation to the physician. The diagnosis of syphilis is aided considerably by locating the characteristic spirochetes in material from a skin lesion. And strep throat may be pinpointed by observing streptococci in bacterial colonies isolated from the throat.

FORMS OF BACTERIA

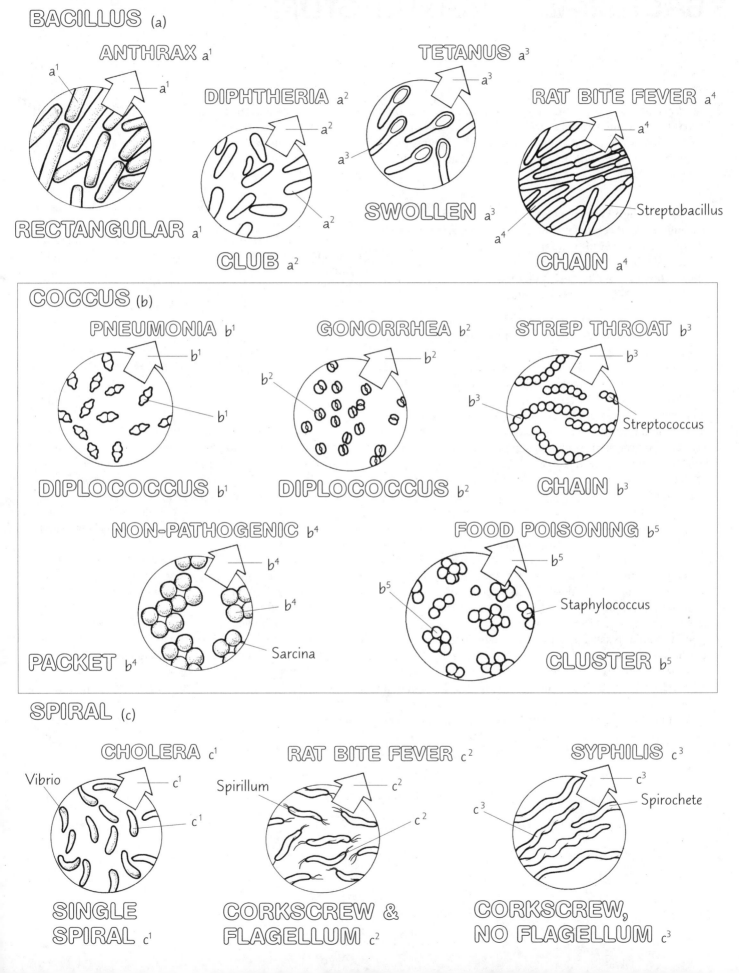

BACILLUS (a)

ANTHRAX a¹
a¹
a¹
RECTANGULAR a¹

DIPHTHERIA a²
a²
a²
CLUB a²

TETANUS a³
a³
a³
SWOLLEN a³

RAT BITE FEVER a⁴
a⁴
Streptobacillus
a⁴
CHAIN a⁴

COCCUS (b)

PNEUMONIA b¹
b¹
b¹
DIPLOCOCCUS b¹

GONORRHEA b²
b²
b²
DIPLOCOCCUS b²

STREP THROAT b³
b³
b³
Streptococcus
CHAIN b³

NON-PATHOGENIC b⁴
b⁴
b⁴
Sarcina
PACKET b⁴

FOOD POISONING b⁵
b⁵
b⁵
Staphylococcus
CLUSTER b⁵

SPIRAL (c)

CHOLERA c¹
Vibrio
c¹
c¹
SINGLE
SPIRAL c¹

RAT BITE FEVER c²
Spirillum
c²
c²
CORKSCREW &
FLAGELLUM c²

SYPHILIS c³
c³
Spirochete
c³
CORKSCREW,
NO FLAGELLUM c³

12
BACTERIAL ULTRASTRUCTURE

The transmission electron microscope magnifies bacteria over 100,000 times and reveals a wealth of detailed structures that can be closely correlated with cellular function. In this plate, bacterial structures will be examined and related to cellular behavior.

Set aside a light yellow color for (i). Color the titles (a) through (k) and related structures. Try to select colors that do not obscure cellular detail. Use darker colors for very small structures. Color over the flagellum/flagella (a) on the bacteria shown in motion on the plate and represented in the illustration by their capsules (b).

Many species of bacilli and spirilla, and a few species of cocci, move by means of one or more flagella (a). The flagellum is composed of thin fibers made of protein. The movement of the bacterial flagellum is rotary thereby creating a propellerlike motion that drives the bacterium forward.

Some species of bacteria, many of them pathogenic, have a capsule (b). The capsule is secreted by the bacterium and adheres to the outer surface of the cell wall. It is composed of complex polysaccharides and small proteins. When the capsule has a looser consistency, it is called a glycocalyx. The capsule is a storehouse for nutrients, a depot for cellular waste products, and a protective shield against dehydration and potentially harmful changes in the external environment. It also retards phagocytosis by white blood cells of the host's immune system.

The cell wall (c) is composed primarily of complex organic acids and is found in virtually all species of bacteria. This structure provides a rigid framework for the organism and helps determine its shape. The molecular construction of the bacterial cell wall is unique, and the nature and significance of this structure is explored in Plate 13.

The cell or plasma membrane (d) is the outermost border of the cytoplasm and internal to the cell wall. The cell membrane (Plate 13) is constructed of protein globules suspended in lipids. It is a boundary layer and a dynamic vehicle for the transport of material into and out of the cell. It holds many of the cell's enzymes, and it is the functional equivalent of the mitochondrion in eukaryotic cells. Antimicrobial substances, such as certain detergents and antibiotics, can interfere with the function of the cell membrane, spelling death to the bacterium.

Certain bacteria, usually Gram-negative bacteria (see Plate 14), have hairlike pili (e; sing. pilus; also called fimbriae) extending from the surface of the cell. They are not to be confused with the larger flagella. Pili help anchor the bacterium to a surface. In some cases, they assist the transfer of genetic material between bacterial cells. Many bacterial species display a coiled inward extension of the cell membrane called the mesosome (f). The function of this structure may be to serve as a site for attachment of deoxyribonucleic acid (DNA) during replication in bacterial reproduction. The mesosome may also be an artifact introduced by chemicals used in fixation for electron microscopy.

As in other prokaryotes, there is no nuclear membrane around the bacterial genetic material. A bacterium exhibits a single long chromosome of DNA arranged as a closed loop folded over itself many times. The chromosome is suspended in the cytoplasm without a covering or membrane. The region of cytoplasm occupied by this chromosome is called the nucleoid (g). The chromosome contains all the hereditary information of the cell, and provides all the necessary "instructions" for producing the proteins essential to the life of the cell. Smaller molecules of DNA, called plasmids (h), form closed loops in the cytoplasm apart from chromosomes. A single bacterium may have several dozen plasmids, each with a few genes. In the plates ahead, the activity of plasmids in genetic recombination will be examined.

The bacterial cytoplasm (i) is a gelatinous mass of proteins, carbohydrates, and other organic and inorganic chemical substances. The cytoplasm is the site of bacterial growth, metabolic reactions, and reproduction. Suspended within it are such bodies as the nucleoid and plasmids, as well as the ribosomes, and inclusions called granules. The ribosomes (j) are composed of ribonucleic acid (RNA) and protein. They are the sites of protein synthesis. They are the places where amino acids, the unit molecules of protein, are bound together by enzymes in the precise sequence that gives each protein its functional character. Granules (k) appear to be protein membrane-lined storage sites for starch, glycogen, lipid, or other essential materials.

BACTERIAL ULTRASTRUCTURE

FLAGELLUM a   CAPSULE b   CELL MEMBRANE d

CELL WALL c   PILUS e

MESOSOME f

NUCLEOID g

PLASMID h

CYTOPLASM i

RIBOSOME j

GRANULE k

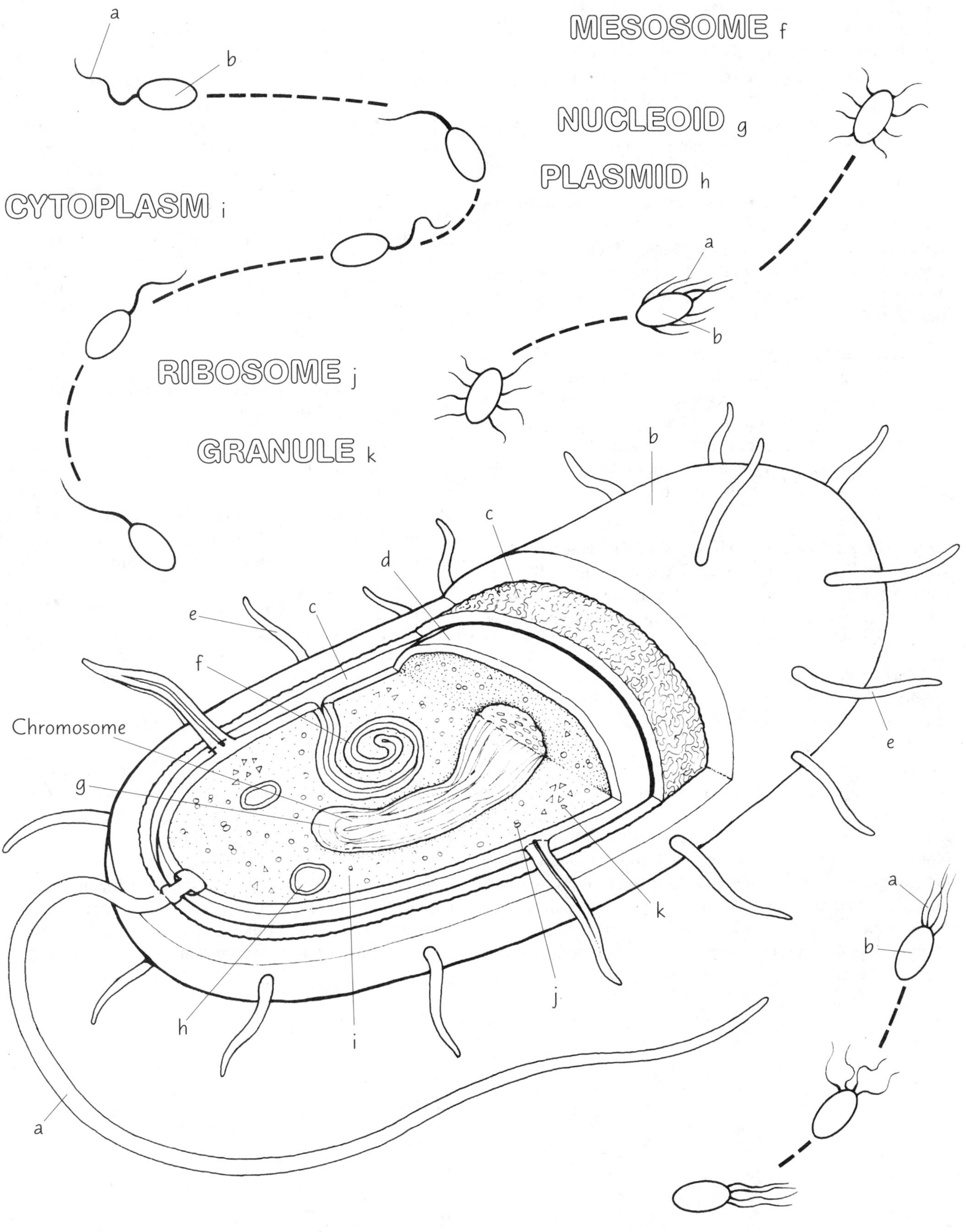

Chromosome

13
BACTERIAL CELL ENVELOPE

The coverings of a bacterial cell are collectively known as the cell envelope. Included in the cell envelope (from the outside moving inward) are the capsule (when present), the cell wall, and the cell membrane. One or more of these structures protect against cell rupture, control transit through the envelope, and provide markers (antigens) for antibody attachment.

Reserve three distinctly separate color groups for (a) through (a²), (b) through (b⁵), and (c) through (c⁴). Note the inset of a bacterium at upper right; color the designated parts of its cell coverings (a), (b), (c) and the related titles. Then color the titles (a¹) and (a²), and the related parts of the capsule.

The outermost layer of the cell envelope of certain bacteria is the capsule (a). The capsule consists of layers of polysaccharides (a¹) interwoven with smaller fibrils of protein (a²). The capsule is both a protective layer and a mechanism for attaching the bacterium to other surfaces. For example, in bacterial species residing on the surface of teeth (enamel), the capsule forms a portion of the "plaque."

Color the subheadings, Gram-Negative Bacteria and Gram-Positive Bacteria, gray. Then color the titles (b¹) through (b⁵) and related structures in the two cell walls. Also color the portion of cell membrane which is illustrated for orientation.

Most bacteria have a cell wall (b). The cell wall of Gram-negative bacteria shows more variety in its component structures (b¹) through (b⁴) than those of Gram-positive bacteria (b¹) and (b⁵). Peptidoglycan (b¹), a complex protein-carbohydrate molecule, is largely responsible for the structural integrity of the cell wall in most bacteria. The cross-linked layers (not shown) of this molecule resist destruction and bursting in the presence of high intracellular osmotic pressures. Note the relatively large amount of peptidoglycan in Gram-positive bacteria in comparison to that of Gram-negative bacteria. External to the relatively thin peptidoglycan layer of Gram-negative bacteria is a layer of lipoprotein (b²). The outer membrane (b³) of the cell wall consists of phospholipids (also seen in the cell membrane) externally attached to lipo-polysaccharide (b⁴; abbreviated LPS). These layers contribute to the control of diffusion through the cell envelope, inhibiting its permeability to chemicals from the external environment, e.g., antibiotics. Lipopolysaccharide may also influence cell membrane protein activity. The LPS of many Gram-negative bacteria are known to be toxic (endotoxins) to humans and other hosts. A well known variety of food poisoning is caused by the LPS of the *Salmonella* cell envelope. In Gram-positive bacteria, internal to the thick peptidoglycan layer, molecules of teichoic acid (b⁵) attach to glycolipid molecules (not shown) on the outer surface of the cell membrane. Teichoic acid is an alcohol polymer concerned with ion transport across the cell envelope. It possesses receptors for antibodies that participate in immune responses.

Color the titles (c¹) through (c⁴), and related structures at the lower part of the plate.

The cell membrane (c; also called plasma membrane) is immediately adjacent to the cell cytoplasm. Chemically it is composed of about 60 percent protein and 40 percent lipid. Most of the lipid is phosphorus-containing phospholipid. The phospholipid molecules are arranged tail to tail in two parallel layers that constitute the phospholipid bilayer. Each phospholipid molecule contains a charged polar head (c¹) which is soluble in water, and an uncharged nonpolar tail (c²) which is insoluble in water. The cell membrane also contains molecules of protein arranged as globules. Some of these globular protein molecules exist at or near the inner or outer membrane surface (c³). Others penetrate the membrane (c⁴). The protein globules as well as the phospholipid molecules appear to move freely within the membrane.

The cell membrane fits the concept of a fluid mosaic model, largely because the protein globules assume different positions, as in a mosaic. The membrane serves as a selective (semipermeable) barrier for materials entering and exiting the cell: materials that normally dissolve in lipid dissolve in the phospholipid bilayer and pass through the membrane, while amino acids and nitrogenous bases, insoluble in lipid, move through passageways in the protein globules. Tiny pores in the membrane also allow certain substances to pass through.

BACTERIAL CELL ENVELOPE

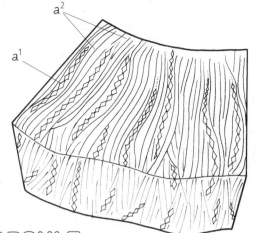

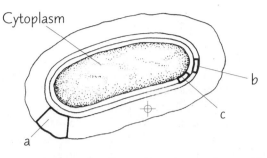

Cytoplasm

Bacterial cell envelope layers (magnified)

CAPSULE a

POLYSACCHARIDE a¹

PROTEIN a²

CELL WALL b

PEPTIDOGLYCAN b¹

LIPOPROTEIN b²

OUTER MEMBRANE b³

LIPOPOLYSACCHARIDE b⁴

TEICHOIC ACID b⁵

CELL MEMBRANE c

PHOSPHOLIPID: POLAR HEAD c¹

PHOSPHOLIPID: NON-POLAR TAIL c²

SURFACE PROTEIN GLOBULE c³

PENETRATING PROTEIN GLOBULE c⁴

GRAM - NEGATIVE BACTERIA ✿

GRAM - POSITIVE BACTERIA ✿

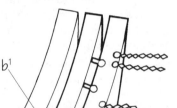

14
STAINING BACTERIA

The cytoplasm of a bacterium lacks color. For that reason, the cytoplasm in unstained bacteria is difficult to see under the light microscope. Therefore, microbiologists commonly stain bacteria to impart color before observing them. Several methods of staining are available, and three are presented here.

Begin by coloring the subheading Simple Stain Technique. Note that the unstained bacteria (a⊕) are not to be colored; however, upon exposure to the dye (b), they are stained the color of the dye (b¹). Use violet for (b) and (b¹). After dye (b) has been applied, note that the bacteria are stained but the background is colorless (for the purpose of distinguishing bacteria from background). Now color the subheading Negative Stain Technique at upper right. Use black for (c•). After dye (c•) has been applied, note that the bacteria remain unstained in a sea of black dye.

The simple stain technique is based on the principle that bacterial cytoplasm generally has a negatively charged cytoplasm which attracts a dye that carries a positive charge. To perform the technique a small amount of unstained bacteria (a⊕) is placed on a slide, dried, and heated. The slide is flooded with drops of a basic dye such as crystal violet (b). The bacterial cytoplasm attracts the dye, and the bacteria become stained (b¹). The stained bacteria are clearly visible under the light microscope.

The negative stain technique works in a manner opposite to that in the simple stain technique. Instead of using a positively charged dye, the technique employs a dye that has a negative charge. The dye is repelled by the negatively charged bacterial cytoplasm. To perform the technique, a small amount of unstained bacteria (a⊕) on a slide is flooded, by dropper, with an acidic dye such as nigrosin (c•). The bacteria remain unstained (a⊕) in a background of dye. The effect is much like seeing a bright moon against the dark sky of night. Since the negative stain involves no chemical reaction within the bacterial cytoplasm, the bacteria often appear less distorted than with the simple stain technique.

Now color the subheading Differential Gram Stain Technique. Note the slide with unstained bacteria (a⊕), one of which is marked positive (+) and the other negative (-). Start with title and dye (b), and color the titles and dyes/bacteria in the sequence indicated by the arrows. Note

carefully that which is to be colored and that which is not. Try to use the colors of the actual dyes used in the laboratory: (b), violet; (c), light brown; (d), blue; (e⊕), uncolored; (f), red. The background dye on the slides is left colorless so as not to obscure the stained bacteria. The sequence is finished with the microscopic view of the Gram-positive and Gram-negative bacteria.

The Gram stain technique is named for Christian Gram, a Danish physician who first devised it in 1884. It differentiates bacteria into two groups: Gram-positive bacteria that take up and retain the crystal violet-iodine complex (d), and Gram-negative bacteria that take up the complex but lose it on decolorization. The different reactivities are based on the different chemical makeup of the Gram-positive and Gram-negative organisms (Plate 13). Most species of bacteria can be differentiated in this way, with significant medical and industrial implications.

The technique begins much like the simple stain technique. Unstained bacteria (a⊕) are dried and heated on a slide and flooded with crystal violet (b), coloring them violet (b). The slide is then washed and flooded with a special iodine solution called Gram's iodine (c). The iodine reacts with the crystal violet and produces a crystal violet-iodine complex (d; here abbreviated CV-I) which appears blue in bacteria (d).

When the bacteria on the slide are washed with a decolorizing agent such as colorless 95% ethyl alcohol (e⊕), the Gram-positive bacteria retain their color (blue) but the crystal violet-iodine complex is washed out of the Gram-negative bacteria.

When the slide is flooded with a red dye, safranine (f), the Gram-negative bacteria pick up the stain (f). The Gram-positive bacteria (d), still stained with crystal violet-iodine complex, remain blue. Examining the slide under the microscope, the microbiologist may assume the blue bacteria are Gram-positive (d) and the red bacteria are Gram-negative (f).

Gram staining is often one of the first steps in identifying a bacterium. Moreover, Gram-positive bacteria are generally susceptible to certain antibiotics such as penicillin while Gram-negative bacteria are killed by other antibiotics. In addition, a particular disinfectant may work against a Gram-positive bacterium but not one that is Gram-negative. Finally, Gram-positive bacteria have certain structures not found in Gram-negative bacteria, and vice versa.

STAINING BACTERIA

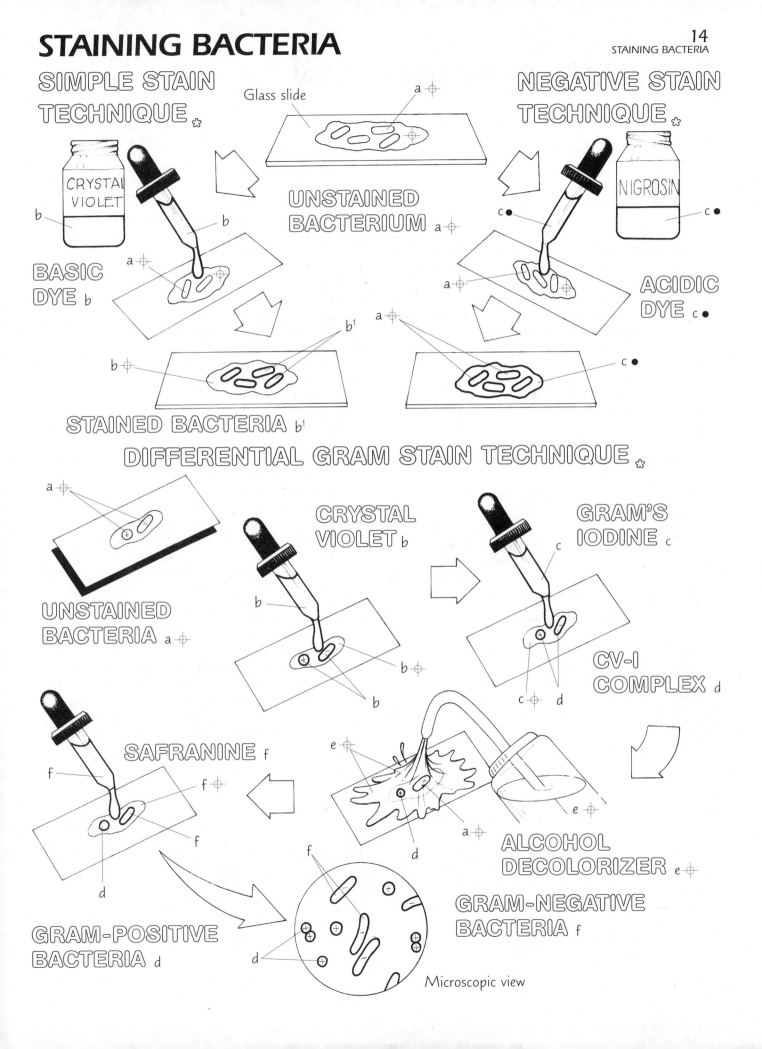

SIMPLE STAIN TECHNIQUE ✿

Glass slide

UNSTAINED BACTERIUM a ⊕

CRYSTAL VIOLET b

BASIC DYE b

STAINED BACTERIA b¹

NEGATIVE STAIN TECHNIQUE ✿

NIGROSIN c

ACIDIC DYE c ●

DIFFERENTIAL GRAM STAIN TECHNIQUE ✿

UNSTAINED BACTERIA a ⊕

CRYSTAL VIOLET b

GRAM'S IODINE c

CV-I COMPLEX d

SAFRANINE f

ALCOHOL DECOLORIZER e ⊕

GRAM-NEGATIVE BACTERIA f

GRAM-POSITIVE BACTERIA d

Microscopic view

15
BACTERIAL REPRODUCTION

In living things, reproduction can occur by two methods: sexual reproduction and asexual reproduction. When sexual reproduction takes place (not shown), parent organisms produce special cells called gametes, each containing one set of chromosomes. The gametes then come together and fuse, thereby reconstituting the two sets of chromosomes in a new cell. Then the cell multiplies until the organism resembles the parent. When asexual reproduction occurs, the chromosome and cell structure replicate. The cell then splits to yield two new cells, and each new cell acquires an equal amount of chromosomal and other cell material.

Asexual reproduction in bacteria is referred to as binary fission; one cell will split (fission) to yield two (binary) cells. The process, though relatively simple, involves several key steps that must take place in the proper sequence if the process is to be successful. In this plate, we shall examine the steps by which bacteria reproduce by binary fission.

Color the title (a), the single bacterium as seen under the microscope, and the arrow labeled (a). Color the titles (b) through (f), and (a[1]), and related structures, starting at the top of the plate and working down. Color the vertically oriented subheading Generation Time gray. As you color through a particular phase of the generation time, color the time period gray. Use a light pastel color for (e).

The process of binary fission begins with the parent cell (a), here taken from a culture of bacteria. The main structures to observe during binary fission are the cell wall (b), the cell membrane (c), the nucleoid that contains DNA (d), and the cytoplasm of the cell (e). Binary fission occurs after the cell has had a period of growth and metabolism.

The process of binary fission begins as the bacterium elongates slightly. While this is taking place, the DNA of the cell replicates (d[1]) to form twice the normal amount of DNA.

A bacterial cell normally has its DNA incorporated into a single chromosome in the form of a closed loop but with the replication of DNA, it now has two chromosomes. The two chromosomes contain identical molecules of DNA.

Protein synthetic activity (e[1]) increases early during binary fission to accommodate the need for added structure. Once the cell has elongated and the DNA replicated, the process continues with organelle replication (e[2]) and cytoplasmic separation into opposite halves of the dividing cell. The separated chromosomes move to either side of the cell. Note that at this point in the process, the cell membrane begins to pinch inward from opposite sides of the cell. In the period of a few minutes, the cell membrane continues to pinch the parent cell into two equal halves, each with an identical chromosome.

It is at about this point in binary fission that penicillin activity (f) can inhibit bacterial reproduction. Penicillin interferes with cell wall synthesis. A defective cell wall is incapable of structurally reinforcing the cell membrane during periods of increased internal pressures associated with metabolic activity. As a result, the cell becomes swollen and eventually bursts and dies.

The process of binary fission concludes with the separation of the two halves of the cell, forming daughter cells (a[1]). The daughter cells have identical DNA and each is chemically identical. With further metabolic activity, these new cells will become mature bacteria, each capable of binary fission. The period of growth and maturation of the bacterial cell will continue for several minutes. Binary fission will take place once again. The interval between the onset of one division and the beginning of the next is the generation time. Generation time is the time it takes the bacterial population to double. The generation time is unique for each bacterial species. For example, in the bacterium shown here, the generation time is a brief 61 minutes: 40 minutes for chromosome doubling followed by a 21 minute period of cell division.

BACTERIAL REPRODUCTION

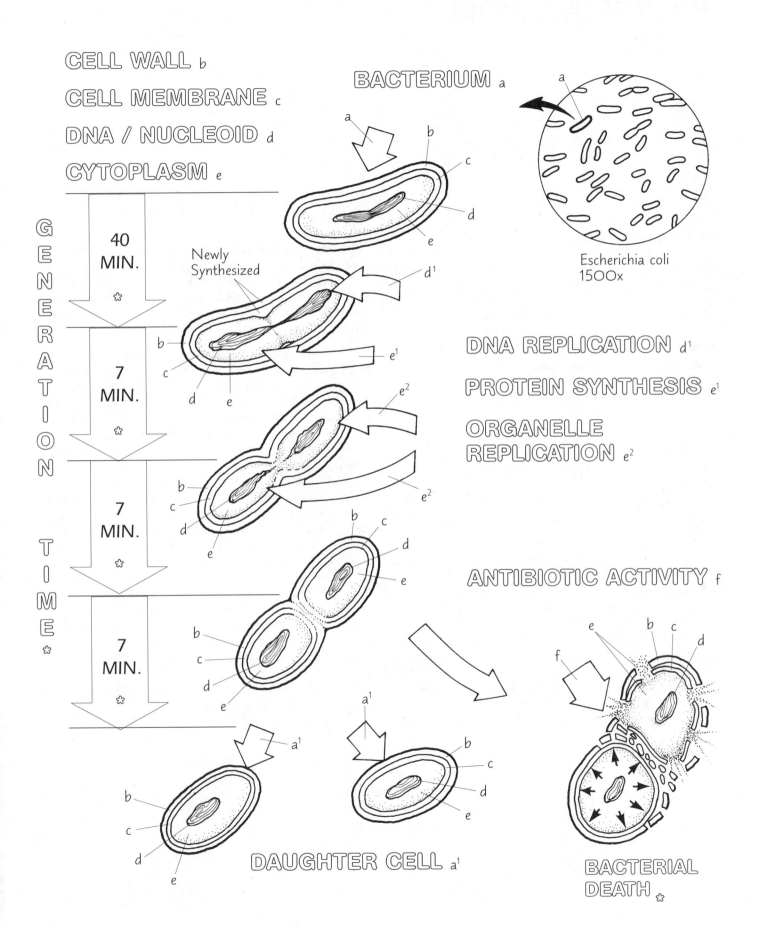

CELL WALL b

CELL MEMBRANE c

DNA / NUCLEOID d

CYTOPLASM e

BACTERIUM a

Escherichia coli
1500x

GENERATION TIME ✿

40 MIN. ✿

7 MIN. ✿

7 MIN. ✿

7 MIN. ✿

Newly Synthesized

DNA REPLICATION d¹

PROTEIN SYNTHESIS e¹

ORGANELLE REPLICATION e²

ANTIBIOTIC ACTIVITY f

DAUGHTER CELL a¹

BACTERIAL DEATH ✿

16
BACTERIAL SPORES

Most bacteria cannot survive extremes in environmental temperature, pressure, and chemistry. For instance, they cannot endure lengthy dehydration or nutritional deprivation. Certain Gram-positive bacteria, however, can do so. Members of the genera *Bacillus* and *Clostridium* have the ability to develop a highly resistant body that can survive such extremes. This body is called an endospore, or simply, spore. The spore is released from the parent bacterial cell on its death.

An endospore is remarkably resistant to severe changes in the immediate environment. For example, most bacteria die quickly when water reaches a temperature of 80° C, but spores can remain alive for up to two hours in water that is boiling (100° C). Bacterial spores have also been recovered alive from the intestines of Egyptian mummies! Bacterial spores carry their own food supply, enzymes, and DNA. When the environment becomes favorable, the spore can convert to a vegetative, reproducing cell. Until that occurs, the spore is dormant, i.e., it has no metabolic activity.

The right side of the plate is to be colored first. Color the subheading, Spore Structure, and the related titles and structures (a, a¹). Then color the titles (b) through (g) and the related structures in the electron microscopic view of the spore at lower right. Light colors for (b), (d), and (g) are recommended. Try to use contrasting colors for (b) through (g).

Spores are round to oval bodies existing free (a) in the environment or within the swollen bacterial cells that produced them (a¹). Note that only one spore forms per bacterial cell. Once their formation is complete, these spores will be freed to the environment as the parent cell disintegrates.

Seen under the transmission electron microscope, spore structure can be appreciated in detail. The core of the spore (b) consists of DNA and dehydrated cytoplasm taken from the parent cell. The core is surrounded by a core membrane or wall (c) taken from the ingrown bacterial cell membrane (spore septum). Outside of and adjacent to the core wall is the cortex

(d). This cortex, the thickest of the spore layers, is composed of an unusual loose arrangement of peptidoglycan. Surrounding the cortex, inner (e) and outer (f) membranes of protein make up the spore coat. This layer confers chemical resistance to the endospore. The outermost layer is the exosporium basal layer (g) composed of glyco- and lipoproteins.

Now color the subheading Sporulation at the upper left side of the plate. Color the titles and structures (b¹) and (b²) in the bacteria at stations 1 and 2, and then color the developing spore structures (b) through (g) in stations 3 through 6. The electron microscopic view is a magnified view of the spore seen in station 6.

The progressive development of the spore within the parent cell is called sporulation. The process begins when the bacterial DNA replicates and half of the DNA (b¹) gathers at one end with a small amount of cytoplasm (b²; station 1). The cell membrane turns inward (spore septum) to form a double membrane enclosing the DNA and cytoplasm (station 2). This double membrane becomes the core wall (c; station 3). Cortex material (d) develops within the core wall and thickens externally (station 3). As the spore forms, the parent cell swells at the spore end (station 4). The protein inner (e) and outer membranes (f) or spore coat appear (station 5), followed by the appearance of the exosporium basal layer (g; station 6). Spore formation is now complete. The cell membrane ruptures (station 6), and the spore is freed.

When the external environment is favorable, spores revert back to vegetative bacilli in a process called germination. In this process (not shown), the spore layers break down and the vegetative bacillus emerges. The bacillus soon undergoes binary fission. The daughter cells begin a new cycle of sporulation (sporulation cycle) when confronted with an unfavorable environment.

At least four serious diseases (anthrax, tetanus, botulism and gas gangrene) are known to be caused by spore formers. These diseases will be discussed in future plates.

BACTERIAL SPORES

SPORULATION ❀

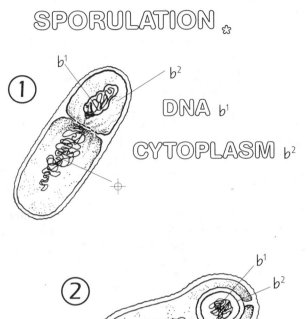

DNA b¹

CYTOPLASM b²

SPORE STRUCTURE ❀

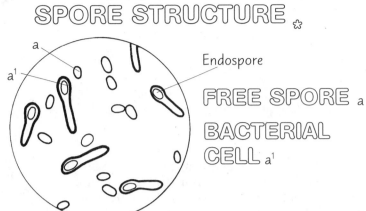

Endospore

FREE SPORE a

BACTERIAL CELL a¹

Light microscopic view

SPORE CORE b

CORE MEMBRANE / WALL c

CORTEX d

INNER MEMBRANE e

OUTER MEMBRANE f

EXOSPORIUM BASAL LAYER g

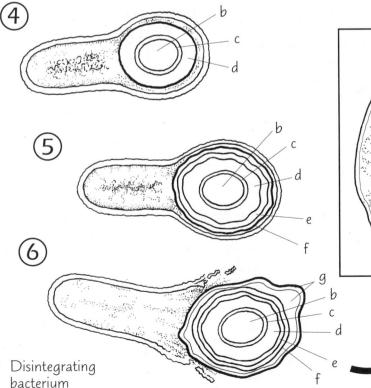

Disintegrating bacterium

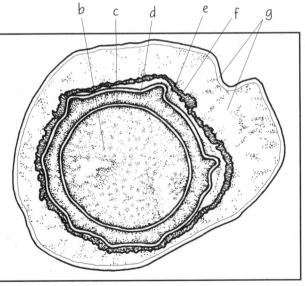

Electron microscopic view

17
BACTERIAL NUTRITION

Bacteria must meet certain nutritional requirements to grow and multiply. This plate is concerned with the means by which bacteria meet their nutrient needs.

Color the subheading Autotrophy at the top of the plate. Plan warm colors for (a) and (b), red for (c), and contrasting colors for (d) and (e). Color the titles and structures (a) through (d) in the upper half of the plate and then read below before going on.

Autotrophy ("self-feeding") is the mechanism by which organisms synthesize their own organic structure from inorganic molecules and carbon dioxide. The energy for such activities comes from the sun (b; solar energy). Energy for cellular work (c) can also come from enzyme-driven chemical reactions taking place in the bacterial cell. Indeed, the biosynthetic pathways in autotrophic organisms are among the most complex known to occur in life. Most autotrophic chemical reactions occur in the cytoplasmic water where enzymes both break down and form the bonds of substances. Start-up materials (a) for synthesis of organic compounds include minerals, such as sodium (a; Na^+), potassium (a; K^+), calcium (a; Ca^{+2}), and magnesium (a; Mg^{+2}). These minerals dissolve in the surrounding water and, along with ammonia (a; NH_3), either diffuse into the cell or are pumped in through a form of active transport. They contribute to membrane structure and are involved in maintaining osmotic equilibrium and control of water volume. More complex molecules, such as nitrates (a; NO_3^-), phosphates (a; PO_4^{-2}), and sulfates (a; SO_4^{-2}) are actively transported into the cell for the synthesis of amino acids (d) and nucleic acids (d). Autotrophic bacteria draw carbon dioxide gas (a; CO_2) from the air or surrounding fluid through the cell envelope into the cell cytoplasm to synthesize organic (carbon-containing) compounds, such as nucleic acids (d; DNA and RNA), amino acids (d; building blocks of protein), phospholipids (not shown; components of cell membranes), and carbohydrates (d). It is believed that autotrophy may have evolved among organisms when heterotrophs (see next section) had used up the more complex, preformed organic molecules in the environment.

Color the subheading Heterotrophy and related titles and structures in the lower half of the plate. Note that the titles (a) and (d) are repeated below, but title (c) is not.

Heterotrophy is the name given to the mechanism by which organisms absorb preformed organic substances (such as lipids, carbohydrates and proteins) and break them down for resynthesis during the formation of specific bacterial components. Such organisms include bacteria (saprobes) that feed on dead or decaying organic matter, such as a rotting logs, and bacteria that feed on living organic matter (parasites). By these means, bacteria ingest and metabolize carbohydrates (a) to obtain energy for life sustaining processes. Glucose (e) is one of the simple sugars resulting from the breakdown of complex carbohydrates. Larger carbohydrate molecules (d) are subsequently synthesized and contribute to cell structure. Lipids (a) are ingested, absorbed and broken down (c; cell energy) into fatty acids (e) for the subsequent synthesis of lipid (d; glycolipids, phospholipids, and others) used in cell structure. Various proteins (a) are ingested and broken down (c; cell energy) in the intracellular environment to release amino acids (e). The amino acids will be used in the synthesis of bacterial proteins (d) that function as enzymes and contribute to cell structure.

BACTERIAL NUTRITION

AUTOTROPHY *
START-UP MATERIALS a
SOLAR ENERGY b
CELL ENERGY c

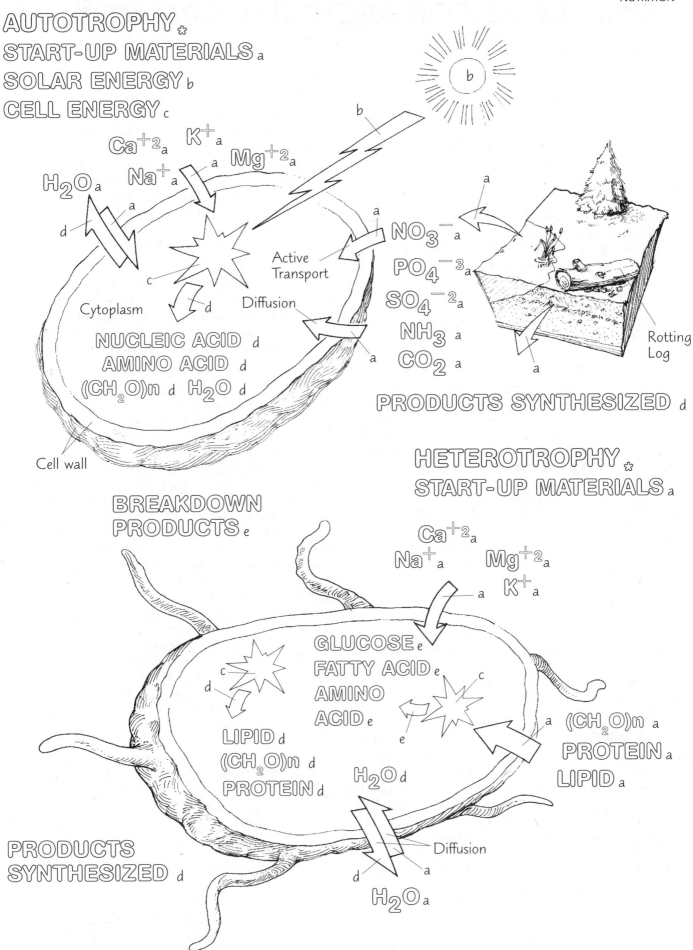

Ca^{+2} a K^+ a
Na^+ a Mg^{+2} a

H_2O a

b

Cytoplasm

Active
Transport

Diffusion

NUCLEIC ACID d
AMINO ACID d
$(CH_2O)n$ d H_2O d

Cell wall

NO_3^- a
PO_4^{-3} a
SO_4^{-2} a
NH_3 a
CO_2 a

Rotting
Log

PRODUCTS SYNTHESIZED d

HETEROTROPHY *
START-UP MATERIALS a

BREAKDOWN
PRODUCTS e

Ca^{+2} a
Na^+ a Mg^{+2} a
K^+ a

GLUCOSE e
FATTY ACID e
AMINO
ACID e

LIPID d
$(CH_2O)n$ d
PROTEIN d

H_2O d

$(CH_2O)n$ a
PROTEIN a
LIPID a

Diffusion

PRODUCTS
SYNTHESIZED d

H_2O a

18
CONDITIONS FOR BACTERIAL GROWTH

Bacteria can grow under a variety of conditions. Depending on the species, bacteria may grow within a wide temperature range, at different pH levels, and at different oxygen levels. These variations permit bacteria to colonize myriad environments.

Start by coloring the subheading Oxygen Level at the upper section of the plate gray. Use dark and light shades of the color used for a to represent high (a) and low (a^1) oxygen content, respectively. Color the titles (b) through (e) and the related bacteria in each tube.

The growth of many bacterial species depends on a plentiful supply of oxygen (a). For example, when bacteria are cultivated in a tube of nutrient agar, the oxygen-requiring bacteria will grow near the top of the tube to obtain oxygen from the atmosphere (high O$_2$ level, a) Microbiologists call these organisms aerobic bacteria (b; obligate aerobes). Other forms, called microaerophilic bacteria (c), require a lesser concentration of oxygen, and maintain a position in the agar some distance below the surface. Another form of bacteria grow either in the presence or the absence of oxygen. These are the facultative bacteria (d), and their colonies are distributed throughout the nutrient agar in the oxygen-rich (a) as well as the oxygen-poor (a^1) environments. Some species of bacteria must have an oxygen-free environment to survive. These bacteria are known as anaerobic bacteria (e). They will grow at the bottom of the tube where oxygen is lacking. Anaerobic bacteria may be so sensitive to oxygen that they die in an oxygen-rich environment. Such bacteria are often located in the mud of swamps and deep within landfills where they produce putrid odors and foul-smelling gases.

In the middle section of the plate, color the subheading pH gray, then color the titles (f), (g) and (h), the related pH levels, and the examples of the different pH environments in which the bacteria can live (f^1) through (h^1).

The term pH refers to the measure of acidity or alkalinity of a solution; a pH of 0 is the most acidic, 14 is the most alkaline, and 7 is neutral. Most bacteria grow best in neutral pH environments, but certain species of bacteria tolerate acid conditions (f) at about pH 5.0. These bacteria commonly inhabit food and dairy products such as cheese and yogurt (f^1). In the stomach of humans (not shown), the pH can be 3.0 or less. Most bacteria have a cytoplasm pH of about 6.5 to 7.5, so survival of such bacteria in the stomach is not likely. However, they will grow in an environment that is chemically neutral (g). One human tissue that offers a generally neutral substrate for bacterial growth is blood (g^1; pH 7.3 - 7.4). Some bacteria exist in harsh alkaline conditions (h), such as alkali lakes (h^1). Here the pH can reach 9 or more.

In the lower section of the plate, color the subheading Temperature gray, then color the indicated parts of the temperature bar, the designations of related bacteria (i) through (k), and the examples of the environments where the temperature variations are found (i^1) through(k^1).

Different species of bacteria have different temperature requirements for the best growth. Some bacteria are cryophilic bacteria (i), growing in cold temperatures that range between 5°C and 20°C. Such temperatures are found in the depths of the oceans and in regions of the Arctic and Antarctic where no other life forms can survive. Cryophilic (psychrophilic) bacteria may also grow in food held in the refrigerator (i^1) where the normal temperature is about 5°C. Here the bacteria may grow and spoil foods and, on occasion, produce toxins that lead to food poisoning. Most bacterial species grow at middle temperature range of 20° to 40°C. Microbiologists call these forms mesophilic bacteria (j). Since the temperature of the human body (j^1) is generally 37°C, the mesophilic bacteria can grow and multiply within many of its fluids and tissues. In the laboratory, many bacterial cultures are incubated at 37°C to provide the proper temperature for growth. Some forms of bacteria multiply at the high temperature range of 40°C to 90°C. These are thermophilic bacteria (k). Since pasteurized milk has been treated at either 63°C or 72°C, the pasteurization process has a negligible effect on thermophilic bacteria. However, the thermophilic bacteria will not infect humans because the 37°C temperature of the body is too cool. Thermophilic bacteria have been isolated from hot water vents in the ocean floor and from hot springs in the western United States (k^1) where the temperature is over 80°C.

CONDITIONS FOR BACTERIAL GROWTH

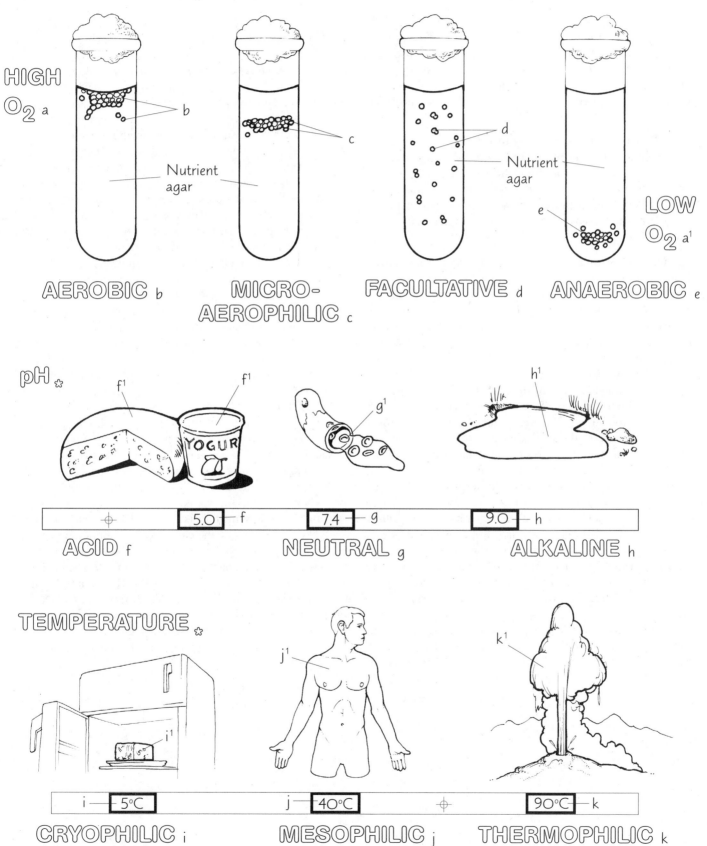

OXYGEN LEVEL ✿

HIGH O₂ a

LOW O₂ a¹

b

c

d

e

Nutrient agar

Nutrient agar

AEROBIC b   MICRO-AEROPHILIC c   FACULTATIVE d   ANAEROBIC e

pH ✿

f¹   f¹   g¹   h¹

YOGUR

ACID f   NEUTRAL g   ALKALINE h

5.0 — f   7.4 — g   9.0 — h

TEMPERATURE ✿

i¹   j¹   k¹

CRYOPHILIC i   MESOPHILIC j   THERMOPHILIC k

i — 5°C   j — 40°C   90°C — k

19
BACTERIAL CULTIVATION

Bacteria can be grown under laboratory conditions for a variety of purposes. In such cases, the cultivated bacteria constitute a culture, and the medium in which they are grown is called a culture medium. In this plate, we present routine and special culture techniques employed in the microbiology laboratory.

Note that five different means of culturing bacteria are represented here and numbered from 1 through 5. Start with station 1 and color titles and materials (a) through (b¹). Then go to station 2 and so on through station 5. Use a light color for large areas, such as (a) and (c); use a darker color for small areas such as (d). Use contrasting colors for (i) through (m).

One kind of liquid medium in which bacteria can be grown is nutrient broth (a; station 1). This medium contains water, beef extract, and peptone, a protein from plant sources such as soy beans. The mixture is sterilized to remove all other microbial contaminants, and the specimen (b; the bacteria to be cultured) is added to the sterilized broth. As the bacteria multiply in the nutrient broth, the broth becomes cloudy with bacterial growth (b¹).

Another important culture medium is nutrient agar (station 2). This medium contains all the components of nutrient broth plus a solidifying agent called agar. Agar is a complex polysaccharide derived from marine plants. It adds no nutrients to the medium and only serves to make the medium solid so that bacteria can be cultivated on its surface. In the laboratory, liquid nutrient agar (c) is prepared and sterilized, then poured carefully into a Petri dish (d), being careful to avoid airborne contamination. The Petri dish (named after its inventor, Julius Petri) is a two-sided plastic or glass dish constructed so that one side fits over and seals the other side. The nutrient agar solidifies in the dish, and the specimen containing the bacteria is inoculated to the surface of the nutrient agar. After several hours, the bacteria multiply sufficiently to form visible colonies (b¹).

Some bacteria are fastidious and require special growth supplements, such as blood or amino acids (not shown). Some bacteria do not grow in culture media at all. These bacteria must be cultivated in live animals, fertilized eggs, or cultures of living cells (e; station 3). An example is the organism of leprosy, a bacterium named *Myobacterium leprae*. This organism must be cultured in the body of an armadillo (not shown). Another example is the rickettsiae which cause typhus. These tiny bacteria (g) will only grow in a fertilized egg or embryo (in the case shown here, the medium is the yolk sac (f) of an early developing chick). Cultivating bacteria in this manner requires an extra measure of laboratory expense, but at present there is no alternative.

Anaerobic bacteria pose special problems in the laboratory because they are often killed by exposure to oxygen. Therefore, microbiologists have devised certain techniques to eliminate the oxygen in the environment. One such technique is the anaerobic container (station 4). Anaerobic bacteria (h) are inoculated on to a culture medium in Petri dishes. A packet of chemicals containing sodium bicarbonate is moistened with water and placed in the jar. The reaction between the water and chemicals produces carbon dioxide and hydrogen. The hydrogen reacts with oxygen in the air, stimulated by a catalyst of palladium pellets located near the top of the container. The oxygen in the container is removed by the water-forming reaction. As carbon dioxide (CO_2) fills the closed container, an anaerobic environment is established. An anaerobic indicator on one of the chemical packets will change color (from blue to colorless) when an oxygen-free environment exists. The high CO_2 concentration also aids the growth of many pathogenic bacteria such as those that cause gonorrhea.

Certain bacteria can be cultivated in a medium where the exact chemical composition is known. Such a medium is called a chemically defined medium (station 5). Using such a medium is advantageous because the chemistry of the microorganism can be studied closely. An example of a chemically defined medium is the one available for *Escherichia coli*. The medium contains glucose (i), ammonium phosphate (j), sodium chloride (k), magnesium sulfate (l), potassium sulfate (m), and water. Such a medium provides an energy source, a nitrogen source for protein, and necessary salts. This medium illustrates the relatively simple growth requirements of many bacteria.

BACTERIAL CULTIVATION

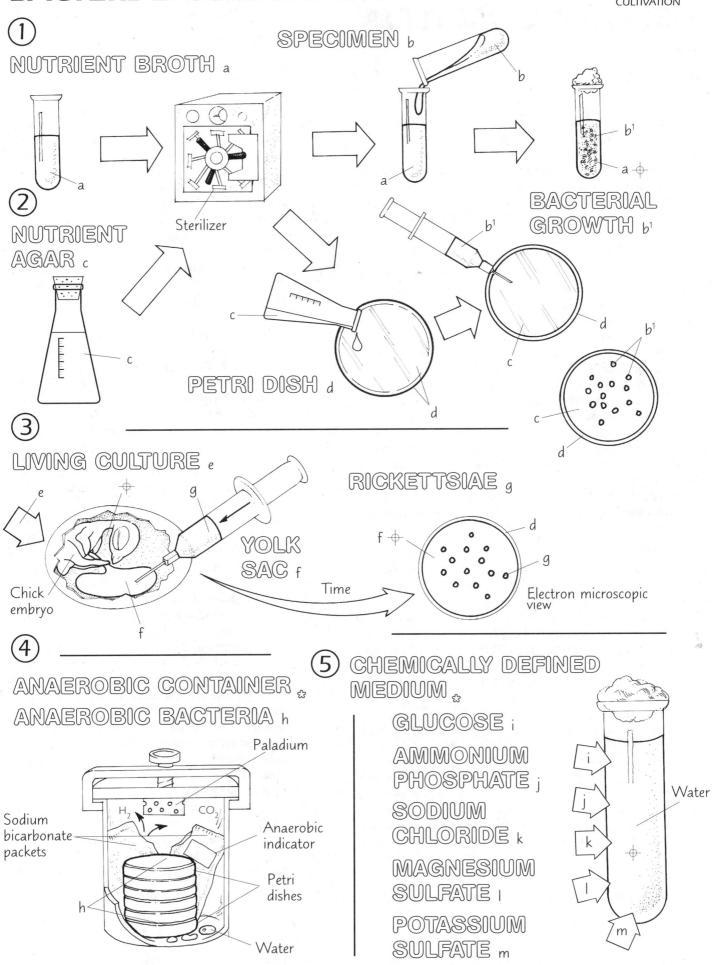

① NUTRIENT BROTH a

SPECIMEN b

b

a

a

b¹

a ⊕

BACTERIAL GROWTH b¹

Sterilizer

② NUTRIENT AGAR c

c

c

b¹

c

d

PETRI DISH d

c

d

c

b¹

d

③ LIVING CULTURE e

RICKETTSIAE g

e

g

d

f ⊕

g

Chick embryo

YOLK SAC f

Time

f

Electron microscopic view

④ ANAEROBIC CONTAINER ✿
ANAEROBIC BACTERIA h

Paladium

Sodium bicarbonate packets

H₂ CO₂

Anaerobic indicator

Petri dishes

h

Water

⑤ CHEMICALLY DEFINED MEDIUM ✿

GLUCOSE i

AMMONIUM PHOSPHATE j

SODIUM CHLORIDE k

MAGNESIUM SULFATE l

POTASSIUM SULFATE m

i

j

k

l

m

Water

20
ISOLATING BACTERIA

Most samples of material, such as a soil suspension, urine sample, or food mixture in fluid, contain several types of bacteria. Such an array of bacteria is referred to as a mixed culture. When a microbiologist wishes to study a particular bacterium in a mixture, it is necessary to isolate the different types. Once isolated, the bacteria can be cultivated in a pure culture. Several methods are available to isolate various bacteria.

Note that a tube of mixed bacteria (a) is being treated by three methods (labeled 1, 2, and 3). Methods are arranged vertically. Color the title (a) and related mixed bacteria. Primary colors for (a) and (b) are recommended, as is the appropriate mixture color for (c), e.g., red for (a) and yellow for (b) makes orange for (c). Color the subheadings gray for each of the methods. Start with the pour plate method (1). Note the title (b¹) in the center of the plate. An identical or similar shade of the color used for (b) is recommended for (b¹). Assume there are three bacterial types in sample (a); when coloring the bacterial colonies (d) at the bottom of the plate, use three contrasting colors to differentiate one colony from another. This applies to each of the three methods shown. It is recommended that you color the solid nutrient agar in each of the dishes first, then overcolor with the colors selected for the bacterial colonies.

One method of isolating bacteria is the pour plate method. To perform this procedure, a sample of the mixed bacteria (a) is transferred to a tube containing melted and cooled (about 50° C) nutrient agar (b). The sample and nutrient agar are mixed thoroughly to form the inoculated agar (c). The agar is poured into a sterile Petri dish, being careful to avoid contamination from the air. The Petri dish is swirled to effect even distribution (here seen from above), and then the agar is allowed to solidify. Next, the plate is placed upside down in an incubator (not shown) and allowed to remain for a period of time (usually 48 hours). During this time individual bacteria grow into their respective colonies (d) within the agar and on its surface. When the agar is observed, these bacterial colonies will usually exhibit different colors, shapes, sizes, and so forth. The microbiologist may then sample the colony desired with a sterile inoculating needle (see needle at the bottom of the spread plate method) and obtain a sample of bacteria for subsequent cultivation in tubes of sterile nutrient agar. In this manner, the various bacteria in the original mixture can be isolated.

Now color the materials in the spread plate method.

In the spread plate procedure, sterile, melted nutrient agar is poured into a Petri dish and allowed to solidify to form solid nutrient agar (b¹). A sample of mixed bacteria (a) is placed on the plate, and a sterilized bent glass rod is used to spread the bacteria evenly over the surface of the plate. The plate is then allowed to incubate at a certain temperature for a specified period of time. Bacterial colonies (d; here differentiated by colors you selected in the previous method) soon appear on the surface of the agar. They appear on the surface only, since this is where they have been placed. The microbiologist may then pick off samples with a sterile inoculating needle and continue the isolation by inoculating fresh nutrient agar.

Now color the materials in the streak plate method to the right. First, color the solid nutrient agar (b¹) in each of the three Petri dishes. Follow/color the streaks made by the inoculating loop with the color used for (a). Streak in one direction, then in the perpendicular direction, then back to the original direction on the other side of the dish. Color the bacterial colonies (d) three different colors as they appear randomly in the streaks at lower right.

In the streak plate method an inoculating loop is used. The loop is a sterile wire wound to form a loop at the end. It is placed into the mixed bacteria (a) and retrieved. The loopful of bacteria is then spread on a plate of solid nutrient agar (b¹) using the pattern shown in the illustration. The inoculating loop is flame sterilized after streaking the first section. The plate is then placed in the incubator as before, using the specified temperature and time at which the bacteria will grow best. At the conclusion of the incubation period, bacterial colonies will appear on the surface of the agar. The place where the first streaks are made contains much mixed bacteria, so much in fact that the colonies usually run together. Where the second set of streaks are made, the amount of mixed bacteria is less, since the loop was sterilized. Distinct colonies are visible because the bacteria have more room to grow. Where the third set of streaks is made, the fewer bacterial colonies are more separated and distinct. The microbiologist can then go to this area and secure samples of the bacteria desired for further study and isolation.

Each of the three isolation methods has its own advantages that suit a particular purpose. For example, the pour plate method is suitable for bacteria that grow in the absence as well as the presence of oxygen. In the spread plate method, fewer materials are needed and the procedure is more rapid, but poor separation may take place. Few materials are needed in the streak plate method and the prospects of isolation are good.

ISOLATING BACTERIA

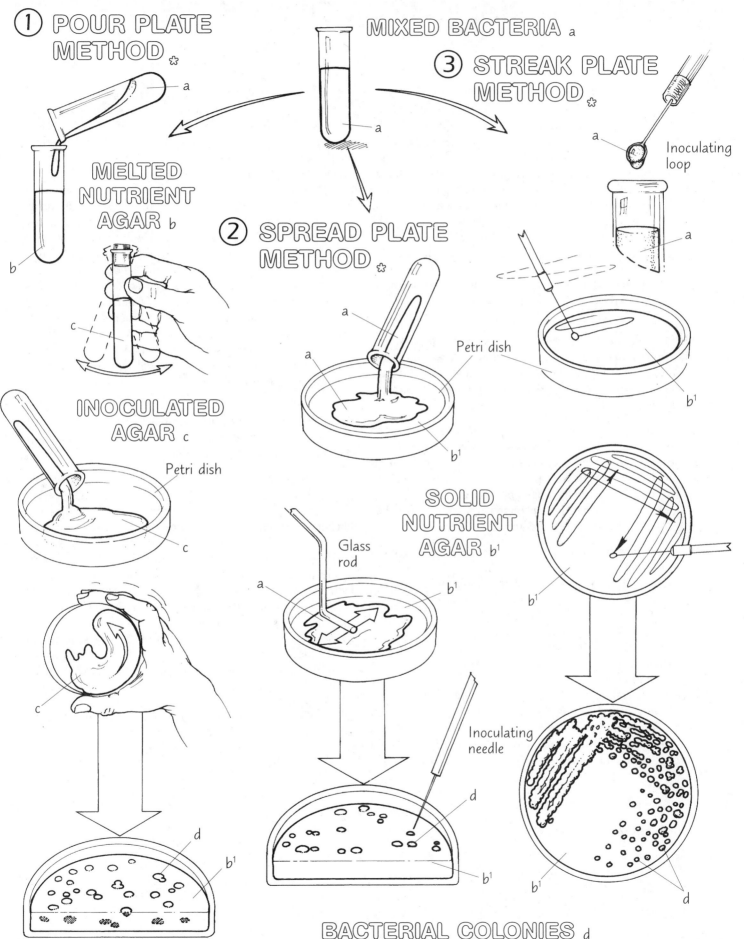

① POUR PLATE METHOD ✿

MIXED BACTERIA a

MELTED NUTRIENT AGAR b

INOCULATED AGAR c

Petri dish

② SPREAD PLATE METHOD ✿

Petri dish

Glass rod

Inoculating needle

SOLID NUTRIENT AGAR b¹

③ STREAK PLATE METHOD ✿

Inoculating loop

BACTERIAL COLONIES d

21
THE BACTERIAL CHROMOSOME

The bacterial chromosome contains genetic information in the form of deoxyribonucleic acid (DNA). DNA directs the synthesis of most of the protein in the organism. Other bacterial structures containing much smaller amounts of DNA include the plasmids. The chromosome consists of a single molecule of DNA. Unlike the DNA in eukaryotic organisms, there is no protein bound to DNA in bacteria and other prokaryotes. The bacterial chromosome is a multi-folded structure in the cytoplasm. It is not membrane bound. It occupies about ten percent of the total volume of a bacterium, and is about 1 mm long when fully extended, a length about 1000 times that of the cell. In this plate, the structure of the chromosome is examined and the chemistry of its DNA is explored.

This and the next three plates share some subscripts, and the titles/structures are color-coordinated, so plan your selection of colors carefully. Note that the subscripts are not in alphabetical order; the subscripts were selected to easily relate to the structures they identify.

Color the title (e), the chromosome coiled in the bacterium at upper left, and the uncoiled chromosome at left. Color the title (s) and related strands a shade of the color used for (e). Color gray the base pairs (b*) above and below the boxed area; color the hydrogen bonds (h) a neutral color. Color the perimeter (e¹) of the boxed area and its title, but wait to color inside the boxed area.

Each bacterium has a single chromosome (e) arranged as a closed loop. The entire chromosome consists of discrete segments of DNA called genes. These genes in prokaryotes are arranged end-to-end, like links of a chain. Each gene formulates the construction of one specific protein. A fraction of a gene (e¹) is enlarged here for coloring.

The precise molecular arrangement within each gene is inherited from the parent cells by a process of genetic recombination during bacterial reproduction. The sum of the inherited structural, functional, and behavioral characteristics of an organism, passed on from generation to generation, is the result of genes instructing cells to make specific proteins.

The total collection of genes in the bacterium (both chromosomal and extra-chromosomal) is called the genome.

The number of genes in an organism's genome varies in proportion to its overall complexity: some viruses have as few as seven genes; the chromosomes of the intestinal bacterium *Escherichia coli* are believed to have over 4000 genes; the 46 human chromosomes contain over 100,000 genes.

The DNA molecule is arranged in the form of a double helix or paired coiled chains in all forms of organisms possessing DNA. This pattern was first reported in 1951 by James D. Watson and Francis H.C. Crick. Each of the two coiled lengths of DNA is called a strand (s). The strands, often referred to as the "backbone" of the DNA molecule, are connected at right angles by nitrogen-containing molecules called bases (b*: their name is derived from the fact that they are alkaline in solution). The bases of each strand are held weakly to bases of the opposite strand by hydrogen bonds (h). Each pair of connecting bases is called a base pair. The construction of the strands and bases is similar to the sides and rungs of a twisted ladder.

Color the structures within the boxed area and the related titles listed center left. The molecules in the strands and bases are magnified for coloring.

Each strand consists of a continuous chain of two molecules: a five-carbon carbohydrate or sugar called deoxyribose (d) and a phosphate group (p). The phosphates bind the deoxyribose molecules together. The bases of the DNA molecule are four: adenine (a), cytosine (c), guanine (g), and thymine (t). Base pairs are bound by hydrogen bonds. The nature of these bonds determines the compatibility of base pairs, and only the pairing of adenine and thymine and the pairing of cytosine and guanine are permitted. These bonds are due to the attraction of oppositely charged areas resulting from the uneven distribution of electrons. It is the sequence of bases along each strand in the DNA molecule that forms the gene; it is the base sequence in each gene that represents a "code" for the synthesis of one protein.

DNA replication occurs just before cell division takes place, and can be explored in the following plate.

THE BACTERIAL CHROMOSOME

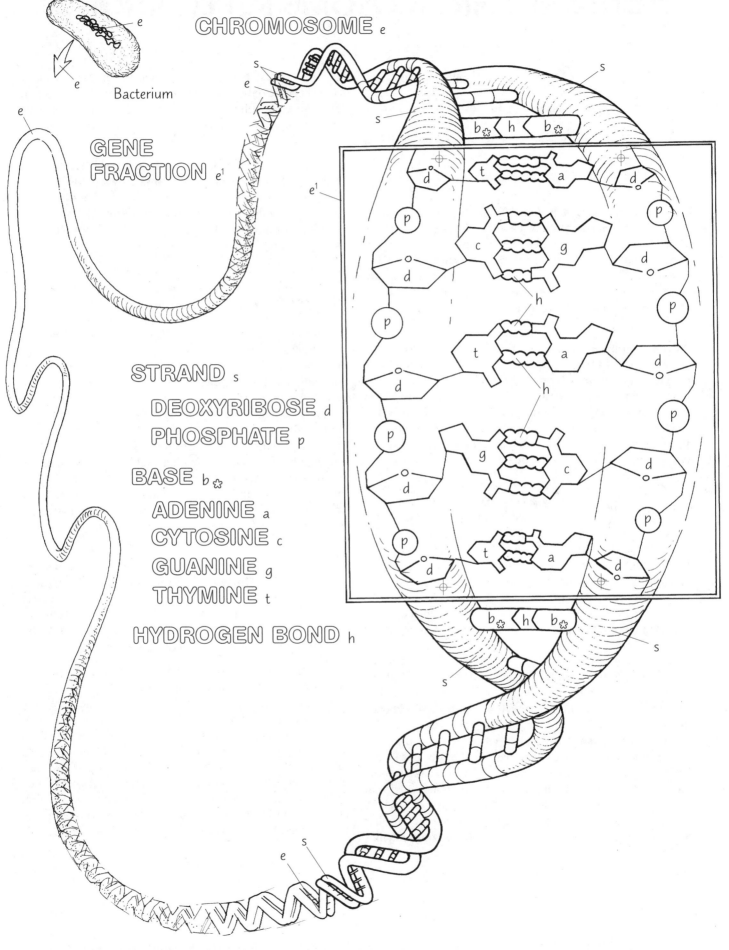

CHROMOSOME e

Bacterium

GENE
FRACTION e¹

STRAND s

DEOXYRIBOSE d

PHOSPHATE p

BASE b✿

ADENINE a
CYTOSINE c
GUANINE g
THYMINE t

HYDROGEN BOND h

BACTERIAL CHROMOSOME REPLICATION

This plate is the second of a sequence, beginning with Plate 21, on bacterial DNA.

When a bacterium undergoes reproduction, the chromosome replicates to form two daughter chromosomes with identical DNA. The replication process ensures that each new bacterium will receive a chromosome with exactly the same genes as the parent cell.

Select the colors used in Plate 21 for the corresponding titles/structures (with the same subscripts) on this plate. Color the title (e) and related parent chromosome, but not in the small bacteria at station 5. At station 1, color the two magnified parent strands (s) emerging from structure (e). Color the bases between the strands gray (b*); color the hydrogen bonds (h) there as well.

Near the upper separated strands, color the arrows (s). Between the upper separated strands, color the three enzyme symbols and their title (z) through (z²) the same color. Color the parts (d) and (p) of the parent strands and related bases (a), (c), (g), and (t). Then color the two free nucleotides (n) and their components. Color the parts of the developing strands just above stations 2 and 3. Do not color further until you read below.

The bacterial chromosome (e) forms a closed loop in the cytoplasm. It consists solely of DNA. The DNA takes the form of a pair of strands (s) intertwined with one another. This construction is called a double helix.

When the process of DNA replication begins, the closed loop opens at one point and the strands separate (station 1), much like a zipper opens or unzips. Note that the separation occurs between base pairs (b) of the strands; specifically, at the hydrogen bonds (h). Hydrolases (z) are responsible for breaking these bonds. The bases of each strand are exposed after the strands have been unzipped.

Repetitive nucleotides form the structure of each of the strands of DNA. A nucleotide consists of a dexoyribose molecule, a phosphate group, and one of four bases. The deoxyribose (d) and phosphate (p) groups form the "backbone" of the strand, and the nitrogenous bases (b) project out at right angles from the strands. In the intact DNA molecule, the bases of one strand are connected to the bases of the opposite strand; each pair of matched or complementary bases is called a base pair. The base pairs are separated after the strands unzip. In the cytoplasm of the bacteria are free (unattached) DNA nucleotides (n). These free nucleotides are bonded to each other and to the parent strand nucleotides by polymerizing enzymes

(z^2; polymerases). Note the shape given to each of the identified bases; because only congruent pairs can form a bond, adenine (a) is only compatible with thymine (t), and thymine only with adenine. Guanine (g) and cytosine (c) are also matched.

Color the arrow (j) pointing to the Okazaki fragment on the left, and its title. Color the arrows showing the direction of replication (i) for each strand. Color the strands (s) and (s¹), and related parts starting at stations 2 and 3 and working down. Color the replicated chromosomes (k) and (k¹) extending down from the strands.

As the nucleotides are aligned along side each parent strand, two new strands (s¹) emerge. The sites of origin of these emerging, replicating strands are called "replicating forks" (stations 2 and 3). The direction of replication (i) and the mechanism of replication is different for each of the two new strands. One strand's development (right side) leads the other (left side). This "lagging" condition on the left is compensated for by attaching free nucleotides in groups of 20 or so (Okasaki fragments, j) to their complementary bases on the parent strand. The nucleotides of the Okazaki fragment are joined by the enzyme DNA ligase (z^1; as in ligature).

As each replicating strand (s¹) takes form, note that it forms a double helix (k) and (k¹) with a strand (s) from the parent chromosome (station 4). Thus, each of the two offspring of the parent bacterium will carry a molecule of DNA in which one of the strands is not replicated, but is retained from the parent chromosome (semiconservative replication).

Color the chromosome (e; station 5) in the small bacterium at upper right. Moving down the right side of the plate, color the parent chromosome (e) and the progressive replacement of (e) with the daughter (replicated) chromosomes (k) and (k¹) in the dividing bacteria. Color the replicated chromosomes in the new (daughter) bacteria.

Each newly-formed double helix constitutes a replicated chromosome (k) and (k¹). Each of these replicated chromosomes progressively replaces the parent chromosome until binary fission is complete. This progression can be seen in the formation of daughter bacteria from a parent bacterium. Prior to DNA replication, the DNA is in the form of a closed loop (e). As replication proceeds, the loop opens and two new chromosomes begin development (k) and (k¹). As replication of the two new DNA molecules progresses, the parent chromosome is gradually replaced. At the completion of fission, each new cell has a closed loop of replicated DNA.

BACTERIAL CHROMOSOME REPLICATION

PARENT CHROMOSOME e
  PARENT STRAND s
  BASE b ✿
  HYDROGEN BOND h

ENZYME z-z²

NUCLEOTIDE n
  DEOXYRIBOSE d
  PHOSPHATE p
  BASE b ✿:
    ADENINE a
    CYTOSINE c
    GUANINE g
    THYMINE t

DIRECTION OF
REPLICATION i

OKAZAKI
FRAGMENT j

REPLICATED
STRAND s¹

REPLICATED
CHROMOSOME 1 k

REPLICATED
CHROMOSOME 2 k¹

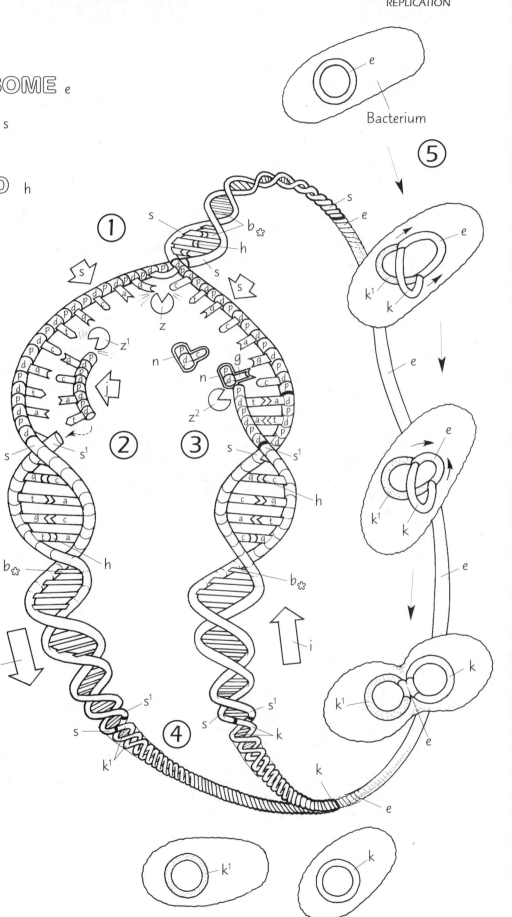

23
PROTEIN SYNTHESIS: TRANSCRIPTION

This plate is the third of a sequence of four plates on DNA structure, DNA replication, and protein synthesis.

The synthesis of protein is a major activity in bacteria and the cells of other organisms. There is protein in the construction of all cell membranes and organelles. Proteins are carriers of large organic molecules through the cell membrane. Proteins form enzymes that catalyze cellular reactions. In larger organisms, proteins contribute to the formation of antibodies and all cells, including blood cells and antibodies. Protein is required for all living things.

Protein synthesis begins with DNA. The overall process is generally divided into two subdivisions called transcription and translation. Transcription is a mechanism for the transfer (transcription) of genetic codes from DNA to RNA. The RNA molecule then carries the code to the ribosomes where the protein is put together (synthesized). This latter process is called translation and is covered in Plate 24.

Set aside the colors used in the previous two plates that will be used for the same structures in this plate. Color the chromosome (e) coiled in the bacterium at upper left and the short chromosome segments at upper left and below. Color the strands (s), but wait to color the segment of the right DNA strand contributing to the formation of the RNA strand. Color the bases (b) gray except for those bases that have subscripts (a), (c), (g), or (t) identifying them. Color the hydrogen bonds (h) except for those connecting RNA and DNA bases. Color the two hydrolase enzyme symbols (z) at the upper and lower parts of the separated strands, and their title.

The chromosome (e) of the bacterium is entirely DNA. The intact DNA molecule consists of two strands (s) connected by complementary base pairs (b) which are held together by weak hydrogen bonds (h). In a process similar to DNA replication, transcription begins with disruption of the hydrogen bonds by hydrolases (z) between the strands along a section of the DNA molecule. The strands separate, bases exposed. This process of detachment is called "unzipping." Whereas free DNA nucleotides formed two new strands of DNA, transcription involves the formation of *one* strand of RNA from *RNA* nucleotides.

Color the free RNA nucleotides (f) shown migrating from the bacterium at upper left toward the developing RNA strand (at arrow f¹) and their title. Color the RNA polymerase symbols (z¹) at each end of the developing RNA strand, and related title. Then color the stacked components of the RNA nucleotides (and related titles at left) forming the RNA strand as well as the arrow (f¹) and its title. Use a new color for uracil (u). Color the arrow (n) and the outline (n) of the DNA nucleotide at right. Color the components of the DNA nucleotides (and their titles).

Free nucleotides abound in the cytoplasm of prokaryotic organisms. There are two kinds: DNA and RNA. When these nucleotides become stacked and aligned in a certain sequence, they form the strands and bases of the DNA and RNA molecules. DNA nucleotides (n) are characterized by deoxyribose sugars (d) attached to phosphate groups (p). Their bases (b) include adenine (a), guanine (g), cytosine (c) and thymine (t). RNA nucleotides (f), on the other hand, consist of ribose sugars (r) attached to phosphate groups (p¹). Further, the base thymine does not exist in RNA nucleotides as it does in DNA nucleotides. Instead, RNA has the base uracil (u). Like thymine, uracil forms base pairs with adenine.

As you color the base pairs between the DNA strand and the developing RNA strand, note the complementarity of specific bases. The synthesis of the RNA molecule continues until well over 1000 nitrogenous bases have been slotted into position and joined together by RNA polymerase (z¹). Note that the RNA molecule being transcribed consists of one strand, unlike the double strand (helix) of DNA.

Color the three types of RNA boxed at the bottom of the plate and related titles. Use light colors to avoid obscuring the illustration detail.

Three kinds of RNA are synthesized in bacteria (as well as in eukaryotic cells). The RNA that carries the genetic code to the ribosomes is called messenger RNA (m; abbreviated mRNA). Messenger RNA provides a sequence of bases that represent the genetic code for synthesis of a specific protein. Its precise role in protein synthesis is presented in the following plate. The ribosome consists of RNA as well. It is abbreviated rRNA (l). The ribosome provides a framework for the mRNA and the protein synthetic process. Its role can be seen in the following plate. Finally, there is an RNA that captures amino acids and brings them to the mRNA at the ribosome. This molecule, called transfer RNA (o; abbreviated tRNA), can also be seen in the next plate.

PROTEIN SYNTHESIS: TRANSCRIPTION

CHROMOSOME e

DNA STRAND s

BASE b ✿

HYDROGEN BOND h

HYDROLASE z

RNA POLYMERASE z^1

DNA NUCLEOTIDE n

DEOXYRIBOSE d
PHOSPHATE p
BASE b ✿ : $A_a C_c G_g T_t$

RNA NUCLEOTIDE f

RIBOSE r
PHOSPHATE p^1
BASE ✿ : $A_a C_c G_g U_u$

(URACIL) u

RNA STRAND f^1

MESSENGER
RNA (mRNA) m

RIBOSOMAL
RNA (rRNA) l

TRANSFER
RNA (tRNA) o

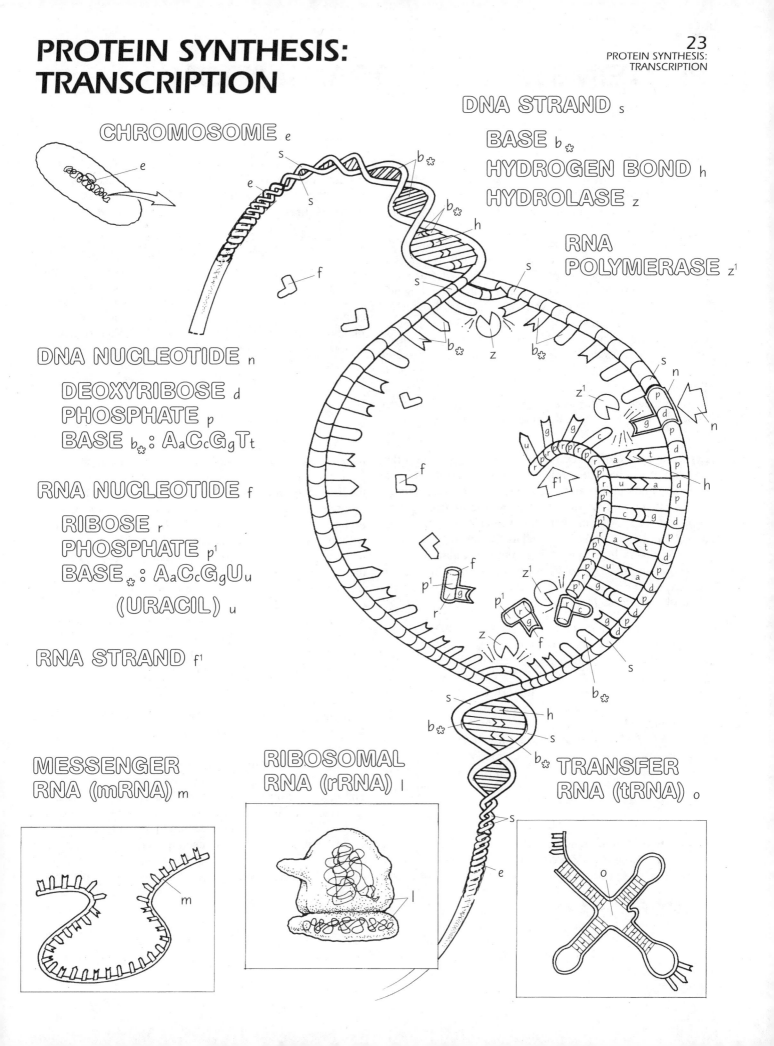

24
PROTEIN SYNTHESIS: TRANSLATION

This plate is the last part of a four plate, color-coordinated series, beginning with Plate 21, on DNA, DNA replication, and protein synthesis (transcription and translation).

Recall that DNA is composed of a double helix consisting of strands and base pairs (Plate 21). The DNA molecule replicates by unzipping the base pairs and forming two complementary strands (Plate 22). The DNA molecule starts the protein synthetic process by unzipping and providing one strand as a template for the synthesis of a messenger RNA molecule. This is called transcription (Plate 23). The newly-derived single strand of RNA carries in the sequence of its bases the genetic code for the synthesis of a specific protein. In the translation phase of protein synthesis, three types of RNA molecules participate in "translating" the genetic code (as represented by sequences of DNA nucleotides) into chains of amino acids that constitute protein molecules.

Set aside the colors used on the previous plate for titles and structures with the same subscript on this plate. Look over the plate and note there are six numbered stations. Begin with station 1. Color the titles and structures (e) and (m), and the arrow (m). Then at station 2 color the titles and structures (l) and (l¹), as well as the arrow (l) at upper left. Use a very light color for (l) and (l¹). Color all of the bases on the mRNA molecule (m) the same color as the strand (m) except those that are identified by a subscript. Wait to color these identified bases.

All of the structures necessary for protein synthesis come from the cytoplasm and the chromosome (e) of the bacterium. Messenger RNA (m; mRNA) is synthesized according to the genetic code in the chromosomal DNA. The mRNA moves away from the chromosome (station 1) toward a ribosome (l). The ribosome is composed of ribosomal RNA (rRNA) and some protein. There are about 30,000 ribosomes in a single bacterium. Each ribosome consists of a large and a small subunit (station 2). In the synthetic process, these subunits embrace the mRNA molecule and grip it. On each of the large subunits are two binding sites (l¹) for the attachment of transfer RNA molecules.

Color the titles and structures at stations 3, 4, 5, and 6 in that order. Note the subscripts of the bases of the tRNA and the mRNA; note that the base pairs that bind at the

binding site are complementary. The colors used for these bases should be the same as those used for the bases in the previous three plates. Color the arrows (o) and (q).

As many as 20 different kinds of amino acids (q; station 3) can be found in abundance throughout the cytoplasm of the bacterial cell. Chains of amino acids characterize proteins, from small peptides with a few amino acids to the larger proteins, such as enzymes, that may consist of tens of thousands of amino acids per molecule. The uniqueness of each protein is a function of the sequence of its amino acids. The sequencing of amino acids is the responsibility of messenger RNA.

Transfer RNA (o; tRNA) consists of short molecules (75-85 nucleotides) that roam the cytoplasm, collect amino acids, and bring them to the ribosomes. The ribosomes are the organelles where the bases of the mRNA bind the complementary bases of the tRNA.

As the tRNA moves through the cell, one end of the molecule binds (v) with a specific amino acid (q; station 4) guided by an enzyme called synthetase (z). At the other end of the tRNA molecule is a loop of three nucleotides. This triplet of three nucleotides is called an anticodon. Note the bases of each anticodon and their shape. These bases are complementary with the bases of three consecutive nucleotides in the mRNA molecule. Each triplet of nucleotides in the mRNA molecule that form base pairs with the anticodon is called a codon.

The tRNA molecule, bonded to an amino acid, migrates to a ribosome and attaches to the ribosomal binding site (station 5). Two tRNA molecules attach in close proximity at separate binding sites. The binding sites permit binding of complementary mRNA codons and tRNA anticodons. A bond forms between the pair of amino acids at the other end of the tRNA molecules. The amino acids thus become connected to the developing chain. The amino acid-tRNA bonds are disrupted enzymatically (z¹). The ribosomal subunits move along the mRNA strand to the next codon, and the process is repeated.

This process will continue until a terminator codon on the mRNA molecule signals the end of the chain formation. This chain of amino acids (station 6) constitutes a protein (q¹) of variable length, and it will move to assume its function in the bacterium (structural part, enzyme, product for export, and so on). This synthesis of a specific protein completed, the mRNA molecule will be dismantled by enzymatic action (not shown) and its nucleotides freed for future syntheses of RNA.

PROTEIN SYNTHESIS: TRANSLATION

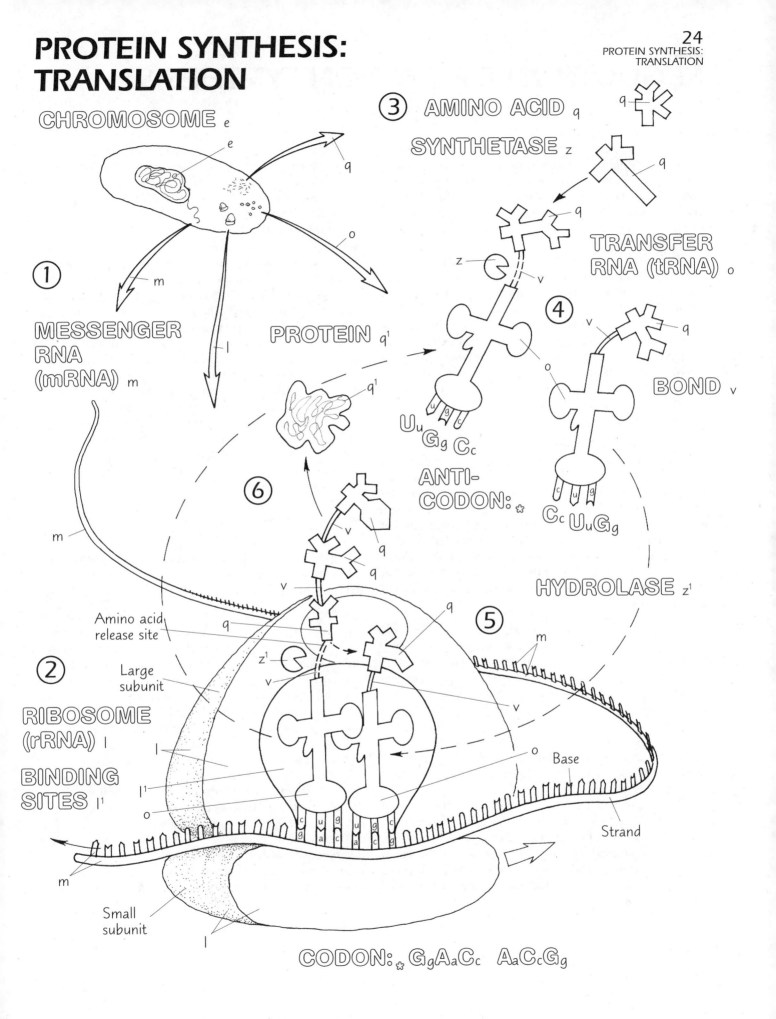

CHROMOSOME e

③ AMINO ACID q
SYNTHETASE z

TRANSFER
RNA (tRNA) o

①

④

MESSENGER
RNA
(mRNA) m

PROTEIN q¹

BOND v

$U_u G_g C_c$
ANTI-
CODON: ✿

HYDROLASE z¹

Amino acid
release site

②

Large
subunit

RIBOSOME
(rRNA) l

BINDING
SITES l¹

Small
subunit

⑤

⑥

Base

Strand

CODON: ✿ $G_g A_a C_c$ $A_a C_c G_g$

$C_c U_u G_g$

REGULATION OF PROTEIN SYNTHESIS

Selected sequences of nucleotides of DNA, responsible for coding the synthesis of proteins, are called genes. Bacterial genes direct protein synthesis through the processes of transcription and translation (Plates 23 and 24). However, such syntheses do not all occur simultaneously. Mechanisms exist that regulate activation and inactivation of DNA transcription of gene sequences into messenger RNA (mRNA). This makes possible control of the synthesis of enzymes. The presence or absence of enzymes can initiate or terminate protein production, making it possible, for example, to turn off a metabolic pathway as the last product has been synthesized. By controlling when metabolic activity occurs, the cell can save a great deal of energy. Two different mechanisms of induction and repression of DNA transcription (synthesis of mRNA) have been identified in *Escherichia coli*, the colon bacillus. Here we show these mechanisms influencing the synthesis of enzymes.

Color the titles and related parts (a) through (d) of the operon in the boxed bacterial chromosome. Use light colors to avoid obscuring the detail of the illustrations. Color the titles and structures at station 1. Start with (b). Note OG OFF is not colored. Note that certain genes in the strand receive no color. For (c^1) and (d^1), use light, pale shades of the colors used for (c) and (d). Color the subheading Induction of Enzyme Synthesis: OG ON and the related titles and parts (i) and (j) of the adjacent DNA strand at station 2. Do not color the illustration in the lower third of the plate for now.

Each of the two mechanisms regulating enzyme synthesis stems from a group of genes called an operon (a). An operon includes three kinds of functional genes. Regulator genes (b) control the activity of the operator genes by coding for a protein called a repressor protein. This repressor protein binds with operator genes. Operator genes (c) stimulate or depress (turn "on," turn "off") the activity of structural genes. If bound by a repressor protein, the operator gene is "turned off," preventing the structural genes from initiating synthesis of enzymes. Structural genes (d) code for a sequence of amino acids that characterize each enzyme or other protein.

There may be a number of different structural genes in each operon (here we show two). Although the operator gene is usually adjacent to the structural genes of the same operon, the regulator gene may be some distance away from the genes of the same operon. The operon constitutes a unit of genes that controls transcription and, ultimately, the stimulation or inhibition of enzyme synthesis.

The first mechanism controlling transcription concerns the induction operon. The induction operon is a group of genes that induces transcription of mRNA by means of an inducer molecule. In the absence of such a molecule, transcription of a specific mRNA molecule cannot occur. The operator gene is turned off (OG off).

Such is shown at station 1: the regulator gene (b) codes for a strand of mRNA (f) that, in concert with a ribosome (g) and an enzyme (e), results in the synthesis of a repressor protein (h). This protein binds with the operator gene (c) of the operon, inactivating it (c^1; OG off); this inactivates the structural genes (d^1) from coding for the synthesis of a particular enzyme.

Induction of enzyme synthesis is shown at station 2. The regulator gene codes for the synthesis of a repressor protein as before, but an inducer molecule (i) attaches to the repressor protein, changing its structure. Now the altered repressor protein cannot interact with the operator gene. Unaffected then, the operator gene stimulates the structural genes to synthesize mRNA resulting in the synthesis of a certain enzyme (j). One example of an inducer substance is allolactose, a conversion molecule from lactose (milk sugar). In lactose metabolism in *E. coli*, allolactose binds with repressor proteins, denaturing them. This permits the synthesis of enzymes necessary for lactose metabolism to take place.

Color the subheading Repression of Enzyme Synthesis and the related titles and structures at station 3. Work from left to right.

The second mechanism of regulating transcription involves a repression operon. Like an induction operon, it consists of structural (b), operator (c), and regulator genes (d). Unlike induction operons, the repression operon inactivates genes that normally code for enzyme synthesis. Similar to the induction operon mechanism, a structural gene (d) codes for an mRNA strand that results in the production of a repressor protein (h). However, the repressor protein does not bind to the operator gene directly. It first must be bound by a co-repressor protein (k). The repressor protein is altered (h^1; enhanced) sufficiently to bind with the operator gene (c). In turn, the operator gene "turns off" the structural genes (d^1), preventing the synthesis of certain strands of mRNA.

Although the models of induction and repression were first observed in bacteria, the models apply equally well to eukaryotic organisms. Although not all enzyme syntheses are so regulated, there are enough that scientists can understand some of the means the bacterium employs to manipulate genes.

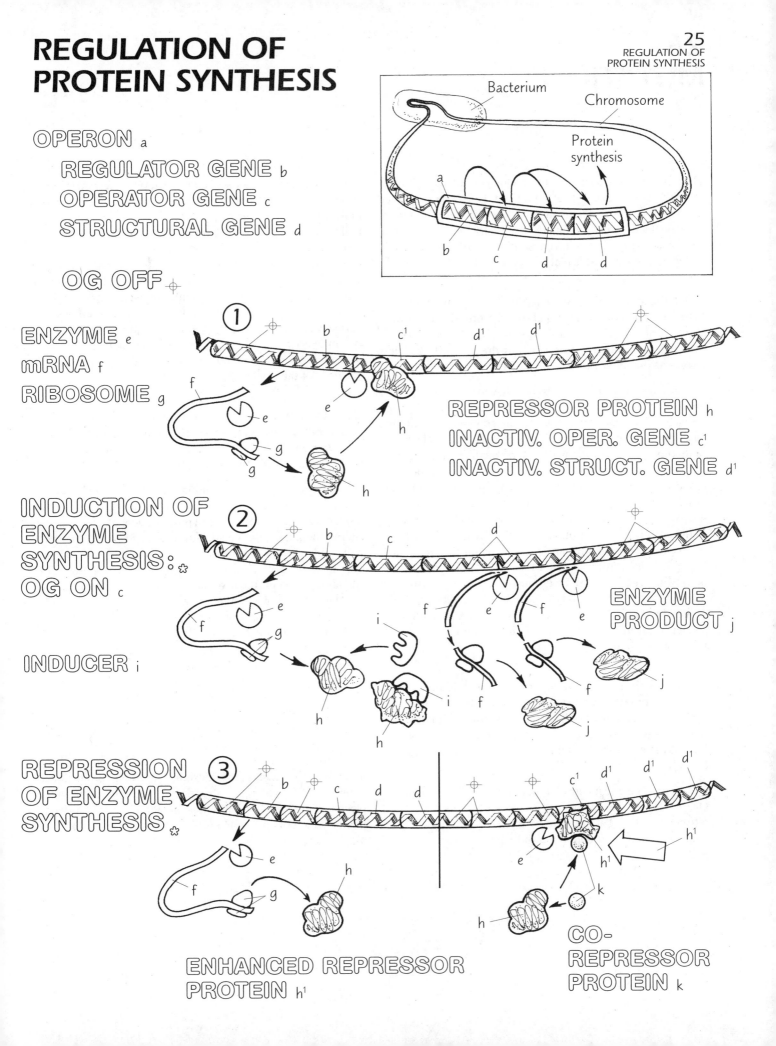

REGULATION OF PROTEIN SYNTHESIS

Bacterium

Chromosome

Protein synthesis

OPERON a

 REGULATOR GENE b
 OPERATOR GENE c
 STRUCTURAL GENE d

 OG OFF

ENZYME e
mRNA f
RIBOSOME g

REPRESSOR PROTEIN h
INACTIV. OPER. GENE c^1
INACTIV. STRUCT. GENE d^1

INDUCTION OF ENZYME SYNTHESIS:
OG ON c

INDUCER i

ENZYME PRODUCT j

REPRESSION OF ENZYME SYNTHESIS

ENHANCED REPRESSOR PROTEIN h^1

CO-REPRESSOR PROTEIN k

26
MUTATION

The DNA in a bacterial chromosome may be altered by any of several methods, natural or engineered. DNA can be altered by mutation, wherein the composition or sequence of one or more nucleotides in a strand of DNA is permanently changed. One means (point mutation), results in a substituted protein and is the subject of this plate.

Color the subheading Normal Condition gray. Color the titles and structures/arrows above the boxed illustration, starting at far left. Use a light color for the large tRNA molecule (e). Note the subscripts for the bases are the first letters of the base names.

You will recall that a strand of DNA consists of a chain of nucleotides characterized by four different bases. The bases of one strand of the DNA molecule are complementary to the bases of the second. Thus, adenine is bound by hydrogen bonds to thymine, and cytosine is bound to guanine. Here a part of one strand of parent DNA is shown (b). Through a process of replication during bacterial binary fission, two daughter strands of DNA are formed with bases complementary to the bases of the parent strand. Part of a daughter DNA strand (b¹) is shown. In the process of transcription, a strand of RNA is formed and the bases are complementary to the bases in the "template" DNA strand. Part of a strand of messenger RNA (d; mRNA) is shown. Recall the base uracil (u) is substituted for thymine in strands of RNA. By way of anticodons on transfer RNA molecules (e; tRNA), the codons in the RNA nucleotide sequence are translated into a chain of amino acids (f) of variable length to form the protein.

Color the title mutagen (h) and the nitrous acid in the flask as well as the titles and structures in the boxed area. The structure of guanine is illustrated to show its similarity to the molecular structure of hypoxanthine, the converted base. A different shade of the color used for (g) is suggested for (g¹).

A substance that causes a mutation is called a mutagen. One example of a mutagen is a chemical compound called nitrous acid (h). This relatively simple, naturally-occurring compound may be found in the soil where bacteria live. The nitrous acid may enter the bacterial cytoplasm and change the structure of the chromosomal DNA by converting adenine molecules to hypoxanthine molecules (g¹). Hypoxanthine molecules have a structure similar to guanine. This conversion of a single base at one (or more) places in a DNA molecule constitutes a point mutation and results in a mutated DNA molecule. The change is permanent and irreversible. Although the hypoxanthine mimics the guanine molecule sufficiently to be integrated with the DNA strand, its presence among the normal bases of nucleotides, both free and within the DNA, can influence enormous alterations in the structure and function of the protein product, and consequently the function and/or behavior of the organism.

Color the subheading Point Mutation and the title (i) and structures at the lower third of the plate. The mutated base in the parent DNA strand is marked by the arrow (h); the mutated base is identified as (g¹).

The presence of hypoxanthine (g¹) has little or no effect on the DNA molecule to which it is attached until replication takes place. At that time, the hypoxanthine molecule, having altered the adenine molecule sufficiently to mimic guanine, will bind with the base that is complementary to guanine: cytosine. The result is an abnormal replicated DNA strand (b¹) containing cytosine (at arrow) where there should be thymine.

The abnormal DNA may not present a problem to the bacterium until transcription of a strand of RNA occurs. In transcription, the mRNA (d) will pick up the base guanine (at arrow), complementary to the abnormally placed cytosine. Compare this with the mRNA strand in the normal condition where adenine (complementary to thymine) is found.

When the abnormal mRNA moves to the ribosome for translation, its code of G-A-A will attract a transfer RNA molecule (e¹) with an anticodon of C-U-U (reading the bases from above to below). Again, contrast this with the normal condition. Since tRNAs with different anticodons bring different amino acids to the ribosome, the wrong or substituted amino acid (i) will be positioned near the mRNA and then placed into the protein. Thus, this mutation in DNA has led to the formation of an erroneous or substituted protein.

MUTATION

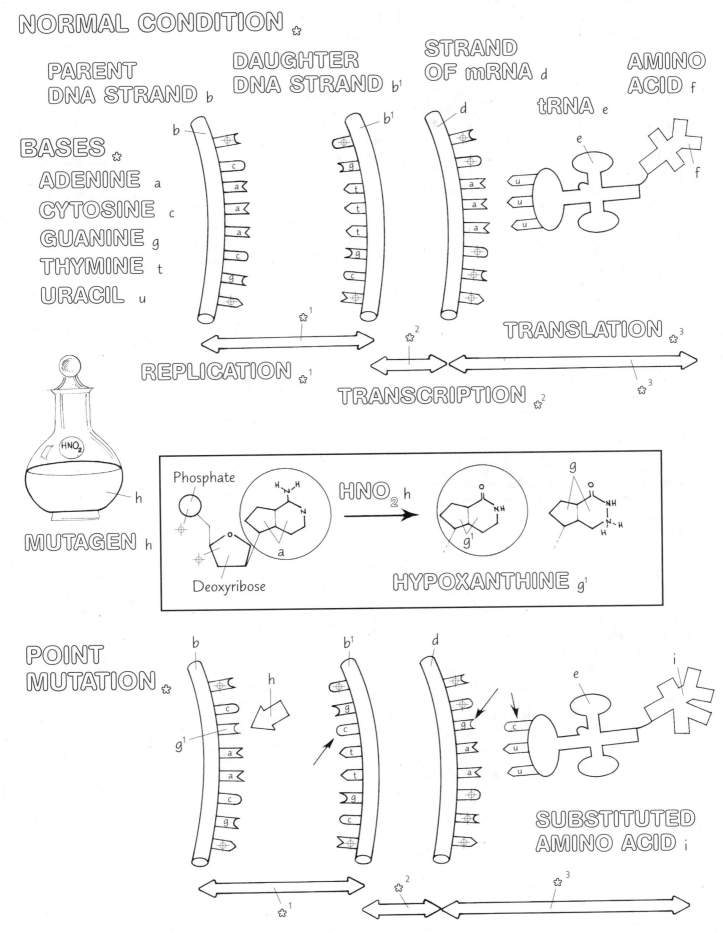

NORMAL CONDITION

PARENT
DNA STRAND b

DAUGHTER
DNA STRAND b¹

STRAND
OF mRNA d

AMINO
ACID f

tRNA e

BASES
  ADENINE a
  CYTOSINE c
  GUANINE g
  THYMINE t
  URACIL u

REPLICATION 1

TRANSCRIPTION 2

TRANSLATION 3

MUTAGEN h

Phosphate

HNO₂ h

HYPOXANTHINE g¹

Deoxyribose

POINT
MUTATION

SUBSTITUTED
AMINO ACID i

27
BACTERIAL TRANSFORMATION

From time to time, microorganisms with new characteristics appear. Frequently, they go unnoticed, but sometimes they present us with new challenges. How new strains of organisms arise in nature is not fully understood. However, research into bacterial genetics has revealed that bacterial DNA can undergo changes. These changes, termed genetic recombinations, introduce new characteristics in a bacterium, and may lead to the development of new species of organisms. One kind of genetic recombination is bacterial transformation.

In 1928, a British medical officer, Frederick Griffith, reported the curious results of a set of experiments with *Streptococcus pneumoniae*, the so-called "pneumococcus." This organism occurs in several strains. A strain is a group of organisms within a species that exhibits certain qualities seen only within that group. In this case, a strain of pneumococcus can be characterized by the presence or absence of a capsule. "S" strain consists of colonies of smooth, encapsulated pneumococcal organisms that are deadly. "R" strain consists of colonies of rough, unencapsulated pneumococcal organisms that are harmless.

Look over the upper half of the plate. Note and color the four stations (experiments), starting with the subheading *Streptococcus pneumoniae* and station 1. Set aside three shades of the color to be used for (s) through (s²); use the lightest color for (s). Set aside two shades of a contrasting color for (r) and (r¹). Select colors light enough to avoid obscuring structural detail.

Griffith found that by injecting deadly "S" strain bacteria (s) into healthy mice (m), pneumonia and death in the mice resulted (station 1). When he injected harmless "R" strain bacteria (r) into healthy mice, no sickness or death resulted from the injections (station 2). In a third experiment, Griffith injected heat-killed "S" strain bacteria into healthy mice, and these animals did not become sick or die (station 3). These results were all expected.

Continuing with his experiments, Griffith injected into a group of mice a mixture of harmless "R" strain bacteria and harmless heat-killed "S" strain bacteria. To his complete surprise, the mice died of pneumonia (station 4). Taking blood samples from the dead mice, he found only "S" strain bacteria– living, encapsulated bacteria! Something had caused the live, harmless "R" strain bacteria to transform into live, deadly "S" strain bacteria.

Griffith's observations of transformation went unexplained for decades. Late in the 1950s, scientists were finally able to piece together most of the details of what was taking place. They termed the process "bacterial transformation." Microbiologists provided evidence that DNA was the basic material of transformation and the basis of all mechanisms of heredity.

Color the subheading Process of Transformation in the lower half of the plate gray. Color the ruptured, heat-killed "S" strain bacterium (s¹) the color used for (s¹) in the upper half of the plate. Color the titles and related bacterial parts in the recipient bacterium as it undergoes transformation and division, and the parts of the offspring as well.

Evidence indicates that the process of transformation begins with the explosive release of the chromosome/DNA from the heat-ruptured bacterium (s¹). Recall that the DNA is packed tightly in the cell's cytoplasm. A rupture in the cell wall will send segments of DNA spilling outside the cell along with the cytoplasm.

A nearby "R" strain bacterium (r) will take one or more segments (s²; about 20 genes) of the extruded "S" strain DNA through its cell membrane and into its cytoplasm. Such a bacterium that possesses this ability is said to be "competent." Enzymes cut out a segment of a single strand of DNA (r¹) from the resident chromosome, dissolve it, and insert the "borrowed" DNA segment (s²) in its place. This segment of DNA from the ruptured bacterium will behave just like the rest of the bacterium's DNA: replicating during the "R" strain bacterium's cell division and passing into the daughter "R" strain cells.

The effect of this chromosomal transformation in once harmless "R" strain bacteria is believed to be reflected in the virulence of the bacterial offspring causing death in the mice (recall station 4). In Griffith's experiments, it is believed that the newly acquired genes of the harmless "R" strain bacteria induced the formation of proteins characteristic of disease-producing "S" strain bacteria, thereby causing the offspring "R" strain bacteria to become deadly.

BACTERIAL TRANSFORMATION

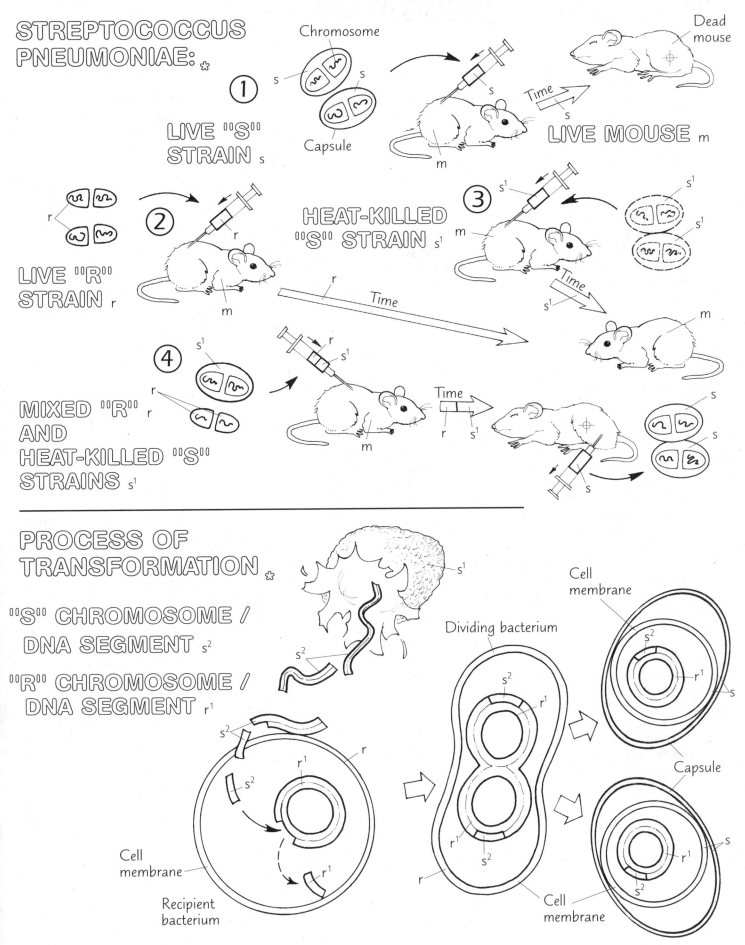

STREPTOCOCCUS PNEUMONIAE: ✿

① Chromosome

LIVE "S" STRAIN s

Capsule

Dead mouse

Time

LIVE MOUSE m

②

LIVE "R" STRAIN r

HEAT-KILLED "S" STRAIN s¹

③

Time

④

MIXED "R" r
AND
HEAT-KILLED "S"
STRAINS s¹

Time

PROCESS OF TRANSFORMATION ✿

"S" CHROMOSOME / DNA SEGMENT s²

"R" CHROMOSOME / DNA SEGMENT r¹

Dividing bacterium

Cell membrane

Cell membrane

Recipient bacterium

Capsule

Cell membrane

28
BACTERIAL CONJUGATION

The process of transformation allows bacterial populations to adapt to their environments through genetic change. Populations can also adapt through another form of recombination called conjugation. Two kinds of bacterial conjugation are shown here.

Set aside three shades of each of the colors selected for (a) and (b), and two shades of the color used for (d). Start by coloring the subheading Plasmid Transfer. Color the titles (a) through (d), and the related structures in the paired bacteria in the upper half of the plate. Begin at left and work to the right.

In the early 1950s, scientists discovered that bacteria were of two mating types: they called these types F (+) and F (-), where F is the abbreviation for "fertility." F (+) bacteria (a) are able to donate DNA to the F (-) bacteria and the F (-) bacteria are capable of receiving it. The double-strand loop of DNA that is transferred in conjugation contains about 20 genes. This loop of DNA is carried in the F (+) bacteria separate from the chromosome (a^1), and is known as a plasmid, fertility factor, or F factor (d).

For plasmid transfer to occur, the F (+) and F (-) cells must contact one another. For a significant number of transfers to take place, there must be an enormous population density—about 10,000,000 cells per cubic millimeter of medium. As an F (+) cell approaches an F (-) cell, it extends a tubular appendage known as a sex pilus (c) toward it, making physical contact. The pilus acts as a conjugation bridge between the F (+) and F (-) cells. As the cells make contact, the F factor in the F (+) cell replicates and the duplicate plasmid, called a daughter F factor (d^1), migrates across the bridge into the cytoplasm of the F (-) cell. The cells then separate. The recipient F (-) cell becomes an F (+) cell upon receipt of the daughter F factor. This new F (+) cell is now capable of producing a pilus and transferring copies of the F factor.

Plasmid transfer between bacteria is important because the genes that confer antibiotic resistance are located largely in plasmids. Antibiotics are chemical substances produced by microorganisms. They have the capacity to inhibit the growth of or to kill other bacteria, and hence are used to treat bacterial infections. Some bacteria can produce enzymes which destroy antibiotics, and the genes regulating the production of these enzymes are generally located on plasmids. Thus, bacteria can pass on antibiotic resistance through transfer of plasmids from F (+) bacteria to F (-) bacteria.

Color the subheading Chromosome Transfer in the lower half of the plate. Use the same colors for structures here as you did above, where applicable. Color the titles and related structures in the lower half of the plate. Begin at the left side, coloring both the F (+) and F (-) bacteria.

In another form of bacterial conjugation, called chromosome transfer, the F (+) bacterium transfers a strand of its chromosomal DNA to the F (-) bacterium. In most cases, the F factor is integrated with the F (+) chromosome before conjugation occurs, and the bacterial cell is called a high frequency of recombination (or Hfr) cell (a^2). In the conjugation process, an Hfr bacterium and an F (-) bacterium approach each other; the Hfr cell makes a conjugation bridge with the F (-) cell using the sex pilus, and the cells contact one another.

The combined F (+) chromosome and plasmid of the Hfr cell then opens, and a single strand of combined DNA (e) is copied. This strand leaves the parent chromosome as the chromosome rotates counterclockwise. The free end of this single DNA strand, containing both F (+) DNA (a^3) and plasmid DNA (d^2), moves out through the conjugation bridge into the cytoplasm of the F (-) bacterium. The migrating strand, almost a complete copy of the Hfr cell's combined chromosome/plasmid, is taken completely into the F (-) cell.

As the combined DNA strand enters the F (-) cell, enzymes displace an equal amount of F (-) DNA (b^2) from the F (-) cell chromosome (b^1), and the combined DNA strand is installed in its place. The F (-) cell is now called a recombinant F (-) cell (b^3).

Both kinds of transfer permit the acquisition of new genetic characteristics in bacteria. Conjugation has been demonstrated among various genera of bacteria, especially among Gram-negative bacteria. It is possible that such processes as conjugation and transformation (Plate 27) are significant mechanisms for the alteration of the genetic constitution, appearance of new bacterial diseases, and bacterial resistance to antibiotic therapy.

BACTERIAL CONJUGATION

PLASMID TRANSFER ✿

F (+) BACTERIUM a
 F (+) CHROMOSOME a¹

F (−) BACTERIUM b
 F (−) CHROMOSOME b¹

SEX PILUS c
F FACTOR / PLASMID d

DAUGHTER
F FACTOR / PLASMID d¹

CHROMOSOME TRANSFER ✿
Hfr BACTERIUM a²
PLASMID DNA d²
F (+) DNA a³

COMBINED DNA STRAND e

DISPLACED F (−) DNA STRAND b²
RECOMBINANT F (−) BACTERIUM b³

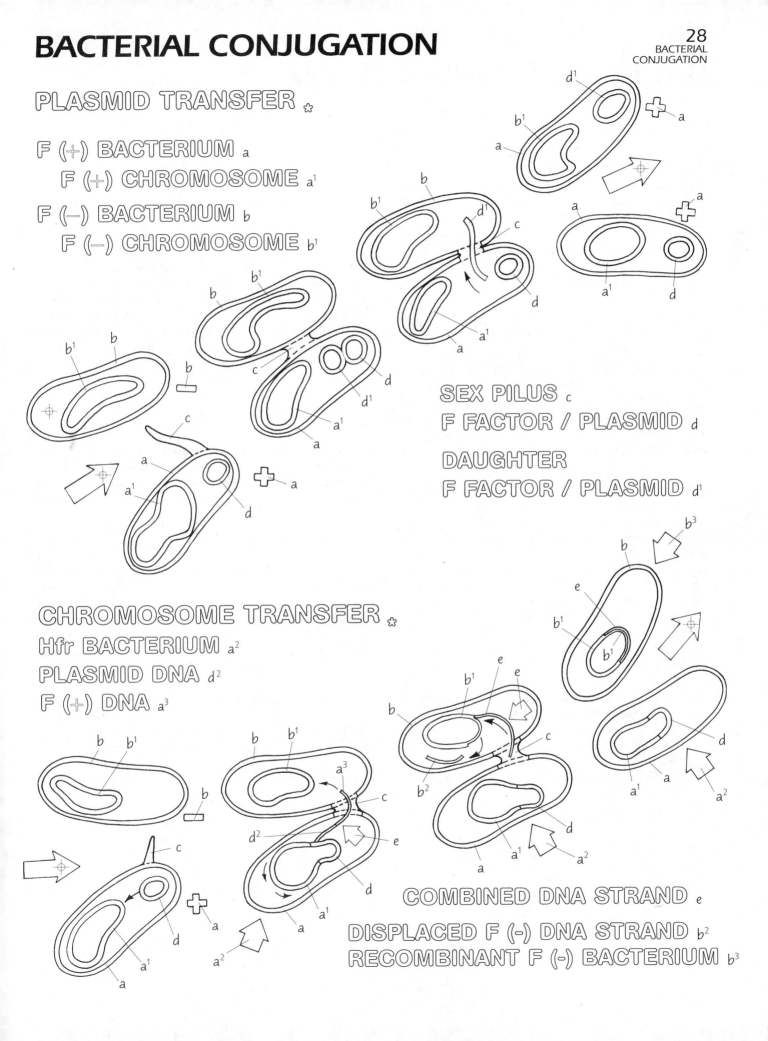

BACTERIAL TRANSDUCTION

Two mechanisms by which bacteria undergo genetic alteration are bacterial transformation and bacterial conjugation (Plate 28). In this plate, the process of bacterial transduction is presented as a third means by which genetic diversity can be accomplished in microorganisms. In transduction, a foreign agent is required to convey DNA from one bacterium to another. This agent is a virus.

Begin by noting the arrangement of illustrations by stations 1 through 6. Color the subheading Generalized Transduction, and the titles and structures (a) through (d¹). Use a light color for (a) and a darker shade of that color for (a¹). The colors selected for (b) through (b²) should sharply contrast with (a) and (a¹). Start at the upper right of the plate (station 1), and continue through station 6. For coloring purposes, the bacteriophage (a) is shown much larger in relation to the bacterium (b) than it actually is.

A bacteriophage (station 1) is a virus that infects bacterial cells. Commonly known as a "phage," the virus penetrates the wall of the bacterial host and alters the cellular metabolism to produce hundreds of new viruses. Essentially, the phage consists of a protein coat (a) enclosing a double strand of DNA (a¹). The process of transduction begins when the phage comes in contact with a bacterium and locks on to a receptor site of the bacterial wall (station 2). The phage then injects its DNA through the cell wall into the bacterial cytoplasm. The phage protein coat remains outside the bacterial wall.

Once inside the bacterial cytoplasm, the phage DNA induces the fragmentation of the bacterial chromosome (b¹; station 2) into DNA segments (b²; station 3). Such fragmentation is undertaken by bacterial enzymes (c) whose synthesis is directed by the phage DNA. Certain bacterial DNA segments (b²) form the starting materials for the replicating molecules of phage DNA (a¹; station 3) and protein coats (a) are formed around these. In addition, other bacterial DNA segments become incorporated into new phages as the entire DNA molecule (b²; stations 3 and 4). Thus, phages in the bacterial cytoplasm have one of two different genetic constitutions: phage or bacterial. In time, the bacterial cell wall ruptures, and both types of phage are released into the external environment (station 4).

Inevitably, many of the released phages will infect other bacteria (station 5). Some of these phages will contain bacterial DNA. As described before, the phage will inject its DNA through the cell wall of another (host) bacterium (d) and into the cytoplasm, leaving the protein coat outside the cell. The injected DNA of bacterial origin (b²) may combine with the DNA of the host bacterial chromosome (d¹; station 6). In this case, formation of phages in the cytoplasm does not occur. Instead, the incorporation of the injected DNA with the host DNA induces new genetic traits in the recipient. Such a process is called generalized bacterial transduction. A more specialized type of transduction is also possible.

Color the subheading Specialized Transduction in the boxed area, and related titles and structures. Begin at upper left within the box, and continue clockwise. Use contrasting colors for (e) and (f), both of which should contrast with the color selected for (a¹).

In generalized transduction, all the genes of the DNA in the donor bacterium are available for transduction. In specialized transduction, only certain genes are available for this process. In illustrating the specialized transduction process, we start with a bacterium (e) in which phage DNA (a¹) has been incorporated with the bacterial DNA (e¹). At some point in the life cycle of the bacterium, in response to a chemical signal, the phage DNA part of the bacterial chromosome detaches, taking some adjacent bacterial DNA (e¹) with it. The phage DNA replicates within the host cell (e) and the bacterial DNA also replicates (not shown). The combined DNA constitutes a hybrid. Following formation of a protein coat, the phage leaves the donor bacterium (e) and attaches to a recipient bacterium (f), penetrating its cell wall and entering the cytoplasm. The hybrid DNA is spliced into the recipient bacterium's DNA (f¹), and transduction has been effected.

BACTERIAL TRANSDUCTION

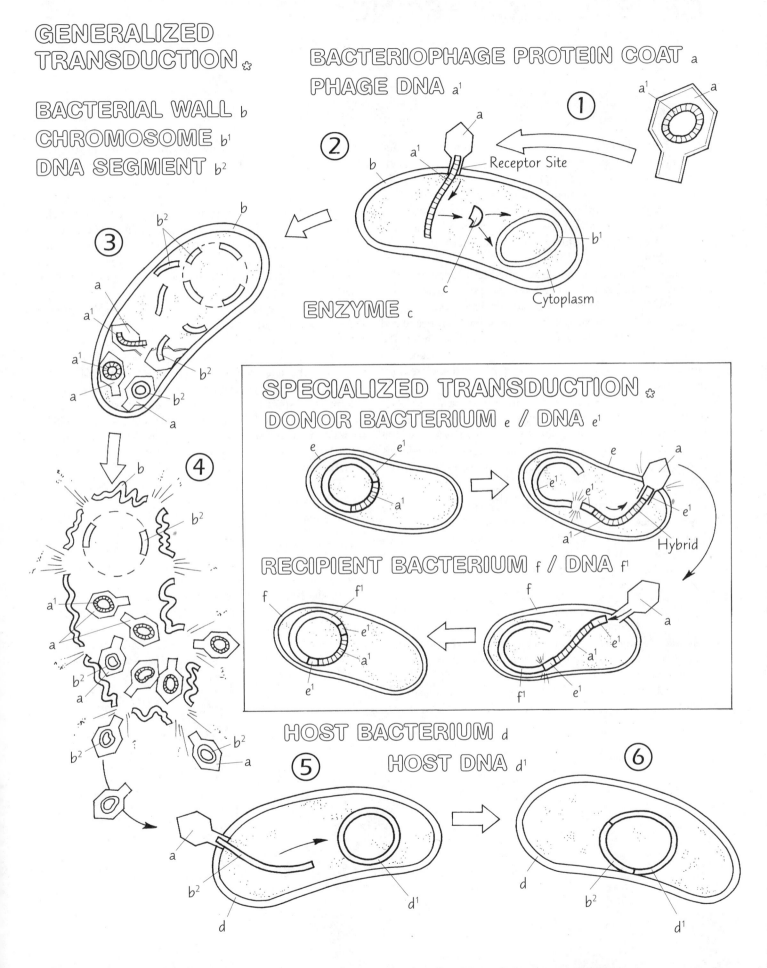

GENERALIZED TRANSDUCTION ✷

BACTERIAL WALL b
CHROMOSOME b¹
DNA SEGMENT b²

BACTERIOPHAGE PROTEIN COAT a
PHAGE DNA a¹

Receptor Site

Cytoplasm

ENZYME c

SPECIALIZED TRANSDUCTION ✷
DONOR BACTERIUM e / DNA e¹

Hybrid

RECIPIENT BACTERIUM f / DNA f¹

HOST BACTERIUM d
HOST DNA d¹

30
GENETIC ENGINEERING

Experiments in bacterial transformation, conjugation, and transduction stimulated interest in the refinement of genetic insertion and deletion techniques. Such techniques are at the core of the process of genetic engineering. Was it possible to insert non-bacterial (foreign) genes into a bacterial chromosome? And would these altered bacteria establish a colony of cells that would produce proteins coded by these foreign genes?

In the 1960s, researchers discovered certain enzymes that would cleave bacterial DNA at designated points (restrictions) of the DNA strand. At these points of cleavage, foreign genes could be inserted. Research workers then found that specialized enzymes, called ligases, would secure the foreign genes to the strands of the bacterial chromosome. This same mechanism operates in DNA replication processes colored in the previous three plates.

As work progressed, scientists found chromosomal DNA difficult to work with, but noted that plasmids were much easier to manipulate. Using these small ringlets of DNA, research workers learned to maneuver genes and "engineer" new forms of bacteria that would synthesize specific proteins on demand.

Look over the arrangement of illustrations; note the counterclockwise progression of events. Set aside three shades of the color selected for (f). Use sharply contrasting colors for (a) and (d). Use a light color for (d). Color the titles and structures (a) through (g), beginning at upper left.

Note in the bacterium *Escherichia coli* the DNA-containing chromosome and the plasmid (a). The plasmid consists of a ring of DNA but a distinctly smaller amount than that found in the chromosome. Genetic engineering techniques begin with the breaking open of the bacterium and isolation of the plasmid (called the "donor plasmid"). The plasmid is cleaved at a desired point by a restriction enzyme (b). This cleaved part is removed and broken down; the plasmid is ready to receive a fragment of foreign DNA.

Foreign DNA (d) can come from a variety of sources, e.g., other bacteria, other animal tissues, even human cells. The foreign DNA is cleaved into fragments. The selected fragment is combined with the plasmid DNA at the restriction point with the aid of the enzyme DNA ligase (c). This combined plasmid is called a chimera (e; mythological lion-goat-serpent monster).

Chimeras are inserted into the bacteria by placing both bacteria and chimeras in calcium chloride ($CaCl_2$) solution and heating them quickly. This process opens the cell walls and membranes, permitting the chimera plasmids to pass through and enter the bacterial cytoplasm. Following cooling, the bacteria soon reproduce, generating a colony of cells each with a chimera plasmid. Soon a significant and measurable volume of bacterial protein (g) is produced. With this volume will be found a "new" protein whose genetic code is carried in the chimera plasmids within the bacterial cytoplasm. This new protein is available for harvesting.

Genetic engineering is the process of manipulating and collecting a known quantity of genes from a known source, opening the recipient DNA at the desired point in the linkage, and combining the strand of a known sequence of DNA with the recipient DNA of living organisms. The protein "factories" then synthesize the anticipated product which can be used for any number of purposes.

By the 1980s, the process of genetic engineering was well-established and biotechnology companies were mass producing proteins that could not be easily obtained otherwise. The process has yielded protein hormones vital to life, such as insulin and growth hormone, as well as vaccines, antiviral interferon, and blood clot-dissolving enzymes.

GENETIC ENGINEERING

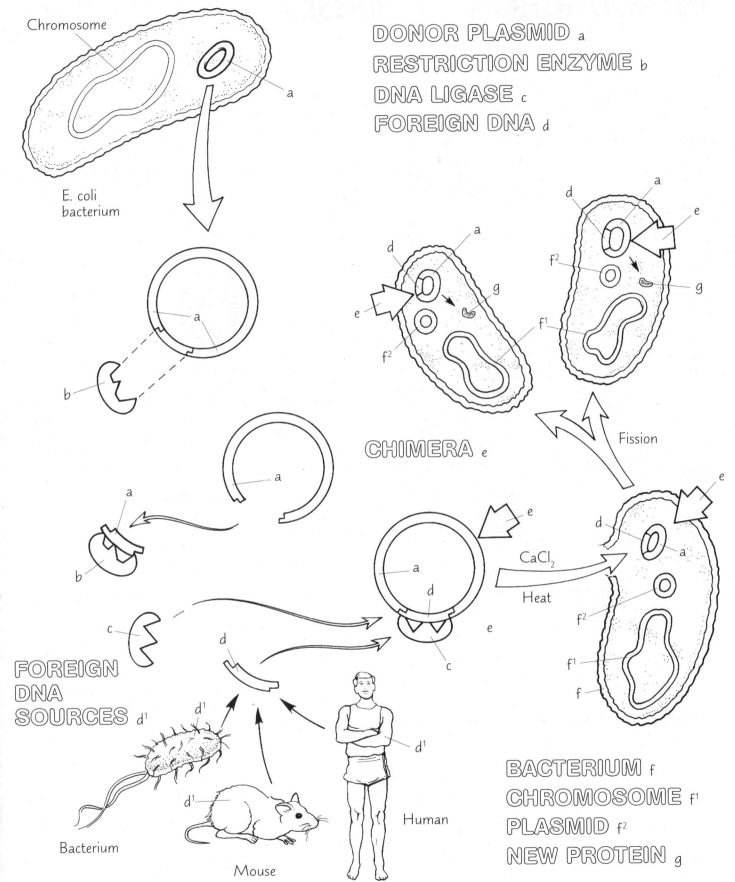

Chromosome

E. coli
bacterium

DONOR PLASMID a
RESTRICTION ENZYME b
DNA LIGASE c
FOREIGN DNA d

CHIMERA e

Fission

$CaCl_2$

Heat

FOREIGN
DNA
SOURCES d^1

Bacterium

Mouse

Human

BACTERIUM f
CHROMOSOME f^1
PLASMID f^2
NEW PROTEIN g

INTRODUCTION TO VIRUSES

The virus is among the smallest and simplest agents of disease. Viruses are so tiny that they cannot be seen with a light microscope and researchers must use the electron microscope to view them. Viruses have a unique chemical structure and a parasitic dependence on other organisms associated with an unusual method of reproduction. Viruses cause such well-known diseases as chicken pox, influenza, hepatitis, and infectious mononucleosis.

During the early 1900s, medical scientists observed that carefully filtered fluids of diseased tissues (disrupted, devitalized tissue caused by the growth and toxins of microorganisms) were capable of inducing disease. They reasoned that tiny microorganisms (viruses) in the diseased fluid passed through the smallest filters, and when this filtered fluid was injected into a living, healthy host, the viruses present induced disease processes to occur.

Color the subheading Size and the titles and viruses (a) through (f) in the upper half of the plate. Note the 300 nm ruler at right. Use light colors for the larger structures; beware of colors that obscure the detail of the illustration.

Viral dimensions are measured and viral structural characteristics are observed with the aid of the electron microscope. Most viruses are substantially smaller than bacteria, but some viruses approximate the sizes of very small bacteria. The average *Escherichia coli* bacterium, used here as a reference bacterium, is about 3000 nm (2000 - 6000 nm) in length (Plate 8). The bacterium *Chlamydia* is about 250 nm long and is tiny in comparison with *E. coli*, but it is about the same size as the smallpox virus (a; about 300 nm long).

Most viruses are about the same size or smaller than the smallpox virus. The tobacco mosaic virus (b), a parasite of tobacco plants; is a mid-size virus with a length of about 300 nm. Note the bacteriophage (c) is only about 300 nm in length, considerably smaller than the *E. coli* bacterium. The rabies virus (d), well known for its catastrophic effect in humans after bites from infected animals, is about 200 nm long. The smaller adenovirus, agents of a number of human miseries, including pneumonia and conjunctivitis, measure about 75 nm in length. One of the smallest of this group of microorganisms

is the polio virus (f) with an average diameter of 25 nm. Polio viruses affect the central nervous system of humans, destroying the motor neurons that supply the skeletal muscles of the body.

Color the subheading Shape, titles (g) and (h), and the related shapes in the boxed area. Then color the subheading Structure, and the related titles and structures (i) through (l) at the lower part of the plate.

Electron microscopy has revealed that viruses generally have one of two shapes. One is the icosahedron (g), a geometric figure characterized by 20 triangular "faces." Icosahedral-shaped viruses include bacteriophages, chicken pox, genital herpes, mononucleosis, and polio. The second shape is that of a helix or tightly wound coil (h), somewhat resembling a corkscrew. Helical viruses include those that cause rabies and tobacco mosaic disease.

Viruses consist of two main components: the outer capsid (i) and the inner genome (j). The capsid is the outer coat, and gives shape to the virus, either icosahedral or helical. In most viruses the capsid is composed of multiple protein subunits called capsomeres (i[1]), the number of which varies among viruses. The genome is found in the core of the virus. It consists of a single or double strand of nucleic acid which is either DNA or RNA, but not both. In some viruses the strand is unbroken; in others it is divided into segments. In icosahedral viruses, the genome is commonly a closed loop folded over itself (not shown); in helical viruses, the genome is coiled in the shape of a helix.

The outermost membrane of some viruses is the flexible, lipoprotein envelope (k) around the capsid. It is usually derived from the host cell during replication. Many envelopes have an array of spikes (l) that contain enzymes that assist in cell penetration. The influenza and human immunodeficiency virus are notable for the presence of spikes coded for by viral genes.

No cytoplasm or organelle has been identified in viruses. The virus is dependent upon a host for metabolic machinery and for reproduction. An apparent inert particle in isolation, it swiftly replicates in the appropriate host, even to the extent of killing the very living entity upon which it depends.

INTRODUCTION TO VIRUSES

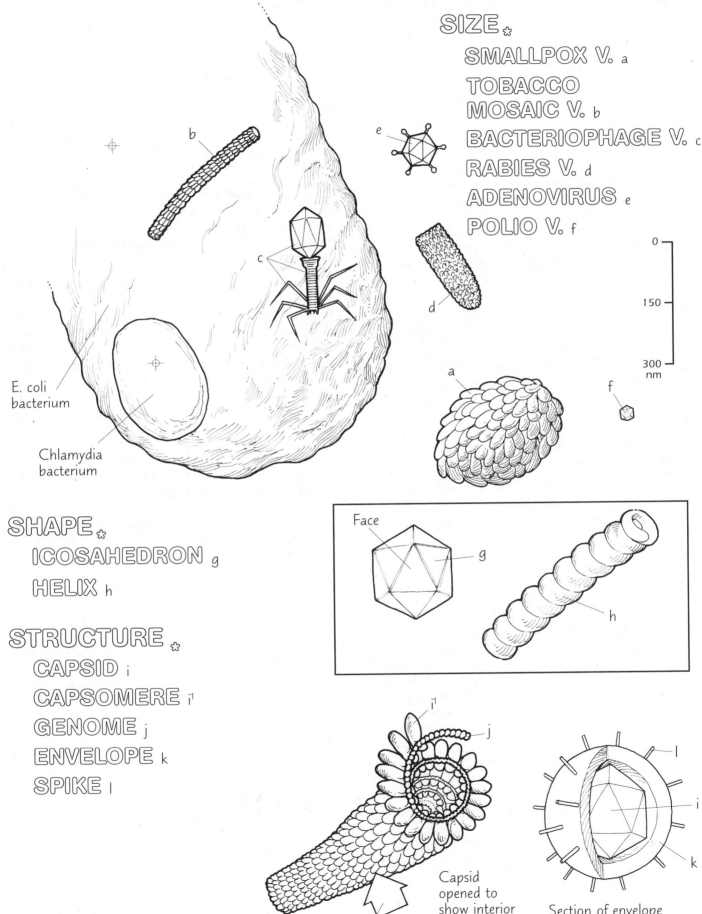

SIZE ✲
SMALLPOX V. a
TOBACCO
MOSAIC V. b
BACTERIOPHAGE V. c
RABIES V. d
ADENOVIRUS e
POLIO V. f

0

150

300
nm

E. coli
bacterium

Chlamydia
bacterium

SHAPE ✲
ICOSAHEDRON g
HELIX h

STRUCTURE ✲
CAPSID i
CAPSOMERE i¹
GENOME j
ENVELOPE k
SPIKE l

Face

Capsid
opened to
show interior

Section of envelope
removed to show interior

Viral replication requires a host cell in which the replication takes place. In the process, the host cell usually dies. In this respect, viral replication is one of the more remarkable phenomena of nature. In general, a virus invades a cell many times its own size and directs the metabolic machinery of the cell to produce copies of the virus. This operation has been studied in plant, animal, and human cells, but it has been most carefully documented in bacteria and the viruses that affect them. Such viruses are called bacteriophages (or simply, phages). In this plate, a bacterium and its phage are employed to explore viral replication.

Note the overall orientation of the plate; there are six stations from above to below. Color the subheading at station 1 (*¹) gray and the related titles and structures (a) through (e) and (g) through (g²). Use a light pastel color for (g) and darker shades of the same color for (g¹) and (g²). Next color the subheading at station 2 (*²) and the titles and structures (f) and (h) and other structures in the related illustration. Read the corresponding text as you color the illustration at each of the two stations.

The phage (a; station 1) is characterized by a capsid (b) having the symmetry of an extended icosahedron. The capsid encloses a strand of DNA (f; station 2). The tail assembly includes a tail sheath (c) surrounding the central core (e) that accommodates the DNA. A pair of pins (d) and multiple fibers (d¹) constitute the lower part of the assembly.

The process of viral replication begins with the attachment of the phage to the bacterial cell wall (g; station 1: union phase). The phage attaches to receptor sites (g¹) of the bacterial wall with its fibers. The central core fits over the penetration site (g²). In the penetration phase (station 2), enzymes from the tail end of the central core break down the bacterial wall. The tail sheath contracts and the core is driven through the bacterial wall. The DNA passes through the core and into the cytoplasm of the bacterium. The viral DNA migrates to the bacterial chromosome and uses the bacterial DNA (h) to synthesize viral DNA. The capsid stays outside the bacterium and disintegrates.

Color the subheadings at stations 3 through 6, in order, and associated titles and structures. Read the corresponding text with each phase you are coloring. The phases illustrated at stations 3 through 6 all take place within the bacterial cell.

Using the viral DNA as a template, a strand of messenger RNA (i; mRNA) is synthesized. This is called the transcription phase (station 3). The mRNA, with instructions for the production of viral protein, passes to the bacterial ribosomes (j) where viral protein (k) is synthesized. Viral proteins soon begin to appear in the bacterial cytoplasm. This is the synthesis phase of viral replication (station 4).

The viral structures are constructed from DNA and the viral protein, and parts of the virus-to-be are soon seen in development. This is the assembly phase (station 5). The assembly of the phage has been observed by researcher workers with the aid of the electron microscope. This assembly is undertaken with enzymes whose production was directed by the mRNA synthesized according to instructions in the viral DNA. As the capsids are put together, viral DNA is enfolded into the interior of these structures. The viral DNA is synthesized from components of the bacterial DNA. As replication activity is completed, all of the bacterial DNA will disappear. The pins attach to the core, and then the tail sheath surrounds the core. The capsid is formed and attached. Attachment of the fibers completes the assembly phase.

As phage construction is completed, the numbers of phages swell within the bacterial wall. A viral-directed enzyme (lysozyme) breaks down the bacterial wall, permitting rupture, and killing the bacterial cell (station 6: release phase). Hundreds of phages are released, each identical in structure and genetic sequence to the one that began the process.

VIRAL REPLICATION

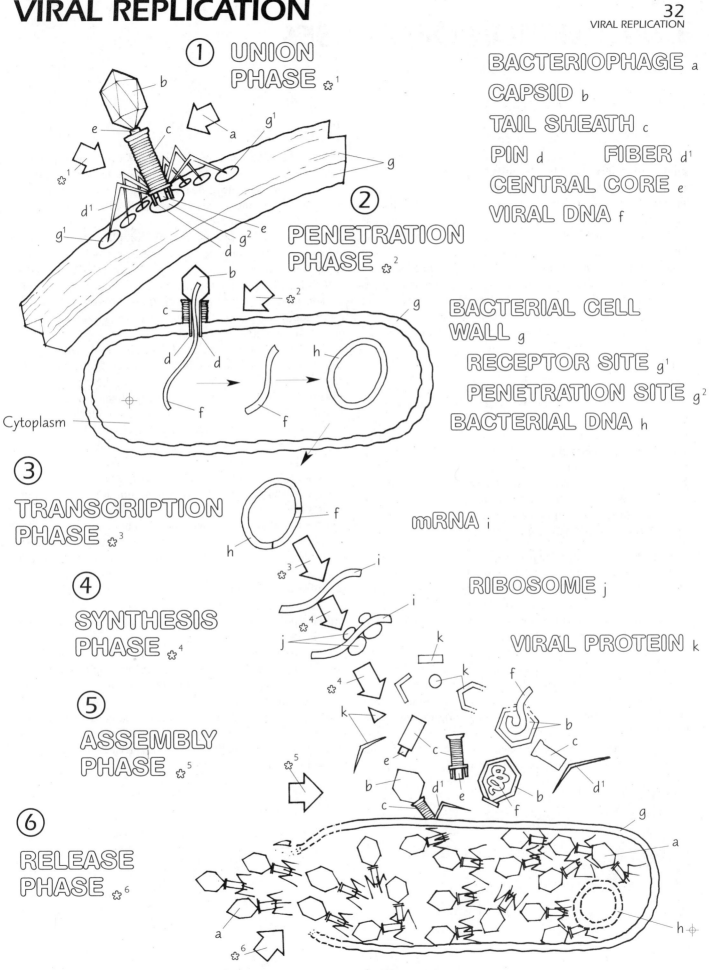

① UNION PHASE ✿¹

② PENETRATION PHASE ✿²

③ TRANSCRIPTION PHASE ✿³

④ SYNTHESIS PHASE ✿⁴

⑤ ASSEMBLY PHASE ✿⁵

⑥ RELEASE PHASE ✿⁶

Cytoplasm

BACTERIOPHAGE a
CAPSID b
TAIL SHEATH c
PIN d   FIBER d¹
CENTRAL CORE e
VIRAL DNA f

BACTERIAL CELL WALL g
RECEPTOR SITE g¹
PENETRATION SITE g²
BACTERIAL DNA h

mRNA i

RIBOSOME j

VIRAL PROTEIN k

INACTIVATION OF VIRUSES

Viruses are susceptible to many of the physical and chemical agents routinely used by the microbiologist to eradicate microorganisms in the external environment. These agents operate on three structures of the virus: the capsid, the nucleic acid, and, when present, the envelope. The chemical components of these viral structures react with these agents and are subjected to processes of decay and disruption (inactivation). Inactivated viruses cannot replicate. Since such viruses cannot increase their numbers, they are rendered inert. Viruses are not alive by present standards, that is, they are not capable of metabolic activity; they do not exhibit growth; and they cannot reproduce without a host cell. In the scientific sense, then, viruses are inactivated, not killed.

Color the subheading Viral Structure and related titles (a) through (c) at the bottom of the plate, and the parts of the icosahedral virus in the center. Use a pale pastel for (a); color the entire envelope, overcoloring the capsid and nucleic acid. Select a slightly darker, contrasting color for (b), and a darker color yet for (c). Color the subheading Inactivating Agents, each of the agents (d) through (i), as well as the arrows leading to the virus. Use light colors for (d), (e), and (f). Note how each arrow terminates on the viral structure with which it reacts.

Functionally significant parts of the virus are the envelope (a), the capsid (b), and the nucleic acid (c) which can be DNA or RNA but not both. The envelope of the virus reacts with ethyl alcohol (d) and other alcohols as well. The envelope is composed of lipid, and the lipid is soluble in alcohol. Thus, exposure of the viral envelope to ethyl alcohol causes its destruction. Since for many viruses, the envelope reacts with the host cell membrane at the initiation of viral replication, exposure to alcohol interrupts the replication process.

Heavy metals (e) are elements in which atoms have large atomic weights and complex electronic configurations. Two heavy metals, silver (Ag) and mercury (Hg) can be used to inactivate viruses. The metals react with the protein in the capsid, disrupting its structure, and effectively destroying the virus. Silver is commonly used in the form of silver nitrate, and mercury is combined with chloride to form mercury chloride. One example of heavy metal use is the silver nitrate often applied to the eyes of newborn to kill the gonorrhea-causing bacteria (if present).

Formaldehyde (f), a relatively simple member of the aldehyde family of organic compounds, reacts with free amine groups on the nucleotide bases of the RNA and DNA molecules in the virus. By binding these groups, formaldehyde prevents nucleic acid activity (replication, transcription, and so on). Formaldehyde is a useful chemical for inactivating viruses during the preparation of vaccines.

Ultraviolet (UV) radiation (g) is a form of energy that is imperceptible to humans. When propagated from a lamp and directed to viruses, the radiation penetrates the capsid and the envelope of the virus, and alters the chemical structure of the cytosine and thymine bases of the nucleic acids. Replication becomes impossible in these irradiated viruses. UV radiation is used in the preparation of some vaccines and as a sterilizing agent to purify the air.

One of the most widely used physical agents in destroying viruses is heat (h). Heat breaks the atomic bonds of organic molecules (proteins, lipids, nucleic acids). The action is nonspecific, that is, heat destroys all components of the virus equally well. Indeed, a direct flame destroys viruses almost instantaneously. Boiling water inactivates most viruses in a few seconds. Hence, boiling water before drinking it is a suitable disinfecting technique.

Phenol (i) is an extremely poisonous extract of coal tar, and is a potent inactivator of viruses. It is, however, rarely used in its pure form. Phenol-containing products, such as hexachlorophene and chlorhexidine are well known antiviral disinfectants. Phenol combines with the protein in the capsid, altering its structure and inactivating the virus.

Thus, the microbiologist has a number of options when viral inactivation is desired outside the body. Inactivation of viruses within the host organism is much more difficult, as reflected by the continuing search for antiviral medications to be used in viral-induced disease states.

INACTIVATION OF VIRUSES

INACTIVATING AGENTS ✿

VIRAL STRUCTURE ✿
ENVELOPE a
CAPSID b
NUCLEIC ACID c

ETHYL ALCOHOL d

FORMALDEHYDE f

HEAVY METAL e

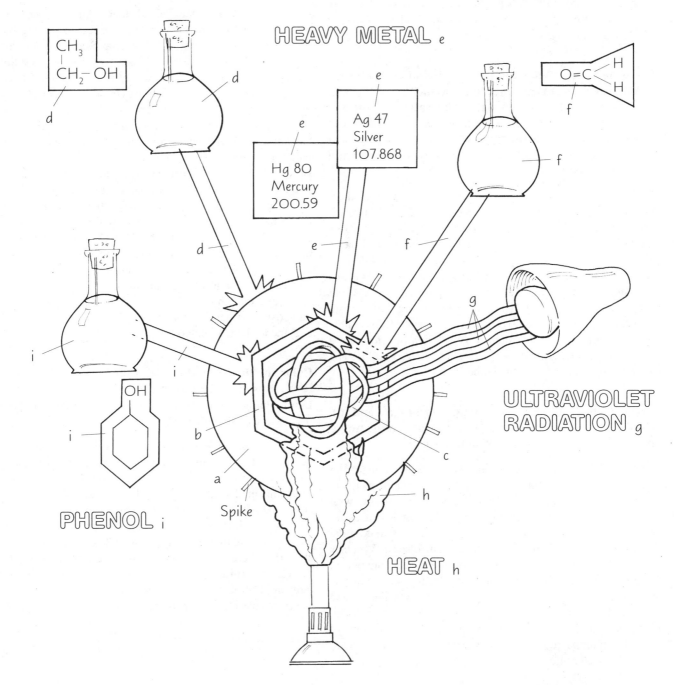

CH₃
CH₂─OH
d

e
Ag 47
Silver
107.868

e
Hg 80
Mercury
200.59

$O=C\begin{smallmatrix}H\\H\end{smallmatrix}$
f

ULTRAVIOLET
RADIATION g

OH
i

b
a
Spike
c
h

PHENOL i

HEAT h

34
ANTIVIRAL VACCINES AND DRUGS

The effort to prevent the spread of viral disease has traditionally been centered on the use of vaccines. Vaccines are agents that stimulate the body's immune system to produce antibodies. Antibodies are proteins that interact with foreign substances in the body. The antibodies provide a type of surveillance by circulating throughout the blood stream and neutralizing viruses when they come upon them.

Set aside five shades of the same or similar colors for (a) through (a⁴). Color the subheading Vaccine Production and note the orientation of the related illustrations: culture of viruses lead to two types of vaccine: inactivated (a²) and attenuated (a⁴). Color the titles, structures, and arrows in the sequence (a) to (a²) first; then read the corresponding text before going on.

One antiviral vaccine in common use is produced by inactivating viruses under controlled conditions. Viruses (a) are cultivated in living cells in a tissue culture (b). The viruses multiply in the culture using the cell's metabolic machinery. As these cultured viruses (a¹) increase in number, they destroy the cells (b¹). In the inactivation process (c), the cultured viruses are subjected to some chemical or physical agent (c¹) that inactivates the viruses, thereby preventing them from replicating. Although inactivated, these viruses are still capable of stimulating antibody production (d) when injected into a host. Such inactivated viruses (a²) are useful in a vaccine. The Salk polio vaccine is manufactured in this way.

Color the titles and related structures (a³) to (a⁴) and (d). Note the clock symbol next to arrow (a³), indicating a process requiring a passage of some time.

A second type of viral vaccine is prepared by transferring viruses from tissue culture to culture over a period of many months or years (a³). During this period, viral variants with significantly slower replication rates evolve. Such viruses are said to be attenuated (a⁴). They are capable of stimulating the host to produce antibody (d) without causing disease. Vaccines containing attenuated viruses are effective over a long period of time and antibody production continues for years. For this reason they are generally preferred to vaccines with inactivated viruses in which production of antibodies is more short-lived. Vaccines containing attenuated viruses are used to prevent measles, mumps, rubella, and influenza. They are also used in the Sabin oral polio vaccine.

Color the subheading Antiviral Drug Production, the subheading Nucleotide, and the titles/symbols (e) through (h). Color the upper two of the three viral DNA strands illustrated; wait to color the third strand for now.

Antibiotics are chemical substances that are produced by microorganisms and have a lethal effect on other living microorganisms, specifically bacteria. Since viruses do not have chemical-producing processes, there is nothing taking place in the virus with which an antibiotic can interfere. Any chemical substance useful against viruses must interfere with viral replication. Such a substance is acyclovir (h).

Acyclovir belongs to a group of drugs called chain terminators. To appreciate its action, it is helpful to review a part of the nucleic acid synthetic process. A nucleoside (e) is a sugar attached to a nitrogenous base. An example is deoxyribose attached to the base guanine. A nucleoside combined with a phosphate group (f) with the aid of an enzyme (g) constitutes a nucleotide (recall Plate 22). Chains of nucleotides form the strands of DNA (or RNA) that make up the genome of the virus. Nucleotides are linked by polymerase enzymes (g) during viral replication; and it is in these linkages that viral replication can be inhibited.

Acyclovir has a chemical constitution similar to a nucleoside. It can be attached to a phosphate group by an enzyme to form a nucleotide.

Color the title (i) and the X in the lowest strand of viral DNA, as well as the rest of the components of the strand. Use a sharp, contrasting color for (i). Finally, color the nucleotide to the left of the X.

The nucleotide formed with acyclovir has one major defect: it lacks a suitable attachment point for the next nucleotide in the production of the viral nucleic acid stand. Thus, this nucleotide is called a "false nucleotide." When it is attached to the nucleic acid strand by a polymerase enzyme, it will not link the next nucleotide into the strand, and the strand "terminates" at that point (i). Viral nucleic acid synthesis ends abruptly wherever false nucleotides are inserted. Acyclovir, the chain terminator, thus slows and ultimately halts the replication process of the viruses. In doing so, it effectively interrupts the infective process.

ANTIVIRAL VACCINES AND DRUGS

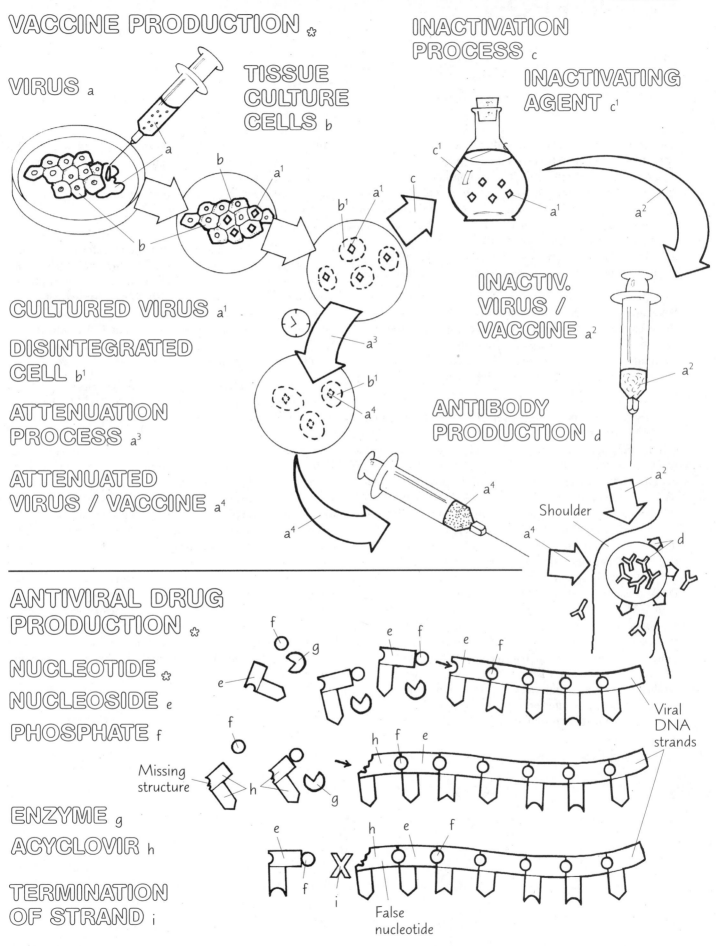

VACCINE PRODUCTION ✲

VIRUS a

TISSUE
CULTURE
CELLS b

INACTIVATION
PROCESS c

INACTIVATING
AGENT c¹

CULTURED VIRUS a¹

DISINTEGRATED
CELL b¹

ATTENUATION
PROCESS a³

ATTENUATED
VIRUS / VACCINE a⁴

INACTIV.
VIRUS /
VACCINE a²

ANTIBODY
PRODUCTION d

Shoulder

ANTIVIRAL DRUG
PRODUCTION ✲

NUCLEOTIDE ✲
NUCLEOSIDE e
PHOSPHATE f

Viral
DNA
strands

Missing
structure

ENZYME g
ACYCLOVIR h

TERMINATION
OF STRAND i

False
nucleotide

35
INTERFERON

Interferon is a complex protein produced by virus-infected cells of vertebrate animals. It is measurable in the cells within 48 hours of the initial infection. It is appropriately named as it "interferes" with viral RNA replication in the infected host cell. Thus, synthesis of viral proteins is inhibited, and replication of virus is prevented in these protected cells. This mechanism of interferon action is an important means by which the body resists viral infection. The sequence of activity in the production and action of interferon is the subject of this plate.

Set aside four shades of one color for (b) through (b³). Color the subheading Unprotected Cell, the titles (a) through (f), and the related structures in the upper drawing; delay coloring the structures below the bacterial cell and the lower illustration for now. Use a light color for (a). Color the arrows (*) gray.

A virus consists of a core of nucleic acid (b) surrounded by a protein coat (a; recall Plate 31). The virus enters the host cell through an interaction with the cell membrane (c). As the virus enters the cell, its nucleic acid is released into the cytoplasm. On reaching the nucleus of the host cell, the viral nucleic acid is integrated into the cell's DNA. Viral mRNA (b¹) is produced, enters the cell's cytoplasm and attaches to the cell's ribosomes (e). Here viral protein is produced. Soon thereafter, new viruses are replicated within the cell (a¹) and (b²) and released into the external environment.

After the host cell is infected by the virus, the cellular DNA is also induced to produce cellular messenger RNA (d¹) with instructions to code a new protein. Available amino acids are bound at the ribosomes (e) and the new protein, interferon (f), is synthesized and exported. It appears that viral RNA is a more potent stimulator of interferon production than viral DNA. The infected cell is unresponsive to its own antiviral product and remains unprotected from viral destruction.

The amount of interferon produced by infected cells is very small. Further, production of effective interferon is generally species-specific; that is, only human cells can produce interferon effective in humans. However, interferon is a useful protective agent against a wide variety of viruses. Large scale genetic recombination techniques can be used to produce significant amounts of interferon.

Color the viral particles (a¹), (b²) and interferon molecules (f) between the two cells. Then color the subheading Protected Cell and the titles and related structures associated with the lower illustration. Start with the cell membrane and the receptor site (c¹), following the interferon and the arrow to the cellular DNA, and thence to (d¹), (e), and (g). Finally color the sequence beginning with (b²) and finishing with (b³).

The newly synthesized interferon is released into the external environment and distributed by the flow of tissue fluids between cells, in lymph vessels, blood vessels, and so on. Molecules of interferon attach to the receptor sites (c¹) of other cells in very precise ways. Once attachment is secured, interferon is transported through the cell membrane into the cytoplasm. Now, interferon induces the cellular DNA to synthesize cellular enzymes (g) that interfere with the formation of viral proteins by destroying the viral mRNA (b³). Degraded viral mRNA particles are discarded, the viruses are unable to synthesize proteins necessary to their structure, and viral replication is terminated.

Since the early 1980s, interferons have been made available for several viral diseases, including hepatitis and herpes zoster ("shingles"). Interferon shrinks various viral-induced cancers, and is used for some forms of leukemia and is being tested for acquired immunodeficiency syndrome (AIDS). At some point in the future, interferons may become the equivalent of the antibacterial drug penicillin.

INTERFERON

UNPROTECTED CELL ✿

VIRAL COAT a
NUCLEIC ACID b

CELL MEMBRANE c
CELL DNA d

VIRAL mRNA b¹
REPLICATED
VIRAL COAT a¹
REPLICATED
NUCLEIC ACID b²

CELL mRNA d¹
RIBOSOME e
INTERFERON f

PROTECTED
CELL ✿

RECEPTOR SITE c¹

ENZYME g

DEGRADED
VIRAL mRNA b³

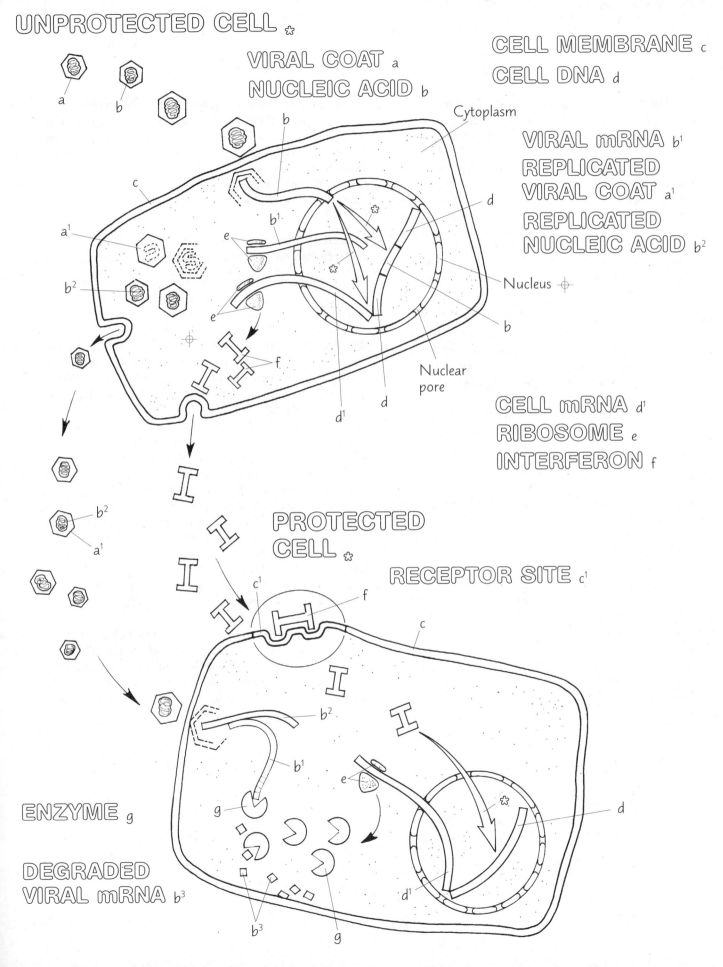

36
RETROVIRUSES

Transcription is a process of transferring the genetic information within a strand of DNA to a strand of messenger RNA (mRNA). This is the case for DNA viruses; that is, viruses that have DNA as their nucleic acid or genome. Some of the RNA viruses have enzymes that permit the transfer of genetic information from a strand of RNA to a strand of DNA. The group of viruses with this capacity are called retroviruses (retro-, backward). The enzyme utilized for the process is called reverse transcriptase. Retroviruses are unique for their ability to remain with the host cells for relatively long periods, integrated with the cell's chromosomes. This plate examines the mechanism by which retroviruses infect a human cell and use the metabolic machinery of the cell to replicate.

Set aside four shades of the color used for (c). Look over the plate and note that the illustrations are arranged into six stations. Each of the subheadings associated with these stations receive the color gray. Starting at station 1, color the titles and structures (a) through (g) at stations 1 through 4. Delay coloring stations 5 and 6 for now. Stay with the text while coloring.

The retrovirus exhibits the icosahedral form of capsid (b) enclosed with an envelope (a) characterized by spikes. Within the capsid is found the nucleic acid genome (c). The genome of the retrovirus consists of two strands of ribonucleic acid (RNA) of which only one strand is shown. The capsid also encloses molecules of the enzyme reverse transcriptase (d).

In the infective process, the envelope of the retrovirus makes contact with and attaches to the cell membrane (e) of the host (station 1: infection). This fusion of envelope to membrane permits the capsid (and the two-strand RNA genome) of the virus to enter the host cell cytoplasm. The envelope disintegrates outside the host cell. This method of cell entry is different than that employed by the bacteriophage (recall Plate 29). The capsid is degraded within the cytoplasm by enzymes (f) and the viral genome is uncoated (station 2: uncoating); that is, the capsid is broken down, exposing the double-strand RNA. The reverse transcriptase brought in with the virus then facilitates the formation of a strand of DNA (c^1) using viral RNA as a template (station 3: reverse transcription). The strand of newly-formed viral DNA enters the cell's nucleus to be incorporated into one of the cell's chromosomes (g). The cell nucleus now has sufficient information to set in motion the synthesis of viral RNA as well as viral protein.

Color the subheadings at stations 5 and 6, and related titles and structures. Read the corresponding text while coloring.

A substantial period of time (days, months, or years; called a latency) may pass before the synthesis of viral RNA is activated. Research workers are currently seeking to determine what stimuli turn the mechanism on. During this latency period, the infected individual may manifest no symptoms or signs of disease. However, when the production of viral protein begins, it consumes the cell's metabolic apparatus. Carried out in thousands of cells virtually simultaneously, the deleterious effects on body functioning are substantial.

Once activated, the viral DNA incorporated in the cell's chromosomes codes for the production of viral messenger RNA (c^2) as well as the genome RNA (c^3; station 4). These nucleic acids pass out through the cell's nuclear membrane. The mRNA moves to the cell's ribosomes (h) where protein for the capsid, envelope, and reverse transcriptase is synthesized (station 5: synthesis/assembly). Fragments of viral RNA in the cytoplasm are enclosed by the developing capsids. As more and more viruses are formed, the cytoplasm swells and the membrane bursts, flooding the external environment with thousands of new retroviruses (station 6: release). The lipoprotein viral envelope is derived from the host cell membrane.

It is believed that retroviruses are responsible for initiating tumors. At least one form of human leukemia is related to a retrovirus. Another retrovirus, the human immunodeficiency virus (HIV) is known to cause AIDS. For these reasons, retroviruses are under intense research investigation.

RETROVIRUSES

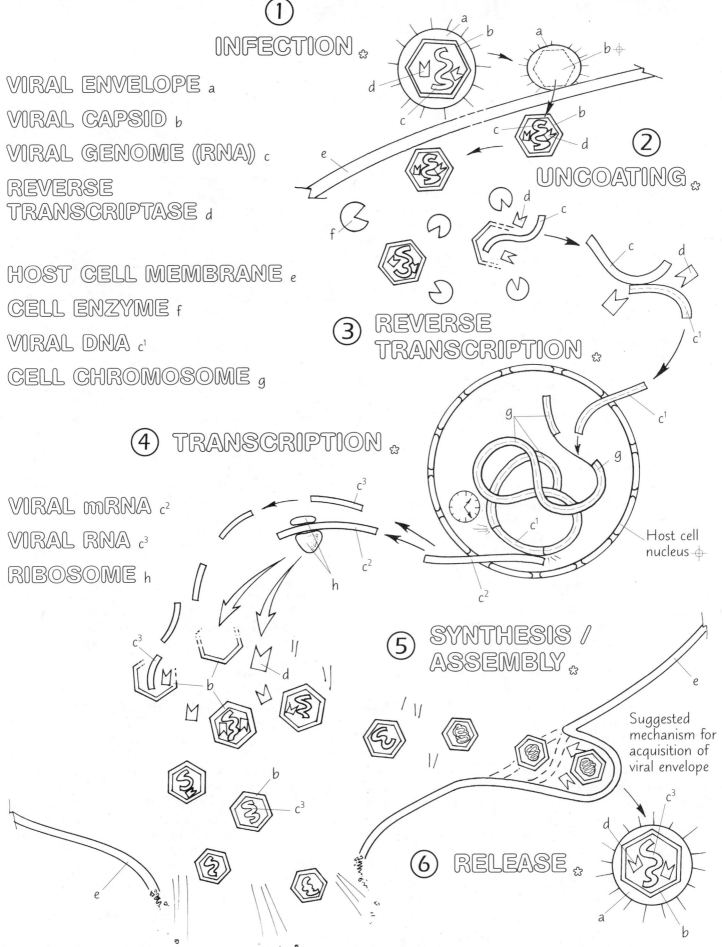

VIRAL ENVELOPE a

VIRAL CAPSID b

VIRAL GENOME (RNA) c

REVERSE
TRANSCRIPTASE d

HOST CELL MEMBRANE e

CELL ENZYME f

VIRAL DNA c¹

CELL CHROMOSOME g

VIRAL mRNA c²

VIRAL RNA c³

RIBOSOME h

① INFECTION

② UNCOATING

③ REVERSE TRANSCRIPTION

④ TRANSCRIPTION

⑤ SYNTHESIS / ASSEMBLY

⑥ RELEASE

Host cell nucleus

Suggested mechanism for acquisition of viral envelope

INTRODUCTION TO FUNGI

Fungi are a diverse group of microorganisms occupying a taxonomic kingdom separate from the plant and animal kingdoms. The group includes fleshy fungi (such as mushrooms), lichens, molds, and yeasts. There are well over 80,000 species of fungi. All have eukaryotic cells; that is, fungi have nuclei containing DNA, nucleoli with RNA, and nuclear membranes. The cytoplasmic organelles include mitochondria, ribosomes, endoplasmic reticulum, and other specialized cellular structures; fungi do not, however, contain chloroplasts or chlorophyll. Fungi are generally nonmotile. They are saprobic (feeding on decomposed organic matter) or parasitic (feeding on living organisms). They absorb nutrients because they have no way of ingesting them. Found largely in the moist habitats of the world, they may be terrestrial or aquatic. Fungi are sporeformers, by either asexual or sexual means. Fungi that reproduce by sexual stages are called perfect. Imperfect fungi reproduce asexually. Many fungi are capable of both sexual and asexual reproduction.

Fungi are taxonomically arranged into two major subdivisions, or phyla: the true fungi (Eumycotina) and the slime molds (Myxomycotina; not shown). Eumycotina has five groups or classes. Examples of four of these groups are presented on this and the next two plates. Examples of imperfect fungi (class Deuteromycetes) are seen on Plate 92.

Set aside five shades of the same or similar colors for (a) through (a⁴). Color the title Mold Structure a dark gray, as well as the mold on the slice of bread (*¹). Color the titles and related structures (a) through (b) in the upper third of the plate. Then color the title Yeast Structure (*²) and the related cells (c).

Fungi have two basic forms: mold (*¹) and yeast (*²), although some fungi may have both forms. Molds are characterized by filamentous hyphae. The furry quality of molds growing on bread or other food is a reflection of these microscopic filaments. Yeasts are round or oval cells, generally do not have hyphae and do not present a furry appearance.

Hyphae (hypha, sing.) are filaments of living substance. They contain nuclei and cytoplasm and a cell wall composed of chitin or cellulose. Hyphae may be continuous streams of cytoplasm; such fungi are said to have nonseptate or coenocytic hyphae (a). Other hyphae may be divided by walls called septa (septum, sing.); these are septate hyphae (a¹). Hyphae grow from single cells. As they develop in the substrate (e.g., old bread), these hyphae form a mass of intertwining filaments called a mycelium (a²). A non-reproductive mycelium is called a vegetative mycelium (a³; also called a thallus). Specialized hyphae that project up into the air from vegetative mycelia are called aerial mycelia (a⁴). These specialized hyphae and their tips have different names according to their structure (condiophores, sporangiophore, sporangia, etc.). The fruiting body (b; sporangia, ascocarp, etc.) contains the spore producing structures.

Yeasts (c) consist of round or oval unicellular structures. They generally do not form true hyphae. Yeasts reproduce asexually; when they reproduce by budding, the newly formed cell may stick to the parent cell giving the appearance of hyphae (pseudohyphae). Bread yeasts belong to the genus *Saccharomyces* of the class Ascomycetes.

Color the title Eumycotina, the class of fungus presented (Oomycetes) and the related structures and titles (d) through (j), including (a⁴), (b) and (b¹). Refer to the titles (a) through (c) at the upper part of the plate for the related structures here. When the sexual cycle is completed, color the subheadings Sexual Reproduction and Asexual Reproduction; then color the asexual cycle.

Fungi of the class Oomycetes consist largely of aquatic organisms called "water molds"; some are terrestrial soil fungi. Species of this class cause downy mildew of grapes (*Plasmopara viticola*) and potato blight disease (*Phytophthora infestans*). The water molds are generally parasitic, feeding on plant material and fish. Recent research on the ultrastructure of Oomycetes organisms casts some doubt on their status as true fungi. It is held by some investigators that Oomycetes qualify as a separate division.

Most species of Oomycetes have mycelia composed of nonseptate hyphae (a). In sexual reproduction, two specialized hyphae develop in one part of the vegetative mycelium. One becomes the male cell or antheridium (e) and the other becomes the female cell or oogonium (f). The oogonium has one or more egg cells called oospheres (g). When the antheridium contacts the oogonium, the nucleus (i) of the antheridium passes through the cell wall into one of the egg cells in the oogonium. This fertilized oosphere is called an oospore (h). It is discharged from the oogonium as a diploid cell (paired chromosomes), and undergoes a reductional (j; meiotic) division to become a haploid cell (with unpaired chromosomes). The oospore germinates to form a hypha (a) which progressively develops into a vegetative mycelium (a³), completing the sexual cycle.

In the asexual mode of reproduction, a specialized hypha grows from the vegetative mycelium. The shaft of this hypha is called a zoosporangiophore (a⁴). The tip of the hypha forms a spore-containing body called a zoosporangium (b). The spores, called zoospores (b¹), are released and ultimately germinate to form new hyphae. Sexual reproduction is preferable because genetic variability is possible due to the mixing of nuclear material, but asexual material generally yields more spores than sexual reproduction.

INTRODUCTION TO FUNGI

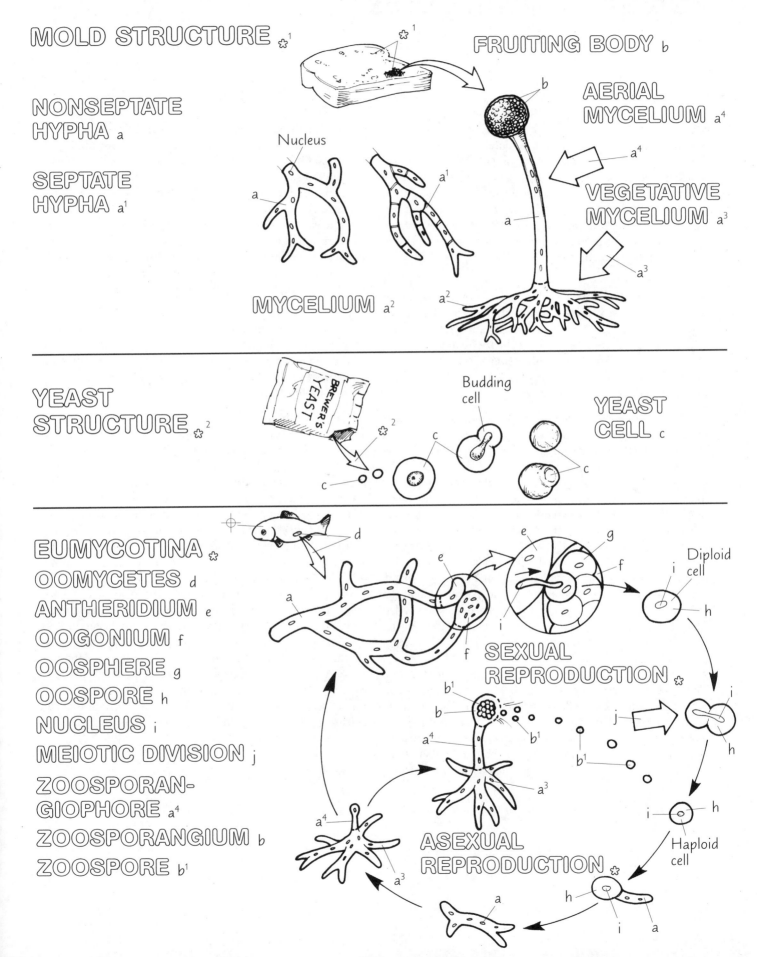

MOLD STRUCTURE ✿¹

FRUITING BODY b

NONSEPTATE
HYPHA a

AERIAL
MYCELIUM a⁴

SEPTATE
HYPHA a¹

VEGETATIVE
MYCELIUM a³

Nucleus

MYCELIUM a²

YEAST
STRUCTURE ✿²

BREWER'S YEAST

Budding
cell

YEAST
CELL c

EUMYCOTINA ✿
OOMYCETES d
ANTHERIDIUM e
OOGONIUM f
OOSPHERE g
OOSPORE h
NUCLEUS i
MEIOTIC DIVISION j
ZOOSPORAN-
GIOPHORE a⁴
ZOOSPORANGIUM b
ZOOSPORE b¹

Diploid
cell

SEXUAL
REPRODUCTION ✿

ASEXUAL
REPRODUCTION

Haploid
cell

FUNGI: ZYGOMYCETES

Rhizopus stolonifer belongs to the class Zygomycetes of the phylum Eumycotina. It is a common fungus whose spores may be seen developing on a slice of bread as black or gray bread mold. It is a terrestrial fungus having nonseptate hyphae. In rare instances, it may be parasitic to animals and plants, but most strains feed on dead organic matter. *Rhizopus* spores travel by air. This fungus can reproduce both sexually and asexually. Both methods of reproduction occur in the same mycelium, and they are illustrated here.

Set aside four shades of the same color for (a) through (a³); plan to use them in both reproductive cycles. Color the subheadings at the top of the plate and the titles and structures (a) through (e), moving clockwise and following the arrows from top left. Consider using shades of the same or similar colors for (b) through (e). Avoid obscuring illustrative detail with dark colors.

Both reproductive cycles of *Rhizopus stolonifer* can occur in the same organism on the same substrate, in this case a slice of contaminated bread. Note that the upper half of the plate shows the sexual reproductive cycle. We begin with the vegetative mycelium (a). Two specialized, sexually-opposite hyphae, called plus (a¹; +) and minus (a²; –) hyphae, come into close contact. Both hyphae may be present in the same mycelium or in different strains and different mycelia of a fungus. A hypha makes contact with a sexually-opposite hypha by chemical attraction. Chemical activity induces the formation of pouchlike outgrowths, called progametangia (b), at the junction of the sexually-opposite hyphae. At some point the progametangia become supported by hyphal extensions called suspensors, and two gametangia (c) are formed. Haploid nuclei from each plus and minus hypha fuse to form diploid nuclei in each gametangium. The paired gametangia fuse to become a single, multinucleated zygote (d) containing diploid nuclei. The zygote develops into a thick, rough-surfaced, heat- and pressure-resistant form called a zygospore (d¹). The zygospore is characteristic of the fungi of this class (Zygomycetes). It enables the fungus to survive environmental fluctuations.

The zygospore remains in the mycelium until conditions are favorable for growth to occur, such as the presence of sufficient nutrients, water, and favorable pH environment. At this time, the diploid nuclei in the zygospore undergo a reduction division, and the chromosome number is halved to the haploid state. These haploid nuclei within the zygospore form meiospores (e). The zygospore soon germinates, characterized by the rupture of its thick wall. The meiospores are released into the environment and blown about by air currents. Eventually, they come to rest on a suitable medium. Each spore germinates to form a nonseptate hypha (a³). Development of the hypha results in the formation of a vegetative mycelium, and the sexual cycle is complete.

Plan on using the same colors in the lower part of the plate as you did on the upper part, as applicable. Color the subheading Asexual Reproduction and the related titles and structures (a) through (j). Begin with (a³) just below the slice of moldy bread.

The mycelium of the fungus *Rhizopus stolonifer* has both sexual elements (plus and minus hyphae) and vegetative elements. The development of the sexual elements has been presented above. Specialized hyphae of the vegetative mycelium (a), called stolons (f) grow across the surface of the bread substrate, and penetrate it, expanding to form extensions of the mycelium or communicate with other mycelia. Short, rootlike hyphae, called rhizoids (g) submerge from the main body of the mycelium deep into the substrate to provide a secure attachment. From each mycelium develops a number of specialized hyphae that grow into the air. These aerial hyphae are called sporangiophores (h). The tips of these structures swell to form saclike structures called sporangia (i; sing. sporangium). Many mitotic divisions occur in each sporangium, forming a single cluster of haploid spores called sporangiospores (j). The collection of these sporangiophores, sporangia, and sporangiospores constitute aerial mycelia. Sporangiospores are cast into the environment following rupture of the sporangia to be carried by air currents to a new substrate. Here they germinate, form hyphae (a³), and the asexual cycle is complete.

FUNGI: ZYGOMYCETES

RHIZOPUS
STOLONIFER ☆[1]

SEXUAL
REPRODUCTION ☆

MYCELIUM a

HYPHA a^3

MEIOSPORE e

(+) HYPHA a^1
(−) HYPHA a^2

PROGAMETANGIA b

GAMETANGIA c

ZYGOSPORE d^1

ZYGOTE d

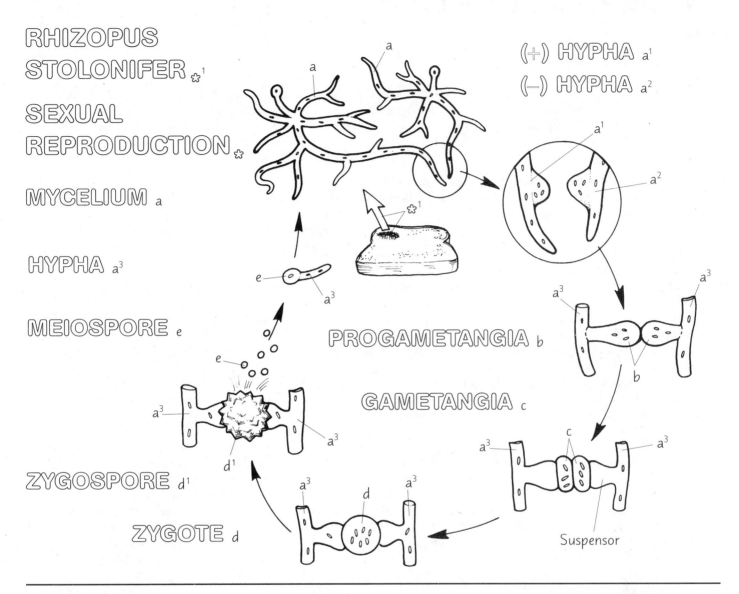

Suspensor

ASEXUAL
REPRODUCTION ☆
VEGETATIVE
MYCELIUM a
STOLON f
RHIZOID g
SPORANGIOPHORE h
SPORANGIUM i
SPORANGIOSPORE j

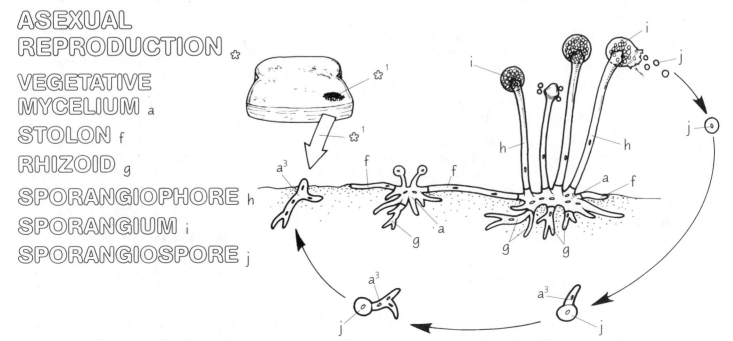

39
COMPLEX FUNGI

The complex classes of fungi include the Ascomycetes and the Basidiomycetes. Members of these classes form complex mycelia that are often compact and visible to the unaided eye. They are commonly found above the ground, as in mushrooms.

Color the subheadings Ascomycetes and Asexual Reproduction gray. Color the titles and structures (a) through (c^1) associated with asexual reproduction.

Ascomycetes, also called sac fungi, include all fungi that form ascospores within sacs called asci during the process of sexual reproduction. Ascomycetes are perfect fungi, capable of both asexual and sexual reproduction. Certain members of this class are known to us in the form of brewing and baker's yeast. Molds of the genus *Penicillium* produce the antibiotic penicillin. Other genera of this class act on milk curds to produce Roquefort and other cheeses. Harmful organisms of this class include the pathogens that cause grape powdery mildew, dutch elm disease, ergot of rye, and root destruction of trees. A member of this class is also responsible for aspergillosis, a fungal disease of the lung. Although many of these organisms are parasitic, some are saprobic as well.

Ascomycete fungi develop large vegetative mycelia (not shown) from septate hyphae (a). During asexual reproduction, specialized hyphae project away from the mycelia to form long stalks called conidiophores (b). The tips of the conidiophores terminate in unprotected chains of spores. The spores are called conidiospores or conidia (c; conidia, dust). Contrast these spores with sporangiospores contained in sporangia in certain Zygomycete fungi (Plate 38). These conidiospores are blown free with air currents, find a substrate, undergo budding (c^1) and germinate to form new hyphae (a).

Color the subheading Sexual Reproduction gray at the upper right of the plate, and then color the titles and structures (a) through (e), following the arrows, associated with sexual reproduction of certain Ascomycetes organisms. Note that (b) and (c) are not used in the sequence from (a) to (e) in this section.

At the ends of certain hyphae in the vegetative mycelium, sexually-opposite hyphae form. The male hypha is called an antheridium (a^1); the female hypha is called an ascogonium (a^2). These specialized hyphae fuse, and after a number of nuclear divisions, the multinucleated terminal cell is called an ascus (d). The nuclei of the ascus are incorporated into the four to eight ascospores (d^1) that form within each ascus. Some asci remain independent; in some species, as shown here, the asci come to be enclosed in a fruiting body called an ascocarp

(e). After a time, during which the ascocarp may become quite large, the ascospores mature, burst from their asci, and escape the ascocarp to the outside environment. These ascospores migrate and germinate in the appropriate substrate to form hyphae (a). This completes the sexual cycle.

Color the subheadings Basidiomycetes and Sexual Reproduction gray. Then color the titles and related structures (a) through (g), starting at the top and following the arrows clockwise, from stations 1 through 6.

Members of the class Basidiomycetes of the true fungi (phylum Eumycotina) include mushrooms, puffballs, shelf fungi, rusts, smuts, and others. Members of the class have septate hyphae. Certain species form large fruiting bodies, best seen as umbrella-shaped structures in the common mushroom (*Agaricus campestris*). Some species of this class are parasitic, causing smut and rust disease in plants. Species of this class are capable of both asexual and sexual reproduction. Hyphal cells can bud or fragment from the parent structure and form new hyphae asexually. Basidiomycetes organisms also produce conidia (a type of asexual spore; not shown) as a form of asexual reproduction. Here we limit the illustration to the sexual cycle of the common mushroom.

Septate hyphae (a) develop under the soil from basidiospores released into the air. Within the developing vegetative mycelium (not shown) there subsequently develop sexually-opposite hyphae (a^1) and (a^2), which fuse (station 1). Through a complex process of nuclear divisions and cell divisions of the fused hyphae, a variably shaped terminal cell, called a basidium (f), is formed. Four spores, called basidiospores (f^2), develop on the surface of each basidium (see the circled illustration at station 5). Progressive growth and reproduction of the hyphae and basidia result in the formation of button-shaped basidiocarps (f^1; station 3). In the case of the mushroom, it will develop a cap-shaped structure growing on a stem called a stipe (g; station 4). On the underside of the cap can be seen linear structures (intertwining hyphae) called gills. From the gills are suspended the conical, spore-bearing basidia (station 5). The basidiospores are released from the basidia into the air. The spores land on the ground and form hyphae, thereby repeating the cycle.

One species of mushrooms of this class forms a large circle on the ground called a fairy ring (station 6). Here exists a wide, circular array of mycelia that have formed from subterranean, radially-branching mycelia. On the edge of this fungal mass, new fruiting bodies (f^1) emerge from the soil, forming the circle. In earlier times, it was believed that fairies danced around this ring of growing mushrooms.

COMPLEX FUNGI

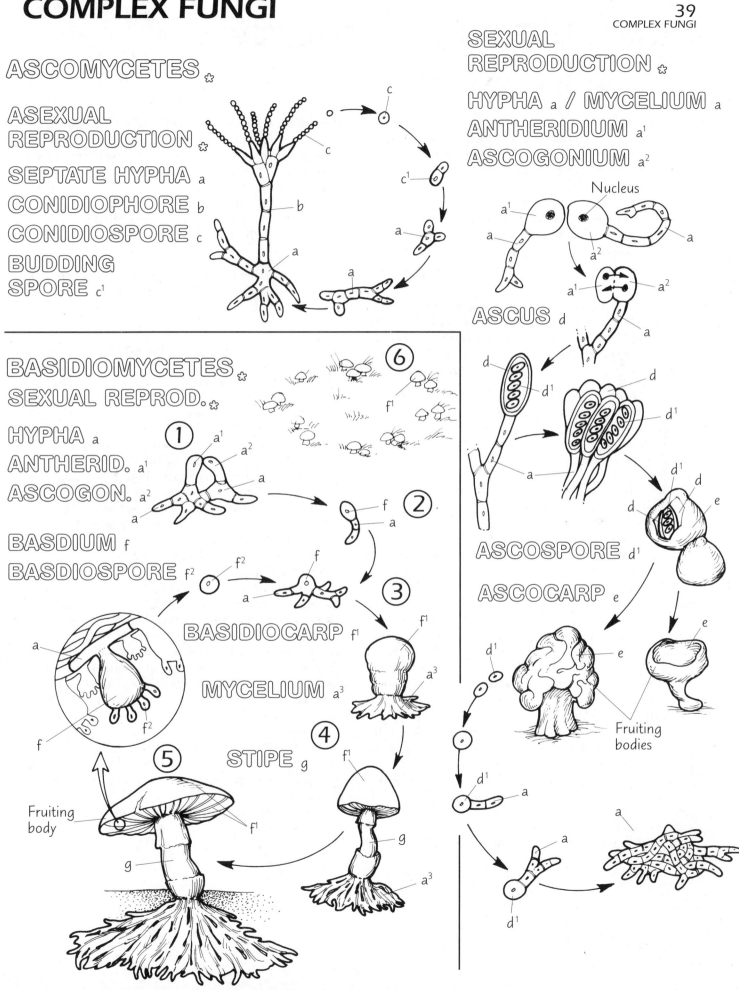

ASCOMYCETES ✿

ASEXUAL REPRODUCTION ✿

SEPTATE HYPHA a
CONIDIOPHORE b
CONIDIOSPORE c
BUDDING SPORE c¹

SEXUAL REPRODUCTION ✿

HYPHA a / MYCELIUM a
ANTHERIDIUM a¹
ASCOGONIUM a²

Nucleus

ASCUS d

BASIDIOMYCETES ✿
SEXUAL REPROD. ✿

HYPHA a
ANTHERID. a¹
ASCOGON. a²

BASDIUM f
BASDIOSPORE f²

BASIDIOCARP f¹

MYCELIUM a³

STIPE g

Fruiting body

ASCOSPORE d¹

ASCOCARP e

Fruiting bodies

The protozoans are a group of about 30,000 species of eukaryotic microorganisms found worldwide and generally considered to be at the first level (proto, first) in the hierarchy of the Animal Kingdom. Protozoa are of particular interest to microbiologists because of their microscopic size and their parasitic role in several important human and animal diseases. Significant diseases caused by protozoans are the topics of Plates 88 to 91. Protozoans have a highly significant impact on aquatic ecology, as they are a principal food source to an incredible range of organisms, from shrimp to baleen whales.

Color the titles and related structures (a) through (k) in the upper part of the plate. The illustration (a) is a longitudinal section through a typical but non-specific protozoan organism. Select a very light color for (c); use light colors generally to avoid obscuring cell detail.

The structure of protozoans varies from gelatinous masses to rigid, skeletal forms. They vary in size from a few micrometers in length (invisible without a microscope) to five centimeters (two inches). Protozoans are generally considered as unicellular animals; their function is as complex or more complex than any other living cell.

Protozoans have a typical cell membrane (b); that is, a membrane composed of a bilipid layer and globular protein. In addition, some protozoans have protective coverings external to the membrane. The cell membrane permits the ingestion and excretion of material by processes of exocytosis and endocytosis. The cytoplasm (c) of protozoans contain numerous organelles, including the energy-generating mitochondria (d), and the usual aggregation of protein-synthesizing structures, such as ribosomes and endoplasmic reticulum (ER; e) and Golgi bodies (f). Associated with these organelles are water and food vacuoles (g), and lysosomes. Protozoans commonly have pigments in their cytoplasm; in some species, these pigments are found in chloroplasts (h) which use sunlight to generate energy (photosynthesis). Each protozoan has a nucleus (i) with a porous nuclear membrane (i^1). The nucleus contains the genetic material (DNA). In certain species, there are two nuclei, one for regulating cellular metabolic activity, the other for reproduction.

Locomotion in protozoans is achieved by flagella (j), pseudopodia (j^1), or cilia (j^2). Flagella may extend in front of the organism, spiral around the organism, or power the protozoan from behind. The whipping action of this tail-like device propels the organism. Pseudopodia are dynamic membrane-lined prolongations of cytoplasm that protrude in front of the animal, like a finger extending from a closed fist. A pseudopodium attaches to the surface, and the rest of the body moves forward, as if it were being pulled by the pseudopodium (see the Amoeba, this plate). Cilia function as collections of tiny oars, beating in unison, power stroking in one direction, resting in the other, resulting in the movement of the animal (see Balantidium, this plate).

Color the titles and related organisms and parts of organisms for each of the four classes shown. Use light colors for the larger areas.

The class Sarcodina consists of protozoans many of which employ pseudopodia. The most well known of this class is the pond water amoeba (k). A cyst-forming type of amoeba (*Entamoeba histolytica*) enters the human gastrointestinal tract by means of impure drinking water. It feeds on the lining tissues, forming inflamed craters or ulcers (amebiasis).

Protozoans of the class Mastigophora constitute an enormous number of organisms which move by means of flagella (j). Examples of this class include the euglena (l), a common inhabitant of pond water. Euglena is capable of photosynthesis due to the presence of chloroplasts in its cytoplasm. This class includes many parasitic protozoans responsible for a number of infections in humans (Plates 88, 90, and 91). One such organism is trichomonas (l^1), responsible for an infection of the female genital tract (trichomoniasis).

Protozoans of the class Sporozoa are non-motile spore producers. These organisms have life cycles in which intermediary forms resemble bacterial or fungal spores (not shown) but are not nearly as resistant to environmental fluctuations as true spores. The toxoplasma species (m) are often crescent-shaped; species of *Toxoplasma gondii* are pathogenic in humans and animals. This organism is transmitted by cats, and manifestations of the disease (toxoplasmosis; refer to Plate 89) include a broad range from mild, short-lived symptoms to severe injury to the brain and other organs.

The class Ciliophora contains the greatest number of species in the phylum Protozoa. Members of this class move by way of cilia (j^2). One representative member of this class is the balantidium (n). *Balantidium coli* is known to cause ulcers of the colon in humans, associated with diarrhea and dysentery (balantidiasis; Plate 91).

INTRODUCTION TO PROTOZOA

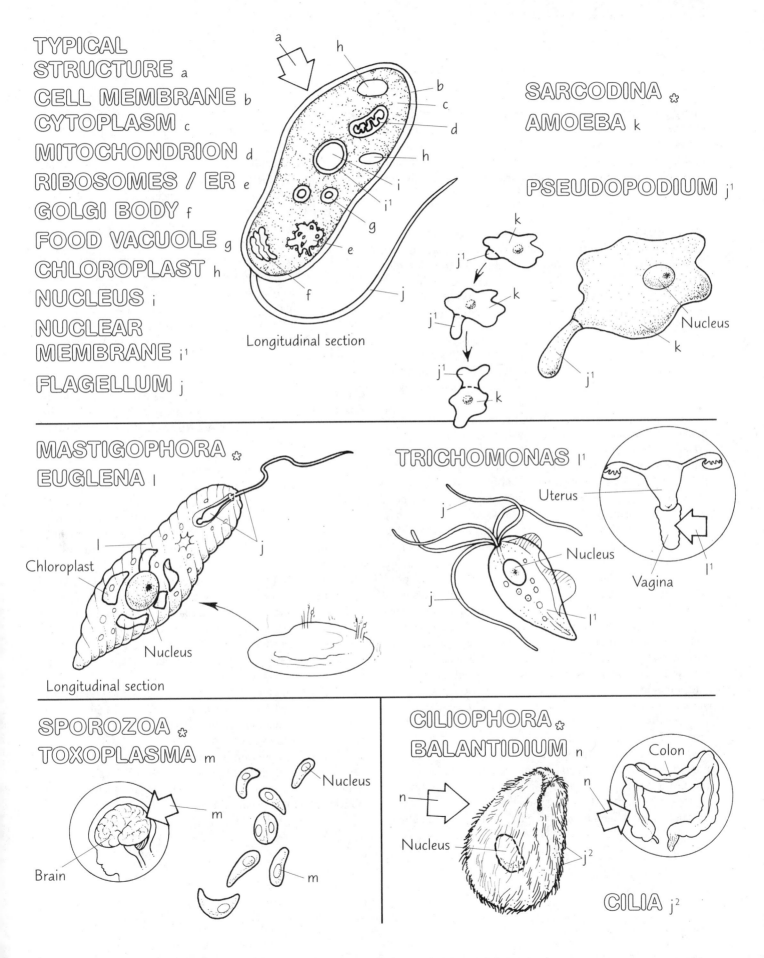

TYPICAL
STRUCTURE a
CELL MEMBRANE b
CYTOPLASM c
MITOCHONDRION d
RIBOSOMES / ER e
GOLGI BODY f
FOOD VACUOLE g
CHLOROPLAST h
NUCLEUS i
NUCLEAR
MEMBRANE i¹
FLAGELLUM j

Longitudinal section

SARCODINA ✳
AMOEBA k

PSEUDOPODIUM j¹

Nucleus

MASTIGOPHORA ✳
EUGLENA l

Chloroplast
Nucleus
Longitudinal section

TRICHOMONAS l¹

Uterus
Nucleus
Vagina

SPOROZOA ✳
TOXOPLASMA m

Nucleus
Brain

CILIOPHORA ✳
BALANTIDIUM n

Colon
Nucleus
CILIA j²

41
PARAMECIUM

Many protozoa are relatively simple, unicellular organisms. Others, however, display a level of complexity observed in multicellular eukaryotes. The paramecium, a representative protozoan of the class Ciliophora, is one of these. In this plate, we explore the cellular structures and their function in this common protozoan.

Start with the main drawing, coloring titles and structures (a) through (l¹). Color the locomotion sequence (a) as you read about it in the text. Use a very light color for (e). Color the ectoplasm (b) with a light color, and then color over the trichocysts (d) with a darker color. Avoid obscuring detail with colors that are too dark. Color over the cilia with the color chosen for (c).

The central illustration here is a slice or section taken through the long axis of the organism. In real life, we would be looking at the organelles within the cytoplasm through a transparent or translucent pellicle.

The outer limiting structure of the paramecium (a) is the gelatinous ectoplasm (b) contained within a flexible, protective covering, the pellicle (a¹; the pellicle is seen cut on edge in the main illustration. Note the large arrow pointing to it). Within the ectoplasm, arise thousands of cilia (c; sing., cilium), projecting out into the external environment. The synchronous beat of the cilia make possible a forward and rotary motion of the paramecium, as illustrated at upper left. Within the ectoplasm, between the bases of the cilia, are arranged a single row of tapered, bottle-shaped trichocysts (d). These organelles discharge long whip-like threads that capture smaller prey, defend the cell against attack, or anchor the cell to a secure surface (as shown) while feeding. The granular, somewhat viscous cytoplasm of the paramecium is called endoplasm (e). Here are found most of the organelles of the paramecium. The genetic material (DNA) of the paramecium is located in both the larger macronucleus (f) and the micronucleus (f¹). The micronucleus is concerned with reproduction; the macronucleus provides the genetic codes for metabolic activities.

The lower surface of the paramecium on one side displays an extended depression called the oral groove (g). This groove is open to the outside. Internally, the groove opens into an oral pore (h; also called a cytostome) that opens into a tubular gullet (i). Cilia around and in the oral groove sweep smaller organisms toward and into the gullet. As the gullet fills with nutrient material, the sac-like end pinches off to form a bubble-like food vacuole (j). The food vacuole circulates around the endoplasmic interior (note arrows) during which food diffuses out of the vesicle. As its contents diminish, the vacuole gets smaller and ultimately disgorges the undigested remains through an anal pore (k).

A unique feature of the paramecium is the contractile vacuole (l). As water diffuses into the organism, a physiologically suitable volume is reached. As excess water enters, the overage diffuses into water vacuoles through feeder canals (l¹). Once full, the vacuole contracts (hence its name), forcing water through an excretory pore to the outside (see the dotted line leading to the outside identified by H_2O; it receives no color). By balancing the content and pressure of the internal water environment, the paramecium maintains acceptable water and salt relationships while ridding itself of dissolved, possibly toxic waste products.

Color the subheading Sexual Reproduction gray, and the related structures at the lower part of the plate. Finally, color the subheading Asexual Reproduction gray, and the duplicating paramecia.

Paramecia display a form of sexual reproduction somewhat similar to that of bacteria (not shown). Begin at station 1 and note that two paramecia come along side one another, make contact, and form a cytoplasmic bridge (conjugation, station 2). The micronucleus of each organism duplicates twice to form four micronuclei. Three of the micronuclei fragment and disappear, leaving one micronucleus per organism (station 3). That micronucleus divides, forming a pair of micronuclei. The conjugated paramecia exchange one of their micronuclei (station 4). Each new pair of micronuclei unites to form a single micronucleus (station 5). The macronucleus in each paramecium, having dissolved and disappeared at station 2 of the reproductive process, reforms in each organism (station 5).

The paramecium will then reproduce by simple mitosis (binary fission, a form of asexual reproduction). In this process, genetic material duplicates, followed by organ duplication. The cell divides, forming two individuals.

Sexual reproduction in paramecia appears to take place during times of environmental stress. The genetic recombination may give rise to an organism that is better adapted to its environment because it is genetically different.

PARAMECIUM

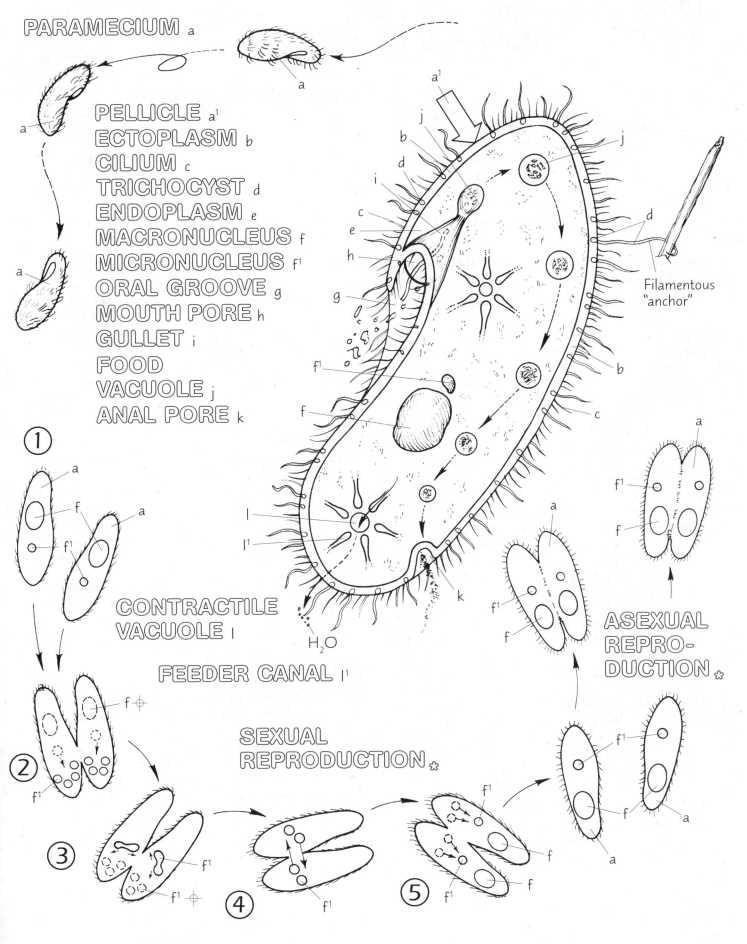

PARAMECIUM a

PELLICLE a¹
ECTOPLASM b
CILIUM c
TRICHOCYST d
ENDOPLASM e
MACRONUCLEUS f
MICRONUCLEUS f¹
ORAL GROOVE g
MOUTH PORE h
GULLET i
FOOD
VACUOLE j
ANAL PORE k

Filamentous
"anchor"

CONTRACTILE
VACUOLE l

FEEDER CANAL l¹

H₂O

SEXUAL
REPRODUCTION ✿

ASEXUAL
REPRO-
DUCTION ✿

42
PHYSICAL CONTROL OF MICROORGANISMS BY HEAT

Control of microorganisms is critical in a laboratory or hospital setting. Two general types of agents are available for limiting microbial populations: physical and chemical. Physical agents include heat, filters, and radiation; antiseptics, germicides, antibiotics, and disinfectants constitute the chemical agents (not shown). Physical agents control microorganism populations by sterilization. Sterilization is a process of destroying all manner of living things, including bacterial spores, the most resistant to destruction of all life forms.

Color the titles and related structures (a) through (f¹). Use light shades of the colors used for (d) and (f) for the killed bacterium (d¹) and the denatured protein (f¹).

The microbiologist often uses direct flame (a) as a means of killing bacterial spores (b). This is achieved by exposing an inoculating loop or the open end of a culture tube to a flame for a few seconds, killing spores and bacteria on the wire or the lip of the tube. The heat kills the spores or bacteria by decomposing the proteins, thereby destroying the enzymes and structure of the organism.

A hot air oven achieves sterilization by the application of dry heat (c). Two hours of dry heat at a temperature of 160° C (320° F) is required for effective destruction of microorganisms (d). The dry heat kills the bacteria by oxidizing their proteins and desiccating (drying) their cytoplasm (d¹). This method of sterilization is used with contaminated dry powders, oily substances, and glass instruments, such as syringes and flasks. Materials must be pre-cleaned before using dry heat. If this is not done, organic matter present on the object to be sterilized will shield the surface against dry heat, and destruction of the microorganisms will be defective.

Moist heat (e) penetrates the organism faster and generally effects sterilization more quickly than dry heat, even though the temperature (100° C) is lower than that of dry heat (160° C). A common moist heat method is simply boiling the contaminated material in water. At normal (sea level) pressure, most bacteria (d), viruses, fungi, and protozoa will die in boiling water within 10 minutes or so. Bacterial spores may survive two hours of boiling water, however. In the moist heat process, the heat disrupts the bonds of protein molecules, causing the protein to coagulate and become non-functional, killing the organism.

Plan to use four shades of the same color for the steam-containing parts (h) in the diagram. Color the subheading Autoclave gray, and color the titles and related parts (g) through (h⁴) in the lower half of the plate. Use contrasting colors for (h) and (i); use the lightest shade of (h) for (h³).

Perhaps the most dependable moist heat method for sterilization in hospitals, clinical laboratories, and research facilities is pressurized steam developed in an instrument called an autoclave. By increasing the pressure environment containing the microorganisms significantly above standard sea level pressure, the temperature of the steam can be raised and sterilization can be effected more quickly. It can sterilize most bacteriological media, instruments, and hospital supplies, but it cannot be used with plastics as they would melt.

In the case illustrated, the materials containing the microorganisms are placed in flasks. The flasks are placed in the sterilizing chamber (j⊕), and the door (g) is sealed by closing it and locking it with a locking device. A valve (h) at the bottom of the autoclave is opened, permitting steam into an entry pipe (h) and then a passageway (h¹, steam jacket) surrounding the chamber. As the steam enters the jacket, the existing air in the sterilizing chamber is eliminated by opening the valves of the air exhaust pipes (i; only the lower air exhaust pipe is illustrated).

Once air is evacuated from the chamber, the air exhaust valve is closed and steam (h³) enters the sterilizing chamber through a steam admit valve (h²) at the back of the chamber. The steam fills the sterilization chamber and builds up a pressure as long as the air exhaust and steam exhaust valves (h⁴) are closed. The pressure is allowed to build up from sea level pressure (14.7 pounds per square inch or psi) to 29.7 psi. If the pressure system reads sea level as 0 psi, then the pressure would be allowed to build to 15 psi. At this increased pressure, the temperature of the steam rises to 121.5° C (or 21.5 degrees above boiling at sea level). At sea level without pressurization, such a temperature could not be achieved.

Under these conditions, the destruction of bacterial spores will occur in 15 minutes or longer. The steam exhaust valve is then opened and the steam is exhausted through the steam exhaust pipes. Once the pressure reaches atmospheric level, the door is opened and the sterilized vessel is removed. The material within the flask is not recontaminated because the flask is sealed with a cap. If instruments are sterilized, then the instruments are wrapped in sterilizing paper or placed in sterilizing envelopes before the process is begun. As long as the sterilized material remains sealed, sterility of the material is assured.

PHYSICAL CONTROL OF MICROORGANISMS BY HEAT

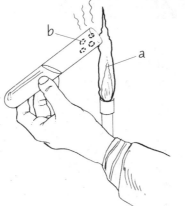

DIRECT
FLAME a
BACTERIAL
SPORE b

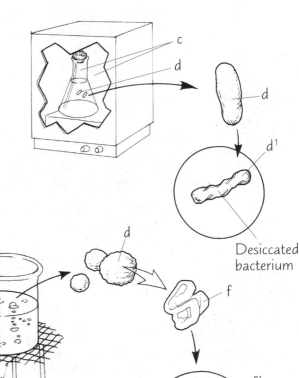

DRY HEAT c
BACTERIUM d
KILLED BACTERIUM d¹

MOIST HEAT e
BACTERIAL PROTEIN f
DENATURED PROTEIN f¹

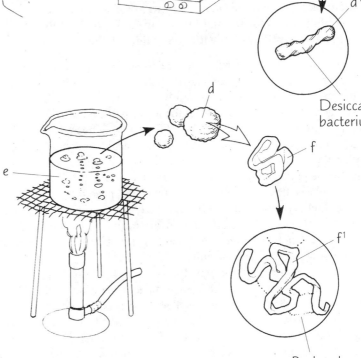

Desiccated
bacterium

Broken bond

AUTOCLAVE ✿

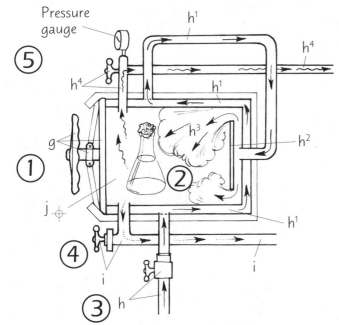

Pressure
gauge

⑤

DOOR / HANDLE g
STEAM ENTRY PIPE / VALVE h
STEAM JACKET h¹
AIR EXHAUST PIPE / VALVE i
STEAM ADMIT VALVE h²
STEAM h³
STERILIZING CHAMBER j
STEAM EXHAUST PIPE / VALVE h⁴

43
PHYSICAL CONTROL
OF MICROORGANISMS

Not all sterilization by physical agents can be accomplished by heat. Some materials to be sterilized are sensitive to heat or would be damaged by heat application. Thus, filtration, ultraviolet light, and ultrasonic vibration are other physical agents available to the microbiologist, though less commonly used than heat, for the control of microorganisms.

Color the subheading Filtration, and the titles and related structures (a) through (e¹) at the upper third of the plate. Select light colors for (b) and (e). Color the Pore Size title and the various filters (e¹) with a very light shade.

A filter is a mechanical device used for separating microorganisms from a contaminated solution (a). Here a depth filter (b) is inserted into the mouth of a flask especially designed for such a purpose. Note the lower part of the flask has a port or tubular opening. When this port is connected by rubber tubing to a vacuum pump (c), a partial vacuum is created within the flask, drawing air through the filter into the flask and out the port. The filtered fluid is called a filtrate (d). With some very fine filters, all of the microorganisms in the solution can be filtered out, leaving the filtrate free of microorganisms (sterile). However, soluble bacterial toxins and other pathogenic chemicals can remain. Therefore, filtration is useful for trapping microorganisms only.

Filters are generally designed in two ways: one to trap microorganisms on the top or front surface of the filter, and another to trap microbes throughout the filter matrix. The depth filter (b) traps microorganisms (a¹) throughout the filter. The depth filter is composed of a matrix of randomly-oriented fibers of cotton, fiberglass, or wool. These fibers are bonded together creating a tortuous maze of flow channels wherein microorganisms can be trapped at any one of many fibrous intersections. The depth filter has the advantage of a high capacity operation, but the pore size is random and much liquid is lost within the filter.

The screen filter (e) is composed of a rigid, uniform mesh of artificial fibers. The pore size is uniform throughout the filter; filters of various pore size (e¹) can be precisely manufactured and selected for a particular operation. Microorganisms are trapped only on the front or top surface. The screen filter is a low capacity filter, clogs easily, and requires frequent replacement.

Color the titles and parts (f) through (g) in the middle set of illustrations concerned with ultraviolet light. Color over the radiation lines (f¹) coming from the overhead light source.

Ultraviolet or "uv" light (f) produces ultraviolet radiation (f¹), a form of invisible energy with a very short wavelength between 100 nm and 400 nm. At a wavelength of about 265 nm, uv light is a valuable sterilizing agent. Ultraviolet light is used to limit contamination in spaces during the absence of human activity, such as operating rooms, toilet facilities, pharmacies, and food service operations. Virtually all microorganisms are susceptible to its effect, including the most resistant of all living things, bacterial spores.

In a hospital setting, for example, uv radiation may be used to de-contaminate the surgical suite during periods of non-use. When the uv radiation strikes living microorganisms (a¹), such as bacteria, the energy penetrates the walls of the organisms and is absorbed by DNA. Ultraviolet radiation interferes with positioning of nucleic acid base pairs during protein synthesis and DNA replication, effectively denaturing the DNA (g) and killing the microorganisms.

Color the titles and structures (h) through (j) in the lowest set of illustrations concerned with ultrasonic vibration. Follow the stations 1 through 4.

Ultrasonic vibrations (h) are high frequency sound waves undetectable by human ears. They can be conducted through a variety of media including living structure and fluids. In the example shown here, ultrasonic waves (h¹) are generated from a sound source and move through a solution containing surgical instruments contaminated with microorganisms (station 1). Movement of the sound waves through the fluid generates vibrations (pressure changes). When the pressure created exceeds the pressure of the fluid medium, the fluid ruptures (cavitation). In cavitation, microscopic bubbles or cavities (i) are formed. In areas of low pressure, the cavities rapidly expand (i¹; station 2); in areas of high pressure, the cavities rapidly collapse (i²; station 3). As a result of the sudden collapse, a tremendous pressure is put on the gas within the cavity (hundreds of times greater than atmospheric pressure at sea level). This pressure is released by the formation of shock waves (j) emanating from the compressed cavities (station 4). The shock waves apply strong forces to the surface of instruments in the fluid medium, cleaning them and rupturing the cell membranes of attached microorganisms, killing them.

Ultrasonic vibrations are not widely used as sterilizing agents because liquid is required and other methods are more efficient. Nevertheless, many laboratories use ultrasonic devices to clean instruments and effect reduction of the microbial population. In these instances, sterilization by heat usually follows. In many research laboratories, ultrasonic vibrations are used to break up the cell walls of selected microorganisms so as to retrieve organelles for further study.

PHYSICAL CONTROL OF MICROORGANISMS

FILTRATION ✿

CONTAMINATED SOLUTION a

VACUUM PUMP c

FILTRATE d

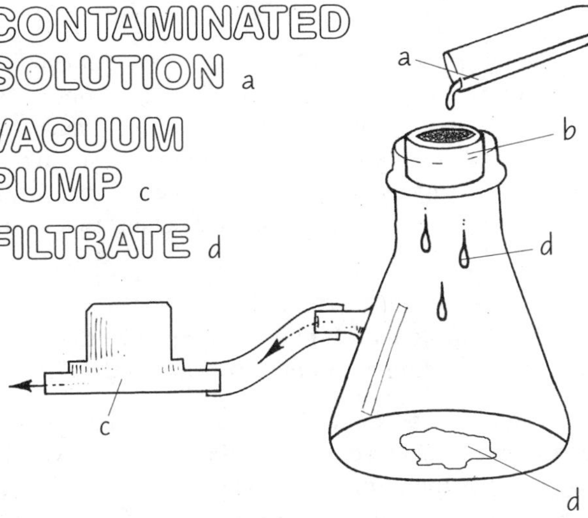

MICROORGANISM a¹

DEPTH FILTER b

SCREEN FILTER e

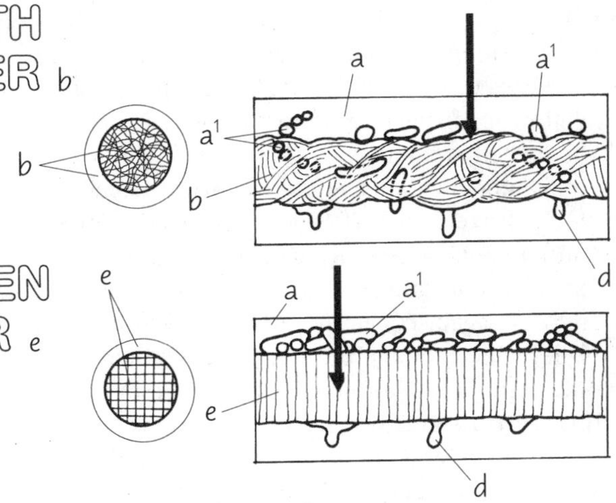

PORE SIZE e¹

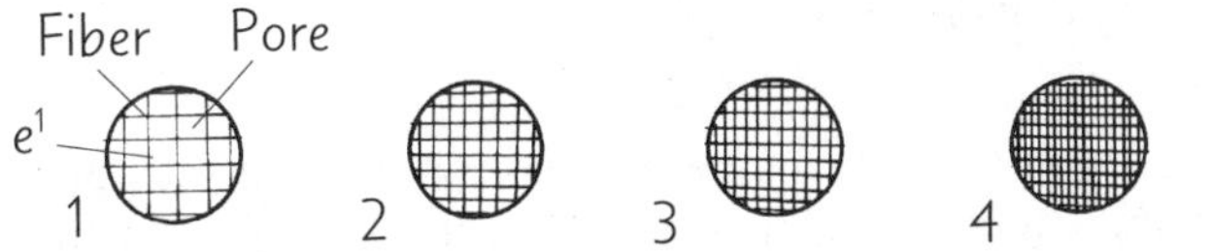

Fiber   Pore

1   2   3   4

ULTRAVIOLET LIGHT f

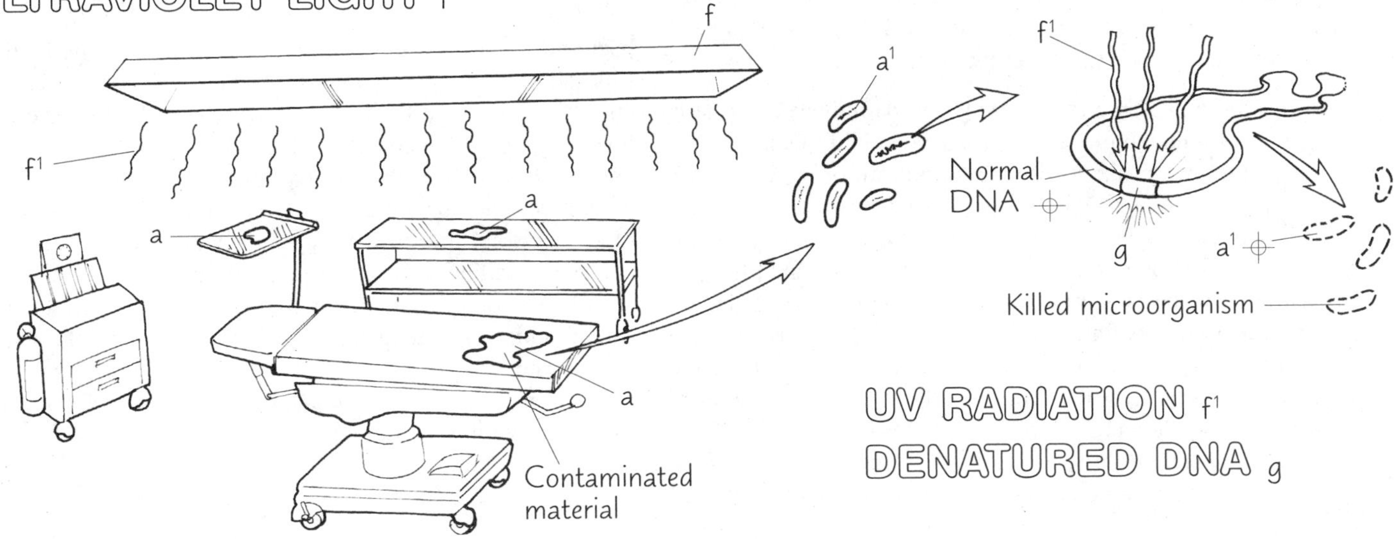

Normal DNA

Killed microorganism

UV RADIATION f¹

DENATURED DNA g

Contaminated material

ULTRASONIC VIBRATION h

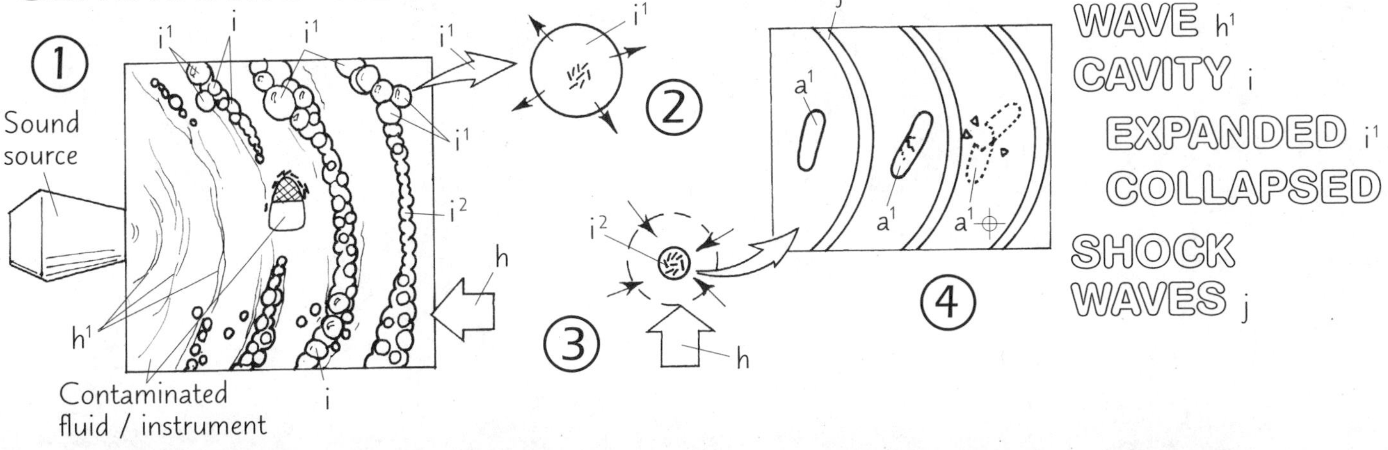

① ② ③ ④

Sound source

Contaminated fluid / instrument

ULTRASONIC WAVE h¹

CAVITY i

EXPANDED i¹

COLLAPSED i²

SHOCK WAVES j

44
CHEMOTHERAPEUTIC AGENTS

One of the most valuable methods of treating infections is by using chemotherapeutic agents (a kind of medication) synthesized in the laboratory. One of the first such agents was synthesized in the laboratories of Gerhard Domagk in 1935. It was called sulfanilamide, and was remarkably effective in combating infections. For its discovery, Domagk received the Nobel Prize in 1939. Since then, over 150 sulfonamides (sulfanilamide altered by attaching a variety of side groups to the basic molecule) have been synthesized. The sulfonamides are active against Gram-positive bacteria, some Gram-negative bacteria, and some protozoa, especially in combination with other mixtures. Sulfonamides are specifically used to treat urinary tract and some other infections. In this plate, we examine the structure of sulfanilamide and how it interferes with the folic acid metabolism of certain infectious bacteria, resulting in their death.

Color the titles and structures (a) through (c) in the upper half of the plate, using light colors for (a) through (c). Color the subheading Synthesis of Folic Acid gray. The structures P, PABA, and G are identified by special subscripts not in numerical order with the other subscripts.

Folic acid (a) is a vitamin synthesized routinely by bacteria as part of a metabolic process associated with nucleic acids. Failure to synthesize it results in the death of the organism. Interestingly, folic acid is not synthesized in humans, and must be obtained in the diet. Folic acid is composed of three parts: pteridine (p), para-aminobenzoic acid (PABA), and glutamic acid (g). Synthesis of folic acid requires the enzyme synthetase (b). Each component of folic acid fits into a structurally-specific site of the enzyme; the arrangement of the three components is specified further by the structural fit between each pair, as shown.

Sulfanilamide (c), a synthetic molecule not found in nature, has a strong resemblance to para-aminobenzoic acid (PABA). There are structural differences, however, as careful examination of the molecule of each reveals. These differences are significant with respect to the synthesis of folic acid: the enzyme synthetase will not accept sulfanilamide as a substitute for PABA in the synthetic process. Herein lies the basis for employing sulfanilamide as a means of blocking folic acid synthesis in bacteria.

Color the titles and structures in the lower part of the plate (b) through (e), starting at left and working to the right.

In our example, a person with a urinary tract infection (d) is being treated with an oral dose of a sulfanilamide derivative (c^1, such as sulfamethoxazole). Following absorption into the blood, the chemotherapeutic agent soon reaches the site of infection where it is taken up by the bacteria (d^1) and achieves a significantly high concentration. Under these conditions, the presence of large concentrations of the sulfanilamide derivative blocks the synthesis of folic acid, resulting in the death of the bacteria, and resolution of the infection.

The mechanism of the folic acid blockade (e) in the bacterial cell is the selection of the sulfanilamide derivative (instead of PABA) by the enzyme folic acid synthetase. The enzyme uses the sulfanilamide derivative because there is a lot more of it immediately available than there is of the naturally occurring PABA. By binding the sulfanilamide derivative, the enzyme can bind glutamic acid but it cannot bind pteridine, due to a lack of structural correspondence between pteridine and the sulfanilamide derivative. As a result, a molecule of folic acid is not synthesized. In effect, the sulfanilamide derivative "competes" with PABA for the active site on the folic acid synthetase enzyme, "wins," and the production of folic acid is inhibited. This process of blockading folic acid synthesis is called competitive inhibition (e).

CHEMOTHERAPEUTIC AGENTS

FOLIC ACID a
    PTERIDINE p
    PABA PABA
    GLUTAMIC ACID g

SYNTHESIS OF
FOLIC ACID ✳

SYNTHETASE b

SULFANILAMIDE
DERIVATIVE c

INFECTION d
S. AUREUS d¹

Bacterium

PABA

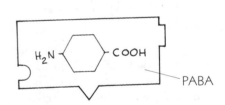

H_2N — SO_2-NH_2

c

H_2N — $COOH$

PABA

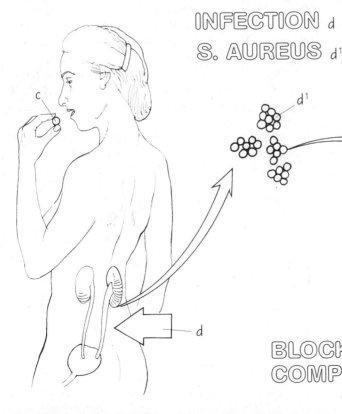

BLOCKED FOLIC ACID SYNTHESIS:
COMPETITIVE INHIBITION e

45
ANTIBIOTICS: PENICILLIN

Antibiotics are naturally occurring metabolic products of primarily soil bacteria and fungi. These substances function to limit microbial growth in the soil environment by killing or inhibiting the growth of competing, generally Gram-positive bacteria. Some antibiotics act by interfering with the synthesis of the bacterial cell wall, weakening it to intracellular pressures, and thereby inducing its rupture, and the subsequent death of the bacterium. Other antibiotics interfere with metabolism of the bacterium, such as protein synthesis. Some antibiotics interfere with DNA synthesis. Those bacteria that are antibiotic producers are species of *Bacillus* and *Streptomyces*. The fungi *Penicillium*, *Cephalosporium*, and *Micromonospora* also produce antibiotics. Certain antibiotics can also be artificially synthesized, at least in part. This results in the production of broad spectrum antibiotics (effective against both Gram-positive and Gram-negative bacteria), and the development of antibiotics against penicillin-resistant bacteria. In this plate, we concentrate on the penicillin antibiotics produced by the mold *Penicillium*.

Color the titles and structures (a) through (e²). Use a light color for (b) through (b²), and contrasting colors for (b), (d), and (e).

Penicillin is a product of the green mold *Penicillium* (a; Ascomycetes; Plate 39). *Penicillium* is a common inhabitant of soil environments. It grows well in fungal culture media and often can be found as a contaminant in bacterial cultures. The naturally occurring penicillins are effective in treating streptococcal infections ("strep throat," tonsillitis, pneumococcal pneumonia, syphilis, gonorrhea, meningitis, and streptococcal endocarditis).

The penicillin molecule (b) consists of two parts. The beta lactam nucleus (b¹) is the key bactericidal element, interacting with the production of amino acids in cell wall synthesis. Lactam is a cyclic amide (ring-shaped molecule) derived from aminocarboxylic acid. The attached group can vary significantly; in this case, we show the group found in the naturally occurring penicillin G, a commonly used antibiotic. Penicillin, however, is a family of antibiotics. Pharmacologists have succeeded in attaching new groups to the beta lactam nucleus, producing synthetic derivatives of penicillin. These broad-spectrum antibiotics are effective against Gram-negative rods,

and include ampicillin, amoxicillin, and others. Other penicillins have been produced in response to bacteria that contain enzymes (beta lactamase or penicillinase) that inactivate the beta lactam ring. These penicillinase-resistant penicillins include nafcillin, cloxacillin, and so on.

Penicillin affects bacteria by interfering with the normal maintenance and synthesis of the bacterial cell wall. Here we show the normal cell wall of a Gram-positive bacterium (recall Plate 13); in this case, a staphylococcal bacterium (c; station 1). These bacteria have a cell membrane enclosing the cytoplasm, and an outer cell wall consisting of a peptidoglycan layer (d). The peptidoglycan layer consists of sheets of carbohydrates with side chains of amino acids (e¹) connected by peptide cross bridges (e²; peptides are short chains of amino acids). On the cell wall surface are receptors called penicillin-binding proteins (e). When penicillin molecules attach to these proteins, synthesis of the cell wall stops. As a result, the cross bridges weaken as synthesis of amino acids in the wall is blocked. As the cross bridges weaken, the entire protein-carbohydrate structure of the cell wall becomes vulnerable to intracellular pressures, and finally ruptures (station 2), killing the bacterium (c¹ ⊕).

Color the subheading Penicillin Resistance at the lower part of the plate, and the titles and structures (b) through (g). Use a light color for (g).

Since the introduction of penicillin in the early 1940s, bacterial species have emerged that are resistant to many antibiotics. Many of these bacteria (c²) produce one of a variety of enzymes called penicillinase (e), probably under the direction of plasmids. Penicillinase converts penicillin to penicilloic acid (g). Penicilloic acid is harmless to bacteria and the bacterial cell wall. Lack of penicillin-binding proteins on the cell wall may also offer resistance to penicillin. The emergence of these penicillin-resistant forms of bacteria reflects a type of evolution wherein bacteria will continue to generate a variety of means to survive anti-bacterial mechanisms. Such evolutionary "tactics" offer significant challenges to the pharmacological industry and practitioners of medicine in their attempt to minimize the negative effects of bacterial-human symbiosis.

ANTIBIOTICS: PENICILLIN

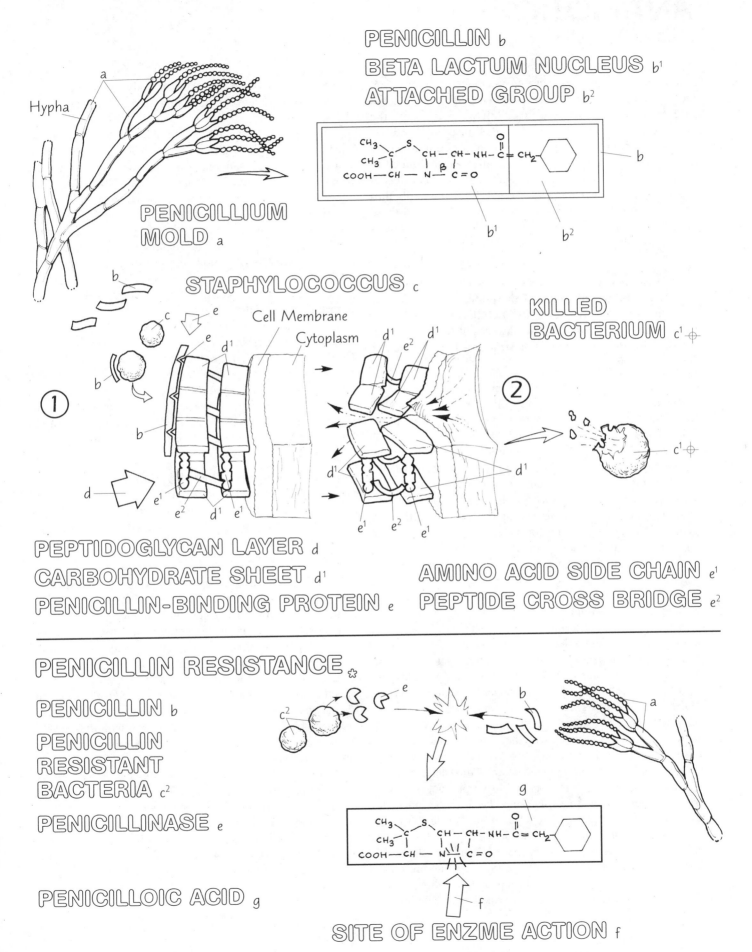

Hypha

PENICILLIUM MOLD a

PENICILLIN b
BETA LACTUM NUCLEUS b¹
ATTACHED GROUP b²

b
b¹
b²

STAPHYLOCOCCUS c

Cell Membrane
Cytoplasm

KILLED
BACTERIUM c¹

① ②

PEPTIDOGLYCAN LAYER d
CARBOHYDRATE SHEET d¹
PENICILLIN-BINDING PROTEIN e

AMINO ACID SIDE CHAIN e¹
PEPTIDE CROSS BRIDGE e²

PENICILLIN RESISTANCE ✿

PENICILLIN b

PENICILLIN
RESISTANT
BACTERIA c²

PENICILLINASE e

g

PENICILLOIC ACID g

SITE OF ENZME ACTION f

Antibiotics are naturally-occurring metabolic products of soil bacteria and fungi. The antibiotics limit the effects of bacterial infection, which could be fatal. Plate 45 explores the characteristics of one antibiotic, penicillin. In this plate we discuss some of the other available antibiotics. Antibiotics are considered narrow-spectrum when their effect is limited to Gram-positive or Gram-negative bacteria; they are considered broad-spectrum when they act on both. While some antibiotics are naturally-occurring, others are semi-synthetic. In these antibiotics, the natural active component in the molecule is combined with a synthetic group.

Color the titles and related structures (a) through (f) in the upper and middle parts of the plate. Use colors light enough to avoid obscuring the details or print in the illustration. Use shades and/or patterns of the same color for (b¹) through (b⁶), and color similarly for (e) through (e⁴).

Many antibiotics are naturally produced by species of moldlike soil bacteria belonging to the genus *Streptomyces* (a). These antibiotics include some aminoglycosides, tetracyclines, chloramphenicol, and others.

Aminoglycosides (b) are antibiotics that contain amino groups (d) bonded to carbohydrate molecules (c; glycosides), which, in turn, are bonded to other carbohydrates. The "R" in the molecule stands for "radical group"; the R group can be a methyl group ($-CH_3$), an amino group ($-NH_2$), or a sulfhydryl group ($-SH$). Among the aminoglycosides in use are streptomycin (b¹), gentamycin (b²; from the mold *Micromonospora*), neomycin (b³), amikacin (b⁴), kanamycin (b⁵), and tobramycin (not shown). In the past, streptomycin has been useful in treating tuberculosis. Gentamycin is effective against staphylococci and Gram-negative bacteria. Neomycin is used for skin infections and conjunctivitis caused by *Haemophilus aegypticus*. Amikacin is derived from kanamycin and is active against serious infections by Gram-negative bacteria as well as Gram-positive staphylococci. Most aminoglycosides are effective against species of Gram-negative bacteria such as *Haemophilus, Bordetella, Salmonella, Yersinia,* and *Pseudomonas*. Aminoglycosides generally have a tendency to cause toxic side effects associated with their use. The site of action of the aminoglycosides are the ribosomes; they interfere with protein synthesis, resulting in the death of the bacterium.

The tetracycline antibiotics are also a product of *Streptomyces*. Tetracycline (e) is so-named because the molecule contains four chemical rings ("tetra," four; "cycles," rings). Aside from tetracycline, the group includes oxytetracycline (e¹), chlortetracycline (e²), doxycycline (e³), minocycline (e⁴), and others. Oxytetracycline and chlortetracycline are older antibiotics, having been developed in the 1960s, while doxycycline and minocycline are more modern antibiotics that are better tolerated by the human body. All tetracyclines interfere with protein synthesis in bacteria, ultimately killing them.

The tetracyclines are broad-spectrum antibiotics, but are used primarily against Gram-negative bacteria such as those causing gonorrhea, salmonellosis, and meningitis. They are also useful against the rickettsiae that cause Rocky Mountain spotted fever. They are valuable against chlamydiae, protozoa, and fungi. In humans, tetracycline tends to bind with certain metals (calcium, magnesium, iron) in the gastrointestinal tract, inhibiting uptake and causing delays in bone growth and yellow-brown discoloration of teeth in young people. In addition, tetracycline can kill the beneficial population of bacteria in the human intestine. Yogurt is often recommended to restore that population.

Chloramphenicol (f) is a broad spectrum antibiotic also derived from species of *Streptomyces*. It is successful against Gram-negative bacteria, rickettsiae, chlamydiae, and fungi. In bacteria, it interferes with the process of protein synthesis, leading to bacterial death. Although a very effective antibiotic, chloramphenicol cannot be used in less than very serious situations because it interferes with bone marrow activity in humans and may lead to a serious (aplastic) form of anemia.

Color the subheading Sites of Antibiotic Activity in the lower half of the plate, and the titles of the antibiotics (g) through (m). As the mechanism of action of each antibiotic is discussed in the text, color the related arrow or part of an arrow pointing to one of the bacterial cell parts at the bottom of the plate.

The sites of antibiotic activity are principally the bacterial cell wall and cellular structures involved in protein synthesis. Antibiotics acting on the cell wall include penicillin (g), cephalosporin (h; similar in structure to the penicillins), vancomycin (i; effective against staphylococci and produced by species of *Streptomyces*), bacitracin (j; produced by *Bacillus subtilis*, and used for skin preparations, and eye and ear infections), and cycloserine (not shown; produced by species of *Streptomyces*). Polymixin (k), a polypeptide antibiotic produced by *Bacillus polymyxa*, interferes with the function of the cell membrane. Protein synthesis is disrupted by aminoglycosides, tetracyclines, chloramphenicol, and erythromycin (l; a product of *Streptomyces*; effective as a penicillin substitute in *Mycoplasma* pneumonia). In some cases, antibiotics combine with and interfere with the activity of a bacterium's nucleic acid. Rifampin (m; derived from rifamycin which is produced by a species of *Streptomyces*) and nalidixic acid (not shown; a synthetic antibiotic especially effective against *Proteus* infections of the urinary tract) are in that group.

ANTIBIOTICS

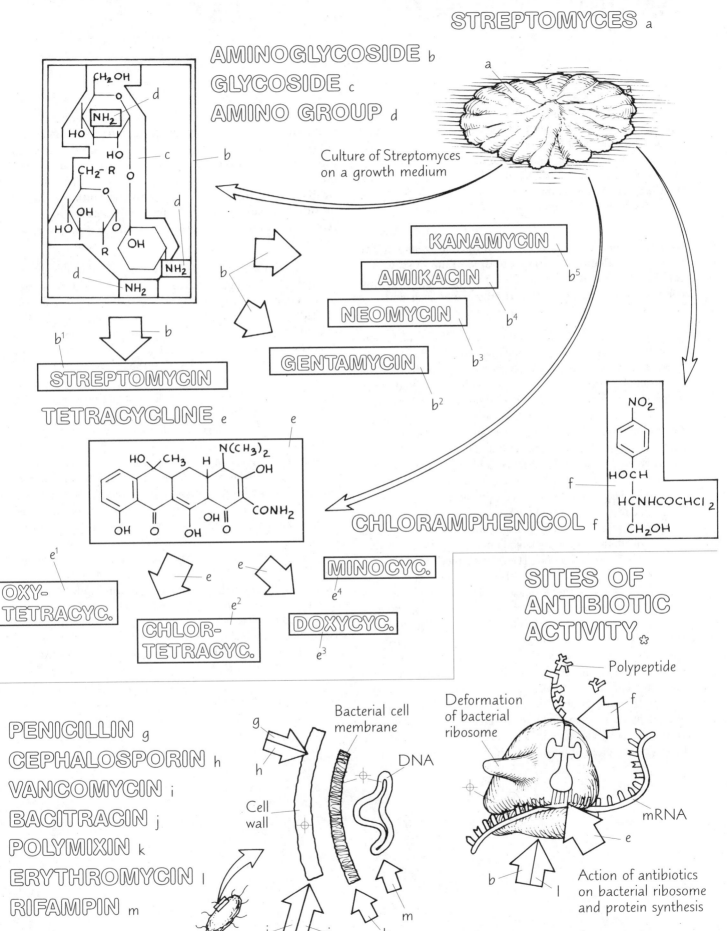

STREPTOMYCES a

AMINOGLYCOSIDE b
GLYCOSIDE c
AMINO GROUP d

Culture of Streptomyces on a growth medium

KANAMYCIN b⁵
AMIKACIN b⁴
NEOMYCIN b³
GENTAMYCIN b²

STREPTOMYCIN b¹

TETRACYCLINE e

CHLORAMPHENICOL f

OXY-TETRACYC. e¹
CHLOR-TETRACYC. e²
DOXYCYC. e³
MINOCYC. e⁴

SITES OF ANTIBIOTIC ACTIVITY

PENICILLIN g
CEPHALOSPORIN h
VANCOMYCIN i
BACITRACIN j
POLYMIXIN k
ERYTHROMYCIN l
RIFAMPIN m

Bacterial cell membrane
DNA
Cell wall
Polypeptide
Deformation of bacterial ribosome
mRNA
Action of antibiotics on bacterial ribosome and protein synthesis

ANTIBIOTIC SUSCEPTIBILITY TEST

Not all bacteria are susceptible to antibiotics. Physicians must select an antibiotic which has a reasonable chance of successfully killing the infectious microorganism (pathogen) in a known body environment, without creating undue side effects, such as generating an allergic response. In this the physician can be assisted by the antibiotic susceptibility test, a laboratory procedure common in any hospital.

Note that the illustrations to be colored are arranged into six stations. Begin at station 1 and continue to station 4, coloring titles and structures (a) through (c). Use a light color for (c). Color over the streaks of bacteria with the color used for (b). Do not color stations 5 and 6 for now.

For an antibiotic to be employed, some idea of the pathogen causing the disease must be acquired. This is done by taking a swab of the infected skin or mucosal surface (lining of the mouth, nose, vagina, and other cavities of the body open to the outside). Alternatively, if no known site of infection exists or is identified, a sample of appropriate body fluids, e.g., blood (a), is taken and placed in an agar-filled Petri dish (station 1). This sample is then cultured (station 2), a process taking about 24 hours or longer. The cultured bacteria (b) are transferred to a liquid medium in a test tube and set aside (station 3). The culture is kept as aseptic as possible by being stored in a test tube plugged with cotton.

A plate of growth medium is prepared to test the cultured bacteria (station 4). A common medium is Mueller-Hinton agar (c), containing sheep blood. The tube of medium is poured into a Petri dish and allowed to set. Within a few minutes, the agar solidifies and is ready for use. When the growth medium is ready, a sterile swab is used to obtain bacteria from the test tube sample, and this bacteria-laden swab is streaked back and forth across the surface of the culture medium in the Petri dish (station 4). A swab is used instead of a wire loop to prevent tearing the fragile growth medium. It is important that the entire plate of medium be streaked evenly to prepare a lawn of bacteria. Airborne contamination must be avoided; this is effected by keeping the medium covered and then lifting the cover only enough to accommodate streaking the medium with the swab. The plate of nutrient medium is now inoculated.

Color the subheading Antibiotic Disc and the names of the antibiotics used on the discs (d) through (h). Use contrasting colors for these five antibiotics. Then color the related discs in the Petri dishes at stations 5 and 6, as well as cultures of bacteria (b) and the blood agar medium (c). In coloring the discs in the dish, take care not to color into the ring of inhibition (read text below if necessary before going on). Leave the rings of inhibition around each of the discs uncolored.

Discs containing selected antibiotics (d) through (h) are commercially prepared and stored in microbiology laboratories for susceptibility tests. For our purposes, five discs have been selected for testing: penicillin G (d), cephalosporin (e), nafcillin (f), chloramphenicol (g), and tetracycline (h). Forceps are used to transfer the discs to the inoculated surface of the growth medium (station 5). Following this, the dish is transferred to an incubator for a period of 16 to 18 hours. During this period, the antibiotic will diffuse from the disc onto the growth medium (station 6). There the bacterial growth may or may not be influenced by the antibiotic. If the bacteria under a disc is susceptible to the antibiotic, the bacteria will be killed under the disc and for a certain distance peripheral to the edge of the disc. Such a clearing is called the ring of inhibition. On the other hand, if the bacteria is resistant to the disc's antibiotic, there will be no ring at all, as the growth of the bacteria has not been affected. In this case, the bacteria are resistant to penicillin G.

The technician may use a rule to measure the width of the ring. This measurement is applied to a chart in the laboratory which gives levels of probability of significance to each measurement (not shown). The results (S or susceptible, R or resistant) are written or checked off on the antibiotic susceptibility test form and sent to the physician. On the basis of this report, and other factors, the physician will order a regimen of antibiotic therapy to which the identified microorganism is known to be susceptible. The patient often makes a dramatic recovery from an acute infectious disease following administration of the appropriate antibiotic.

ANTIBIOTIC SUSCEPTIBILITY TEST

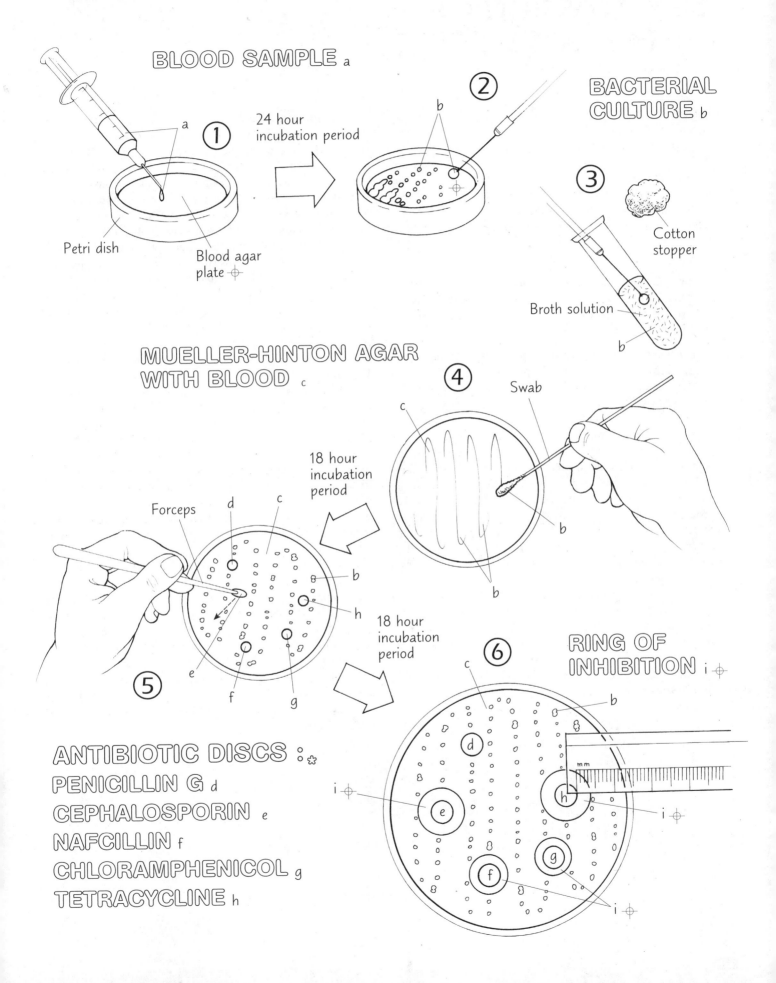

BLOOD SAMPLE a

① 24 hour incubation period

Petri dish

Blood agar plate

② BACTERIAL CULTURE b

③ Cotton stopper

Broth solution

b

MUELLER-HINTON AGAR WITH BLOOD c

④ Swab

b

b

c

18 hour incubation period

Forceps

d

c

b

h

⑤ e f g

18 hour incubation period

⑥ RING OF INHIBITION i

c

b

d

i e

h i

f g i

ANTIBIOTIC DISCS :

PENICILLIN G d

CEPHALOSPORIN e

NAFCILLIN f

CHLORAMPHENICOL g

TETRACYCLINE h

mm

48
TRANSMISSION OF INFECTIOUS DISEASE

Disease may be conceptualized as any change from a general state of good health. In practice, disease disrupts the normal structure and functioning of the tissue affected. Many types of disease exist, including those caused by inadequate nutrition, genetic abnormalities, trauma, and structural or functional disorders. Illness induced by microorganisms (bacteria, viruses, protozoa, fungi) constitutes infectious disease, a subject of interest to medical microbiologists. In this plate, modes of transmission of infection are surveyed.

Color the titles and related structures associated with stations 1 through 4. Use light colors throughout except for (b). In coloring the droplet cloud, create your own pattern to simulate a cloud.

Droplets (a¹) are tiny particles of mucus and saliva expelled from the mouth, nose or pharynx during a cough or sneeze (station 1). These droplets contain microorganisms (b). Some of these microorganisms are pathogenic, and when swept into the nose and mouth of the recipient, they can penetrate the lining tissues or mucosa (c). Microorganisms invade and proliferate in the underlying vascular connective tissues, disrupting normal structure and function. The result is redness and swelling in the affected mucosa, active secretion of mucus, and a sore throat. Among the bacterial diseases transmitted by droplet transmission (a) are strep throat (*Streptococcus*), pertussis or whooping cough (*Bordetella pertussis*), and tuberculosis (*Mycobacterium tuberculosis*). Viral diseases communicated by droplet transmission include influenza, measles, and chicken pox.

The most common arthropods that transmit infectious diseases to humans are insects, characterized by six jointed appendages and segmented bodies (station 2). The common housefly (d¹) and mosquito are examples of insects. These insects carry various microorganisms (b) on their appendages (d²) and body parts, transferring the microbes to the landing site, in this case, fresh food. When ingested, the infective microorganisms can invade the mucosa of the gastrointestinal tract, proliferate, and cause disease. In some cases, insects may deliver the offending microorganism by penetrating the skin ("bite"). The mosquito, for example, becomes infected while consuming the blood of an infected person, then transmits the malaria-causing protozoan with subsequent penetrations of the host skin (not shown). Arachnids are arthropods with eight legs. One arachnid, the tick, can transmit Lyme disease or Rocky Mountain spotted fever disease (not shown).

Contaminated water (e) contains matter presumed to have pathogenic microorganisms (b). Microorganisms in the fecal matter of animals (f) can contaminate streams (station 3) and other bodies of water, especially if the water is stagnant. Ingestion of these microorganisms during the drinking of water transports pathogens to the gastrointestinal tract where they proliferate. The release of toxins may then induce nausea, vomiting, dehydration, fever, and other symptoms and signs. Bacteria associated with typhoid fever (*Salmonella typhi*), the hepatitis A virus, and the protozoan associated with amebiasis (*Entamoeba histolytica*) are transmitted in this way.

Injection of pathogens deep to the skin (i) of humans by an animal bite (g) is another means of transmission of infection (station 4). Once introduced into the tissues by the teeth (h), the pathogenic microorganisms (b) can rapidly reproduce, irritating and destroying local structure and disturbing life functions. Spread of the organisms through the blood and lymph (bacteremia, viremia, septicemia) can involve distant organ systems. Bites of animals infected with the rabies virus introduce the pathogen with the saliva. Other diseases transmitted by animal bites include leptospirosis, brucellosis, rat bite fever, and Q fever.

Human-to-human contact (not shown) of infected mucosal surfaces is also a well known source of infectious diseases. The microorganisms of disease may also be transmitted to humans by any device or substance that contacts the human. Human wounds, for example, can be contaminated by tetanus bacilli from the soil. Finally, humans themselves are a primary source of contamination; carriers of disease-producing microorganisms are often responsible for contaminating food.

TRANSMISSION OF INFECTIOUS DISEASE

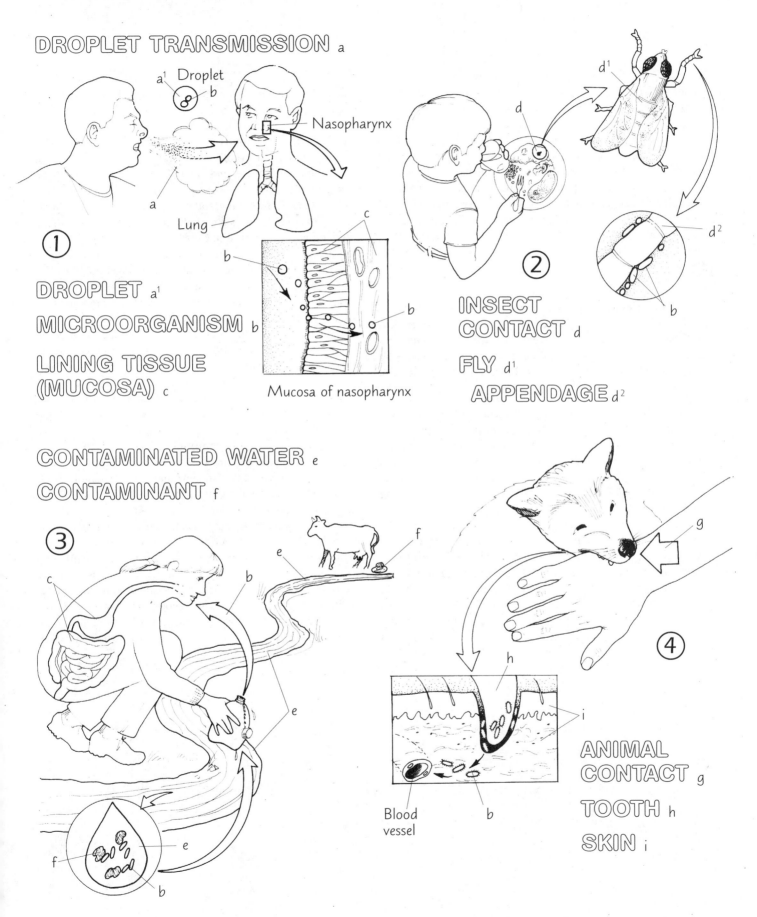

DROPLET TRANSMISSION a

a¹ Droplet
b
Nasopharynx
Lung
a

①

DROPLET a¹

MICROORGANISM b

LINING TISSUE
(MUCOSA) c

Mucosa of nasopharynx

b
c
b

②

INSECT
CONTACT d

FLY d¹
APPENDAGE d²

d¹
d
d²
b

CONTAMINATED WATER e
CONTAMINANT f

③

c
b
e
f
e
b

Blood
vessel
h
b
i
g

④

ANIMAL
CONTACT g

TOOTH h

SKIN i

49
ESTABLISHMENT OF DISEASE

When infectious organisms enter the human body, they compete with the body defenses for supremacy. In a rapid, dramatic, and dynamic series of events collectively termed the "inflammatory response," the body defenses work to destroy the micro-organisms. In some instances, however, the functional tissue is ruptured, and the signs and symptoms of disease manifest themselves. This plate illustrates mechanisms that micro-organisms use to establish themselves in the tissues. For simplicity, three different bacteria are illustrated at the same inflammatory event. Due to limitations of space, only a few of the cells present at an inflammatory site are shown. See also Plate 53.

Color the titles (a) through (c²) and the related parts in the central illustration (station 1) representing an area of skin penetrated by a splinter. Then color the titles and structures at upper left (station 2), and continue with stations 3, 4, and 5. Use light colors for (a) and (b), red for (c) and (c¹), and purple for (c²).

The skin consists of an upper, epithelial part (a) composed of multiple layers of cells (station 1). Deep to the epithelium is a layer of connective tissue (b) that contains a number of different kinds of cells, a dense array of fibers, and some small blood vessels (c). Here a splinter, carrying numerous bacteria, has entered the skin, perforating one of the vessels, and creating a pool of localized blood (c¹) and clotted material (c²).

The introduction of staphylococci (e) in the vascular tissue (station 2) prompts a rapid migration of phagocytes (f) to the scene. As the phagocytes move toward the bacteria, the staphylococci may secrete an enzyme called coagulase which stimulates the formation of blood clots (coagulase activity; d). As the staphylococci become encircled by the sticky mass of fibrin and platelets, the microbes are shielded from the phagocytic cells. These bacteria proliferate within the clot and over a short time form a painful lesion (abscess) of swollen,

reddened (inflamed) tissue, filled with staphylococci and dead phagocytes (not shown). Staphylococci which do not produce coagulase are rapidly phagocytosed and do not represent a threat (i.e., they are not pathogenic).

Epithelial (lining) cells are bound together by a variety of intercellular substances and fibers (i; station 3). One of the intercellular substances is hyaluronic acid (not shown). This tough "cement" limits the spread of bacteria in the cellular tissue. Infection through skin without a break or rupture in the cellular complex is unusual. Certain species of staphylococci, streptococci, and clostridia (h) produce an enzyme termed hyaluronidase which digests hyaluronic acid among the cells (hyaluronidase activity; g). As a result of this activity, the intercellular structure (i) is broken down, inviting enhanced movement of the pathogens through the tissues.

Certain species of streptococcus (k) interact with blood clots in a different way (station 4). Although clots may be somewhat protective to bacteria, they also limit their proliferation. Certain streptococci produce and secrete an enzyme called streptokinase. Streptokinase acts on the clot and dissolves it (k). The bacteria can then spread out beyond the initial infected area.

Certain species of staphylococci, streptococci, and clostridia produce other enzymes to assist their establishment in the tissues they invade. One enzyme, termed hemolysin (l; station 5), is directed at red blood corpuscles (RBCs; c³). The enzyme ruptures the membranes of the RBCs (c⁴), releasing hemoglobin, and diminishing the oxygen-carrying capacity in the body. Hemolysin functions to lower the oxygen levels sufficiently to decrease the metabolic activity of the infected region of cells. Consequently, the bacteria are able to multiply in the local tissues without much resistance. This anaerobic condition is particularly advantageous for clostridia, which grow anaerobically. Clostridia cause gas gangrene, and an associated sign of that condition is anemia (reduced number of red blood corpuscles).

ESTABLISHMENT OF DISEASE

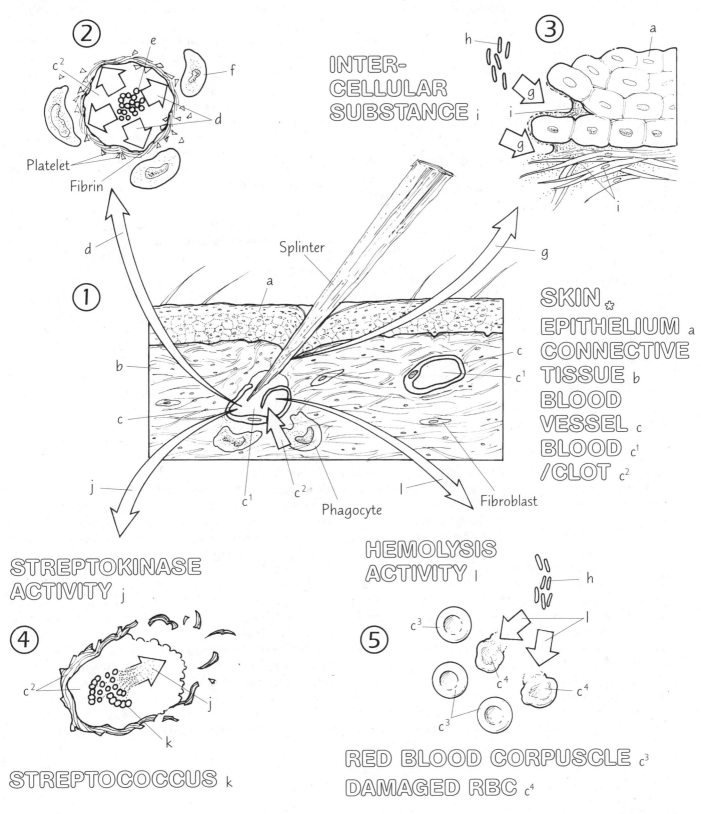

COAGULASE ACTIVITY d
STAPHYLOCOCCUS e
PHAGOCYTE f

② c² e f d Platelet Fibrin

HYALURONIDASE
ACTIVITY g
CLOSTRIDIUM h

③ h g i g a i

INTER-
CELLULAR
SUBSTANCE i

① Splinter a b c c c¹ c² Phagocyte Fibroblast d g j l

SKIN *
EPITHELIUM a
CONNECTIVE
TISSUE b
BLOOD
VESSEL c
BLOOD c¹
/CLOT c²

STREPTOKINASE
ACTIVITY j

④ c² j k

STREPTOCOCCUS k

HEMOLYSIS
ACTIVITY l

⑤ h l c³ c⁴ c⁴ c³ c³

RED BLOOD CORPUSCLE c³
DAMAGED RBC c⁴

50
TOXINS

The establishment and course of infectious disease can be profoundly influenced by the presence of toxins. Toxins are poisonous substances produced by certain microorganisms. They increase the damaging effect (pathogenicity) of microorganisms by altering the metabolism of cells, tissues, and organs of the host's body. In some cases, toxins may merely disturb bodily functions, but in other cases the effects of the toxins may be deadly. Two types of toxins are generally recognized: exotoxins and endotoxins.

Color the titles and structures (a) through (e) in the upper part of the illustration, including the subheading Intestine. Then color the subheading Exotoxin Source/Name, and the related titles and arrows (f) through (i).

Exotoxins are proteins that are produced by certain species of largely Gram-positive bacteria (a). The bacteria manufacture these proteins as part of their normal metabolism. These proteins are not toxic to the bacteria; they are, however, toxic to the host of the bacteria. These proteins are accumulated within the cytoplasm of the bacteria, and then released with waste products of metabolism to the surrounding environment. For this reason, they are called exotoxins (exo-; out). They are soluble in body fluids. Here we show circular bacteria in the lumen (cavity) of the host's intestine releasing exotoxins into the lumen and into the lining cells.

Within the host, exotoxins (b) are released into the tissue fluids, taken into the blood (e), and dispersed to organs throughout the body. Exotoxins cause direct damage to tissues (e.g., disrupt cell membranes and membranes of organelles; here showing damage to a mitochondrion; c) and they interrupt metabolic functions (d), interfering with cell functions. Exotoxins may also be produced and released by bacteria in food and other substances. These exotoxins must be consumed by the host before they can initiate their damaging effect.

Some sources and names of exotoxin are listed (f) through (i); the site of tissue damage is shown in the human figure at left. *Staphylococcus aureus* (f) produces exotoxins which are absorbed by and damage the intestinal lining cells, resulting in nausea, vomiting, abdominal cramps, and diarrhea. For this reason the exotoxins are called enterotoxins (f; entero-; intestine). Toxic shock syndrome, a condition largely affecting populations of menstruating women using tampons, is caused by exotoxins of *S. aureus*. The exotoxin of *Clostridium botulinum* (g) gains access to the nervous system by way of the blood stream. This exotoxin inhibits the release of acetylcholine at neuromuscular junctions. The absence of this neurotransmitter prevents the passage of nerve impulses, causing paralysis. For this reason, this exotoxin is called a neurotoxin (g).

The neurotoxin (h) of *Clostridium tetani* (h) interferes with neuromuscular function by binding to nerve cells that are inhibitory to skeletal muscle contraction. Muscles contracting without inhibition remain in a state of constant contraction (tetany) and become spastic and rigid (lock up). The endotoxin (i) of *Corynebacterium diphtheriae* (i) kills cells in the respiratory tract by interfering with protein synthesis. Dead cells, cellular remnants and debris, collected in mucus, block the small air passageways, making breathing difficult. Such an exotoxin is called a cytotoxin (i).

Color the titles and structures (j) through (m); then color the subheading Endotoxin Shock gray, and the titles listed below it, (n) through (r). Note there are no corresponding structures for these titles.

Endotoxins (l) are produced chiefly by Gram-negative bacteria. They are complexes of protein, polysaccharide, and lipid (collectively called LPS; k) that are part of the bacterial cell wall (j). They are released during bacterial disintegration caused by antibiotics or phagocytosis.

Endotoxins are not tissue or organ specific; they rapidly diffuse into the extracellular fluid and blood and are distributed by blood vessels throughout the body. Their effects are systemic; that is, they damage or effect a wide range of tissues which disrupt the functional stability of the host. They cause blood vessels to dilate (m), resulting in lowered blood pressure (hypotension; n) Tissues fail to receive nutrients and oxygen due to inadequate tissue perfusion pressures, and shock occurs. Endotoxins induce increased body temperature or fever (o) and generalized muscular weakness (p). Populations of white blood cells are depressed (leukopenia; q). Blood vessels may disintegrate, prompting uncontrolled hemorrhage (r; disseminated intravascular coagulopathy, or DIC). These conditions are collectively called endotoxin shock. Some degree of endotoxin-induced sickness is seen in *Salmonella* infection (salmonellosis), certain types of meningitis, and urinary tract infections.

Both exotoxins and endotoxins can contribute substantially to the pathogenicity of a particular microorganism; indeed, organisms producing toxins are more likely to cause disease than organisms which cannot. There are nonspecific and specific forms of resistance to such organisms and these mechanisms will be illustrated in the following plates.

TOXINS

INTESTINE ✿

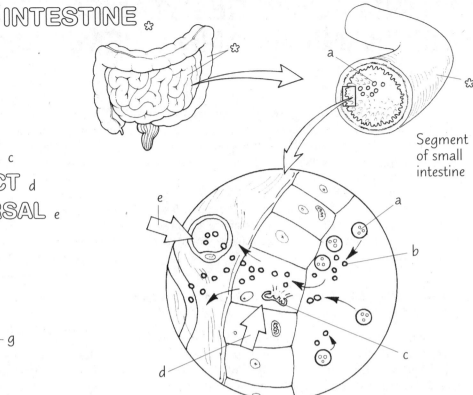

Segment of small intestine

GRAM-POSITIVE BACTERIUM a
EXOTOXIN b
DAMAGED TISSUE c
METABOLIC EFFECT d
SYSTEMIC DISPERSAL e

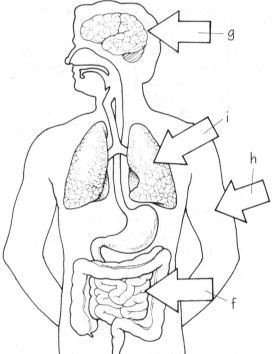

EXOTOXIN SOURCES / NAME: ✿

S. AUREUS ⟹ ENTEROTOXIN f

C. BOTULINUM ⟹ NEUROTOXIN g

C. TETANI ⟹ NEUROTOXIN h

C. DIPHTHERIAE ⟹ CYTOTOXIN i

ENDOTOXIN SHOCK: ✿

HYPOTENSION n
FEVER o
WEAKNESS p
LEUKOPENIA q
DIC r

GRAM-NEGATIVE BACTERIAL WALL j
LPS k
ENDOTOXIN l
VASODILATION m

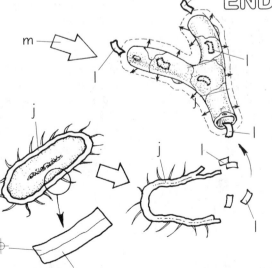

Peptidoglycan ⊕

NONSPECIFIC RESISTANCE TO DISEASE: MECHANICAL AND CHEMICAL BARRIERS

The body's resistance to disease takes two major forms: nonspecific and specific. Nonspecific resistance provides protection in the form of inborn structural and chemical barriers, and patterned responses. Identification of invading microorganisms is not a prerequisite to nonspecific resistance. No particular organism triggers this kind of resistance. In addition to the four kinds of nonspecific resistance illustrated here, other examples of nonspecific resistance are presented in Plates 35 (antiviral interferon), 52 (phagocytosis), and 53 (inflammation).

Color the titles and related structures (a) and (b) at upper left. Use a light color for (a) to avoid obscuring details.

An important nonspecific defense mechanism is the skin barrier (a). Skin consists of tightly bound layers of cells (epidermis) and an underlying layer of dense vascular connective tissue (dermis). This skin barrier is impenetrable to most microorganisms (b). However, the skin is vulnerable to burns and penetrating forces. Microorganisms can enter the subcutaneous tissues and deeper with such exposure (not shown). Bacteria can accompany penetrating objects such as the proboscis of mosquitoes and splinters of foreign material (not shown). Insect bites may inject microorganisms through the skin barrier (not shown). Malaria, Lyme disease, plague, and Rocky Mountain spotted fever may pass from person to person by this mode. Although it resists microorganisms, the skin barrier can be overcome and disease may be established.

Color the subheading Gastrointestinal Barrier gray and related titles and structures (c) through (e), as well as (b¹), at upper right.

Microorganisms entering the body through the mouth are swallowed and enter a hostile environment in the stomach and small intestine (gastrointestinal barrier). Hydrochloric acid (HCl) is secreted by cells lining the stomach, creating an acid pH of about 2.0 in the stomach. Most bacteria (b¹) are killed by the gastric acidity (c). There are microorganisms that are not killed, notably acid-resistant and pathogenic microorganisms such as *Salmonella* species (the causes of typhoid fever and salmonellosis) and the virus that causes hepatitis A.

These microorganisms can pass through the stomach into the intestines where they can penetrate the lining cells and establish disease (not shown). Further nonspecific resistance is provided in the second and third parts of the duodenum (part of the small intestine) in the form of bile (d) and enzymes (e). Bile is produced in the liver, stored in the gall bladder, and discharged to the duodenum via the common bile duct. Enzymes are released into the intestinal lumen from the pancreas (via the pancreatic duct) and cells lining the intestine walls. Both bile and enzymes have the capacity to disrupt the cell membranes of certain microorganisms.

Color the subheading Lysozyme, and the titles and related structures (f) and (g) at lower left.

Lysozyme is a small protein present in human lacrimal secretions (f; tears) and saliva (g). It is an enzyme that affects Gram-positive bacteria (b) by digesting the peptidoglycan in their cell walls when the bacteria come in contact with saliva in the mouth cavity and lacrimal secretions coating the conjunctiva (the transparent lining of the eye). This action kills the bacteria (b¹). Not surprisingly, most bacterial infections of the eyes are caused by Gram-negative bacteria, which are resistant to the effects of lysozyme.

Color the subheading Ciliary Trap and Motion, and the titles and related structures (h) through (j) at lower right. Use a light color for (h).

Microorganisms (b) entering the respiratory tract are carried through the pharynx, larynx, trachea, and bronchial tree by mucus droplets. Nonspecific resistance is provided by the lining tissue of the respiratory tract (h; respiratory epithelium). Here glandular secretions (i) released into the respiratory tract trap and kill the microorganisms (b¹). The cilia on the surface of the lining cells, undulating rhythmically, sweep the microorganisms toward the mouth to be swallowed or expectorated. Ciliary motion (j) is characterized by power strokes and resting strokes (not shown); the power strokes move the substances trapped in mucus toward the mouth and away from the lungs.

NONSPECIFIC RESISTANCE TO DISEASE: MECHANICAL AND CHEMICAL BARRIERS

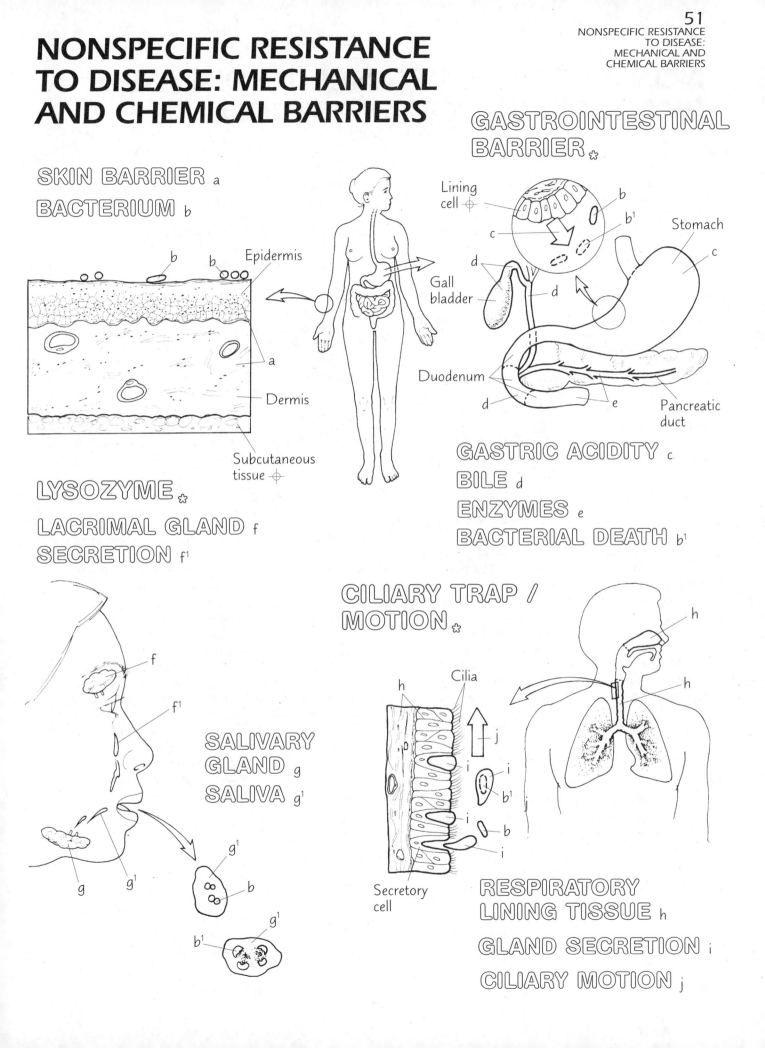

GASTROINTESTINAL BARRIER ✿

SKIN BARRIER a
BACTERIUM b

Epidermis

b

Lining cell ⊕

b

b^1

c

Stomach

d

c

a

d

Dermis

Gall bladder

Duodenum

d

e

Pancreatic duct

Subcutaneous tissue ⊕

GASTRIC ACIDITY c
BILE d
ENZYMES e
BACTERIAL DEATH b^1

LYSOZYME ✿
LACRIMAL GLAND f
SECRETION f^1

CILIARY TRAP / MOTION ✿

f

Cilia

h

f^1

h

j

SALIVARY GLAND g
SALIVA g^1

i

i

b^1

i

b

g^1

i

b

g

g^1

Secretory cell

RESPIRATORY LINING TISSUE h

b^1

g^1

GLAND SECRETION i

CILIARY MOTION j

52
NONSPECIFIC RESISTANCE TO DISEASE: PHAGOCYTOSIS/OPSONIZATION

The mechanism of nonspecific resistance includes a process at the cellular level called phagocytosis. This process is undertaken by certain cells, which engulf and digest microorganisms, cellular debris, and foreign particles. Opsonization is another form of nonspecific resistance combining complement or antibodies with the process of phagocytosis.

Color the subheading Phagocytes, and the titles and related structures (a) through (a²) in the upper right of the illustration. Use light colors for these cells.

Phagocytosis is carried on by a number of different cells, called phagocytes. There are three types of phagocytes: neutrophils, monocytes, and tissue macrophages. Neutrophils (a) make up about 65% of the total white blood cell population, and are found in both the blood and in the extracellular (connective) tissues. They are ameboid cells, assuming countless shapes as they move about, projecting pseudopods toward "prey," and squeezing between the cells of the capillaries as they move into and out of the blood. A distinctive feature of the neutrophil is the multi-lobed nucleus. Neutrophils migrate from local capillaries in an area of injured tissue, swarm around microorganisms, and phagocytose them.

Monocytes (a¹) make up about 4% of the total white blood cell population and are found only in the blood. They are larger than neutrophils, and each is characterized by a large, kidney-shaped nucleus taking up almost the entire cell interior. Like the neutrophils, they contain digestive enzymes capable of destroying ingested particles and microorganisms.

Macrophages (a²; called tissue macrophages) are found in many tissues throughout the body, including primarily the brain, spleen, bone marrow, liver, kidney, and general connective tissues. They are not found in the blood; however, they are probably derived from monocytes in the blood. They may be fixed (as specialized cells lining the blood vessels or lymphoid networks) or free (in the body tissue fluids).

Color the subheading Phagocytosis, and related titles and structures (b) through (g) in the central part of the illustration. Color only the cell membrane of the macrophage (a²).

Phagocytes are attracted to microorganisms and nonliving cellular debris or foreign material by means of chemical attraction (chemotaxis). Once close to the phagocytic cells, microorganisms can be surrounded by pseudopods or rapidly undulating sections of the phagocyte's cell membrane. As the bacteria (b) contact the surface of the macrophage (a²), they are drawn into the cell interior by means of a rapid encirclement by a portion of the cell membrane (c; endocytosis). This newly formed bacteria-containing vesicle (d; phagosome) moves from the cell membrane through the cytoplasm. It soon merges with a lysosome (e) containing digestive enzymes (e¹). The combined vesicles, called a phagolysosome (f), bring enzymes together with the bacteria and there begins the disruption and breakdown of the bacterial cells into fragments (b¹). The phagolysosome migrates towards the cell membrane, fuses with it, and ejects the digested contents to the outside in a process called exocytosis (g). These fragments are no longer a threat to the host organism.

Color the subheading Opsonization, and the titles and related structures at the lower part of the plate.

Phagocytosis can be enhanced by certain protein molecules that are soluble in the body fluids. Such proteins are called complement (to be presented in more detail in Plate 61) and antibodies. This process of phagocytic enhancement is called opsonization, and the participating complement or antibody is called an opsonin. Circulating complement (h) binds with the polysaccharides in the cell membranes of microorganisms; such microorganisms are said to be opsonized (i). Phagocytes of the blood (neutrophils) have specific receptors (j) for these opsonins and they phagocytose the attached microorganisms with greater rapidity than in the absence of opsonins.

In opsonization, we see the merging of specific (complement, antibody) and nonspecific (phagocytosis) mechanisms in the body's defense against infection. Both specific and nonspecific forms of resistance operate in the body simultaneously.

NONSPECIFIC RESISTANCE TO DISEASE: PHAGOCYTOSIS/ OPSONIZATION

PHAGOCYTES *

NEUTROPHIL a
MONOCYTE a^1
MACROPHAGE a^2

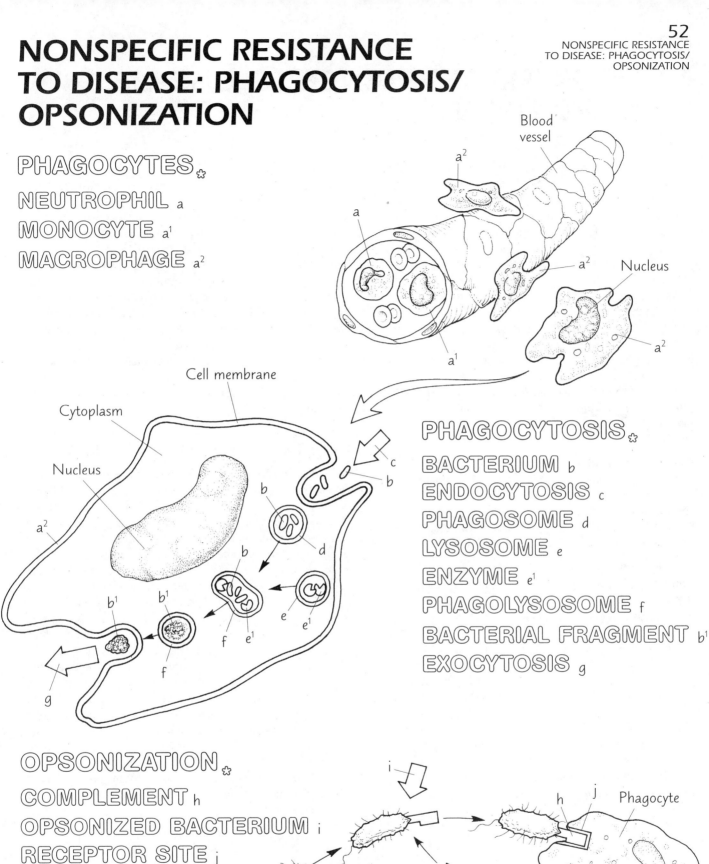

Blood vessel

a^2

a

a^2

Nucleus

a^1

a^2

Cell membrane

Cytoplasm

Nucleus

a^2

b

d

b

b^1

b^1

f

e^1

e

e^1

f

g

PHAGOCYTOSIS *

BACTERIUM b
ENDOCYTOSIS c
PHAGOSOME d
LYSOSOME e
ENZYME e^1
PHAGOLYSOSOME f
BACTERIAL FRAGMENT b^1
EXOCYTOSIS g

c

b

OPSONIZATION *

COMPLEMENT h
OPSONIZED BACTERIUM i
RECEPTOR SITE j

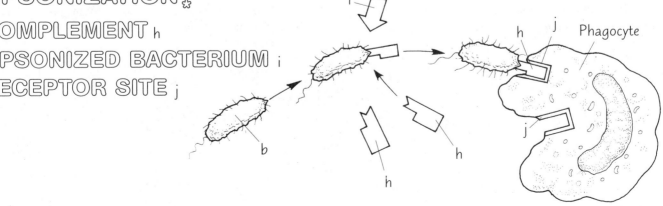

i

h

j

Phagocyte

b

h

h

j

53
NONSPECIFIC RESISTANCE TO DISEASE: INFLAMMATION

Inflammation is a nonspecific response to irritation that occurs in the presence of blood vessels. The dilatation of blood vessels initiates and characterizes the inflammatory response; without blood vessels, inflammation cannot occur. It is an extremely common event, and can be observed following any cut in or penetration of the skin. Inflammation is generally effective in (1) controlling proliferation of microorganisms when they are thrust into the tissues and (2) restoring the normal state of the body (homeostasis).

This plate examines the inflammatory response of the connective tissues just beneath the skin following penetration by a bacteria-contaminated nail or tack (a). The four illustrations (stations) are microscopic views of changes that occur at the cellular and tissue level during the inflammatory process. The structures within each station are not to scale, but are enlarged to show detail.

Color the subheading Inflammation gray. Color the skin of the finger at upper right; use a color consistent with your normal skin for (a), but add a reddish tone to reflect the inflammation at (b). Then color the structures and related titles (a) through (j) in stations 1 and 2. Be careful not to use colors so dark as to obscure detail.

The normal skin (a; station 1) consists of a firm, non-swollen, warm, pain-free, cellular epidermis and a fibrous dermis. The normal fibrous (connective) tissue is generally characterized by a network of blood vessels (c) containing red blood cells (c^1; RBCs) and occasional neutrophils (d) and other white blood cells (not shown), and a loose irregular weave of connective tissue fibers (f^1). The fibers support a variety of cells, including macrophages (e) and fibroblasts (f; fiber-producing cell). A ground substance, including intercellular fluid, forms the matrix of the dermis.

Disruption of tissue in the presence of blood vessels induces an inflammatory response (station 2, inflamed skin). Inflammation is characterized by *redness* (due to dilatation of blood vessels close to the surface of the skin), *swelling* (due to release of plasma or blood fluid into the intercellular spaces), *heat* (as a consequence of the infusion of warm plasma into

the tissues), and *pain* (due to release of pain receptor stimulating chemicals).

At the microscopic level (station 2), there is a great increase in activity in the tissues injured or irritated by the presence of the foreign object (the tack). Clotting in torn vessels occurs. Bacteria (g), having entered into the tissues with the invading tack, begin to move and proliferate, releasing chemicals that are dispersed throughout the local tissue fluids. Local blood vessels dilate (h). Within a few hours, new blood vessels are formed in the injured area. White blood cells, predominately neutrophils, adhere to the margins of the vessels, then push out between lining cells (i; diapedesis) to enter the connective tissue. Great numbers of these cells congregate in the tissues, swarm toward the injured site (j; chemotaxis), and begin to phagocytose the microorganisms.

Continue by coloring the parts of stations 3 and 4.

The neutrophils undertake phagocytosis of all foreign elements, such as bacteria, tissue debris, and clotted blood (station 3). In the case of bacteria, opsonins are often employed (not shown). Having phagocytosed large numbers of particles, many neutrophils rupture their cell membranes and die. This accumulation of dead cells is called pus. Macrophages soon arrive to phagocytose (k) all residual debris, including dead neutrophils and any remaining microorganisms. Under certain conditions, specific immune responses may occur (presented in the following plates).

When phagocytosis of microorganisms is completed, local blood vessels diminish in number and size, and neutrophil populations return to a more normal state (the healing stage, station 4). Fibroblasts secrete new fibers (f^2; fibrosis) in the area of tissue disruption. This activity restabilizes the area. Such fibrous deposition may be significant in shape and size, and as such it is generally called scar tissue. Most scar tissue reorganizes along the lines of tension, disappearing into the background of fibrous tissue; larger masses may remain visible or palpable near the skin surface. Generally, the inflamed tissue returns to a normal state and merges inconspicuously with adjacent tissues.

NONSPECIFIC RESISTANCE TO DISEASE: INFLAMMATION

INFLAMMATION ⁎

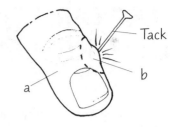

Tack
a
b

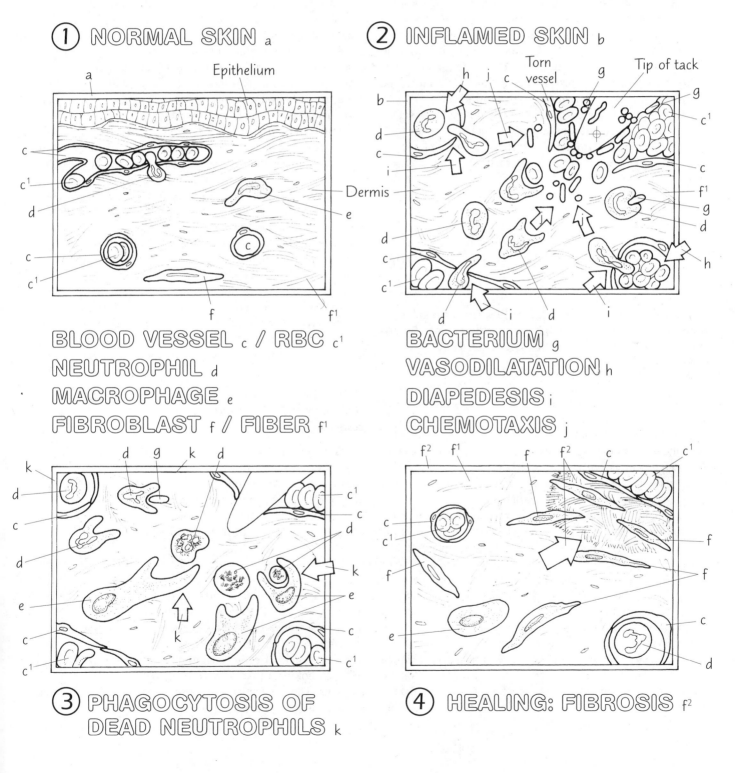

① NORMAL SKIN a

a
Epithelium
c
c¹
d
c
c¹
f
Dermis
e
f¹

BLOOD VESSEL c / RBC c¹
NEUTROPHIL d
MACROPHAGE e
FIBROBLAST f / FIBER f¹

② INFLAMED SKIN b

h j c Torn vessel g Tip of tack
b
d
c
i
g
c¹
c
f¹
g
d
h
d
c
c¹
d i d i

BACTERIUM g
VASODILATATION h
DIAPEDESIS i
CHEMOTAXIS j

③ PHAGOCYTOSIS OF DEAD NEUTROPHILS k

k
d
d g k d
c
c¹
d
c
e
c
c¹
c¹
d
k
e
c

④ HEALING: FIBROSIS f²

f² f¹ f f² c c¹
c
c¹
f
f
f
e
c
d

54
SPECIFIC RESISTANCE TO DISEASE: CHARACTERISTICS OF THE IMMUNE RESPONSE

Specific resistance to microorganisms is characterized by an interaction between cells or cellular products of the host organism and a part of the structure of the invading microorganism. The interaction is specific because it involves establishment of a structural connection between reciprocal parts of the two interacting components. The interaction is called an immune response. It is a response induced by the microorganism when it comes into the host's body.

Immunity is a condition of the body in which resistance is offered against microorganisms. Specific resistance is characterized by (1) structural specificity between interacting elements, (2) diversity of receptor structure and mechanisms of response, (3) the capacity to recall immune responses over time, and (4) to recognize what is "self" and what is not, and to engage the "non-self" molecule in an immune response.

In the previous three plates, nonspecific forms of resistance have been presented. The next several plates examine several aspects of the key activity of specific resistance: the immune response.

Color the structures and related titles (a) through (f) in stations 1 through 3. Use light contrasting colors for (b) and (c), and darker contrasting colors for (b²) and (c¹).

A cardinal feature of the immune response is specificity (station 1). Specificity (a) refers to those sites on the surface of the host cell membrane and the microorganism's cell wall or membrane that are reciprocally structured, permitting attachment, and the elicitation of the immune response. That part of the microorganism (b) that induces an immune response is called an antigen (b¹). The specific part of the antigen that interacts with the host cell is the antigenic determinant (b²). Antigenic determinants are the subject of Plate 56.

The cell that interacts with the antigenic determinant to set off the immune response is the lymphocyte (c), one of the white blood cells. That specific part of the lymphocyte that interacts with the antigenic determinant is called a receptor (c¹). It is impossible for an immune response to occur between lymphocytes and antigens lacking structures that are reciprocally shaped to one another. Products of certain lymphocytes (antibodies) also have receptors for antigenic determinants.

There are millions of antigenic determinants, each with a specific structural form. In turn, lymphocytes have or are capable of developing millions of receptors for these antigenic determinants, reflecting great diversity (station 2) in dealing with antigens. Some simple molecules are not antigenic until they are incorporated into a carrier molecule. Such simple molecules are called haptens (b³). An example of a hapten is a penicillin molecule. This molecule is not antigenic until it combines with a carrier molecule (d), usually a protein. The combination of a carrier and the hapten create an antigen that can initiate an immune response. Not only is there diversity in the array of lymphocyte receptors for antigen, there is great diversity in the mechanisms by which immune response is generated (see station 4).

Following a first encounter with an antigen, the lymphocyte is activated. By this activation, it achieves an ability to respond to that same antigen again some time (days to years) later, and with enhanced responsivity. Such is called immunological memory (e; station 3). This memory is transferred to the offspring of activated lymphocytes; these offspring retain the same receptors as the parent cells. Given the fact that encounters with antigen induce proliferation of lymphocytes, subsequent exposure of antigen is met with increased numbers of antigen-specific lymphocytes. Cells which retain the receptors for specific determinants after an immune response are called memory cells. They have long lives and are capable of eliciting an immune response with an antigen they "remember" from a previous encounter.

Color the subheading Immune Response gray, and structures and related titles concerned with the immune response in station 4.

When microorganisms enter the body, lymphocytes and/ or certain phagocytes recognize the "non-self" signature of these cells, and initiate an immune response. Certain phagocytes act as antigen presenter cells to lymphocytes. That is, the phagocytes ingest the microorganisms and activate lymphocytes. Activation of the lymphocytes to secrete substances or directly attack specific microorganisms constitutes the immune response (*¹). Antibodies (c²) on the surface of certain lymphocytes (c) or free in the tissue fluids or blood may attach to the microorganism, facilitating phagocytosis (g). Certain lymphocytes are induced to secrete toxic substances (c³) that kill the microorganism. Some lymphocytes (c⁴; called killer cells) directly attack the microorganism. Certain lymphocytes secrete substances (c⁵; lymphokines) that activate phagocytes (g) to ingest the antigens. The mechanisms of these responses are considered in the following plates.

SPECIFIC RESISTANCE TO DISEASE: CHARACTERISTICS OF THE IMMUNE RESPONSE

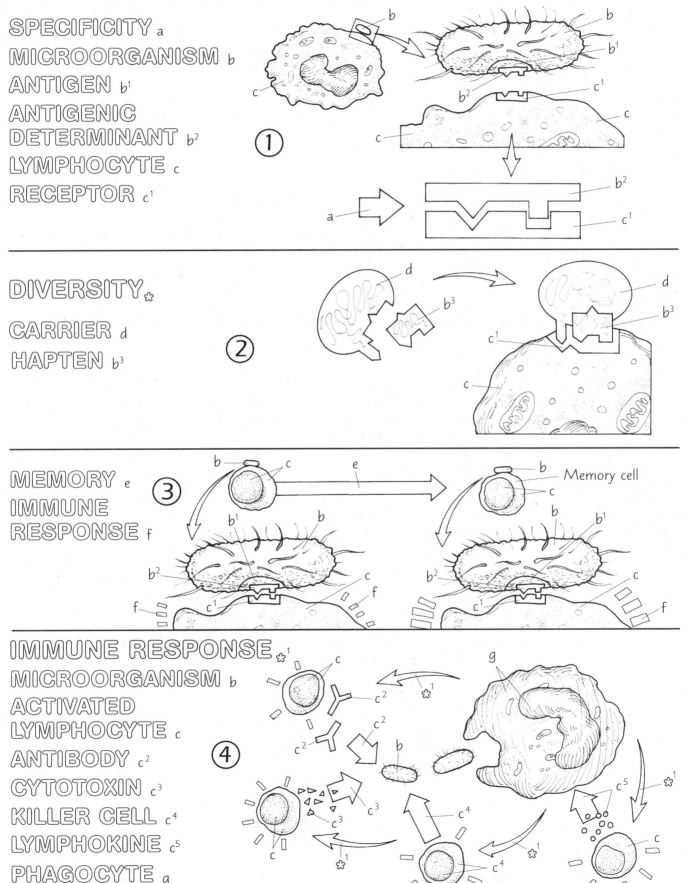

SPECIFICITY a
MICROORGANISM b
ANTIGEN b¹
ANTIGENIC DETERMINANT b²
LYMPHOCYTE c
RECEPTOR c¹

DIVERSITY ✿
CARRIER d
HAPTEN b³

MEMORY e
IMMUNE RESPONSE f

Memory cell

IMMUNE RESPONSE ✿¹
MICROORGANISM b
ACTIVATED LYMPHOCYTE c
ANTIBODY c²
CYTOTOXIN c³
KILLER CELL c⁴
LYMPHOKINE c⁵
PHAGOCYTE g

DEVELOPMENT AND ORGANIZATION OF THE IMMUNE SYSTEM

The immune system consists of a complex network of cells and cell products interacting in the immune response. One function of this response is to provide resistance to infectious diseases. The immune system develops and resides largely in lymphoid tissues and organs.

Color the titles and related structures (a) through (d¹). Use shades of the same color for (b), (b¹), and (b²). Use contrasting colors for (c) and (d).

The components of the immune system arise in the fetal bone marrow (a) during the third to sixth months of fetal development. Here precursor cells known as stem cells (b) arise and multiply to yield two major cell lines. One cell line composed of hematopoietic cells (b¹) develops red and most white blood cells in a process called hematopoiesis. Red blood cells (RBCs) are not involved in the immune system. The second developmental line, consisting of lymphopoietic cells (b²) produces lymphocytes in a process called lymphopoiesis.

Cells developed along the lymphopoietic line may differentiate by bursa processing (c²) or thymus processing (d¹). Each process involves a modifying organ; one which lends a distinct functional characteristic to the developing lymphocyte. A bursa (c) near the cloacal opening of the bird is lined with lymphoid tissue known to process the B lymphocyte line. In humans, there is no bursa. The bursa-equivalent (c¹) in humans is not known with certainty, but is probably the bone marrow or the fetal liver. Bursa or bursa-equivalent processing results in the formation of B lymph-ocytes (c³; B for bursa; also called B cells). Most B lymphocytes migrate from the bone marrow or liver to lymphoid tissues and organs. Perhaps 20% of them remain in the circulation. B lymphocytes can transform into plasma cells; both B lymphocytes and plasma cells can secrete antibodies (Plate 59). B lymphocytes are responsible for humoral immunity; that is, immunity associated with soluble antibodies and body fluids.

Cells processed along the thymus line develop in the thymus (d). The thymus is a two-lobed organ overlying the upper part of the heart. It is large in children. Lymphopoietic

cells are modified here to form T lymphocytes (d²; T for thymus; also called T cells). T lymphocytes make up about 75% of the blood lymphocytes. They have different receptor sites than B cells, and they do not produce antibodies. T lymphocytes are responsible for cell-mediated immunity; that is, immunity associated with cellular interactions. After the child reaches the age of about sixteen, the thymus undergoes progressive deterioration.

Color the subheading Lymphoid Tissue, related titles, and the structures in the boxed inset at lower right. Then color the subheading Lymphoid Organs, and related titles at lower left, and the structures in the human figure.

Lymph is fluid derived from the intercellular tissues and is found in lymph (lymphatic) vessels. Cells characteristically found in lymph are lymphocytes. Lymphoid tissue consists of a network of slender reticular fibers (e¹) produced by reticular cells (e) supporting masses of T and B lymphocytes (c³, d²), varied numbers of plasma cells (f), and phagocytes (g). Lymphoid tissue is the basic structural matrix of lymphoid organs.

Lymphoid organs are generally encapsulated, organized masses of lymphoid tissue. The primary lymphoid organs functioning as sources of lymphocytes are the bone marrow (c¹) and the thymus (d). Secondary lymphoid organs, functioning as satellites of lymphocyte production and activity, are the spleen (h) and lymph nodes (i). Unencapsulated lymphoid tissue includes the tonsils and adenoids in the wall of the pharynx, the vermiform appendix, and diffusely-disseminated lymphoid tissue in the lining membranes (mucosae) of cavity-containing organs. These latter structures constitute mucosal-associated lymphoid tissue or M.A.L.T (j).

B lymphocytes and T lymphocytes occupy central roles in the immune system. From their location in the lymphoid tissues and organs, they encounter most antigens that enter the body tissues and fluids. The nature of antigens and cell-mediated and humoral immunity are presented in the succeeding plates.

DEVELOPMENT AND ORGANIZATION OF THE IMMUNE SYSTEM

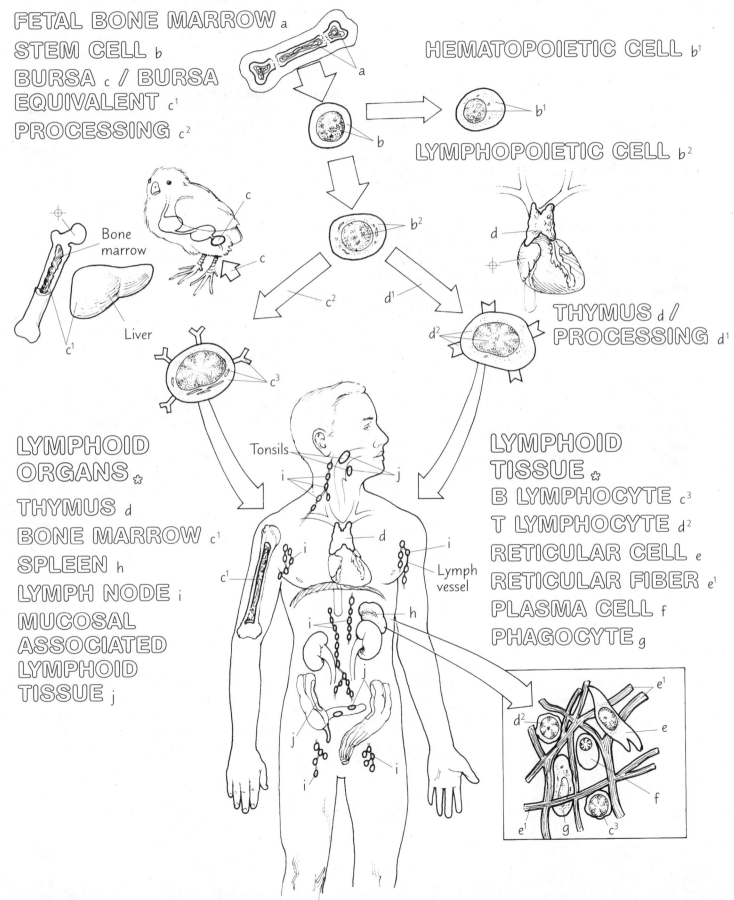

FETAL BONE MARROW a
STEM CELL b
BURSA c / BURSA EQUIVALENT c¹
PROCESSING c²

HEMATOPOIETIC CELL b¹

LYMPHOPOIETIC CELL b²

Bone marrow
Liver

THYMUS d / PROCESSING d¹

LYMPHOID ORGANS ✱
THYMUS d
BONE MARROW c¹
SPLEEN h
LYMPH NODE i
MUCOSAL ASSOCIATED LYMPHOID TISSUE j

Tonsils
Lymph vessel

LYMPHOID TISSUE ✱
B LYMPHOCYTE c³
T LYMPHOCYTE d²
RETICULAR CELL e
RETICULAR FIBER e¹
PLASMA CELL f
PHAGOCYTE g

56
ANTIGENS

Chemical substances that elicit a specific (immune) response are known as antigens (a). Most antigens are proteins; some are polysaccharides; very few are lipids or nucleic acids. In the older scientific literature, antigens were defined as foreign substances, but scientists now know that the body's own tissues can be antigens if the immune system interprets them as "non-self."

Color the bacterial antigens (a¹) through (a⁶), the antigen molecules (b) on the magnified portion of the flagellum, and the antigenic determinants (b¹), and related titles. Use shades of the same color or closely related colors for (a¹) through (a⁶). Note that the subheading Bacterial Antigens receives an (a) color but has no structure to be colored.

A bacterium may be considered to be an antigen, but in fact the single organism contains hundreds of different antigens (a¹) through (a⁶). The key feature of an antigen molecule (b) is the site called the antigenic determinant (b¹). The antigenic determinant is that part of the antigen that interacts with receptor sites on lymphocytes or antibodies to create an immune response. A single antigen molecule may have multiple antigenic determinants. In the case of the antigen molecule from the flagellum, a number of antigenic determinants in the chain of amino acids can exist. Each one of them is capable of eliciting an immune response.

Color the examples of exogenous antigens and related title, as well as the structures and related titles associated with endogenous antigens, at the lower part of the plate.

The list of possible antigens is enormously diverse. Exogenous antigens (c) exist outside the body but can come in contact with the skin or internal surfaces by a number of different means. Such antigens include microorganisms, foreign protein from the skin of domestic animals, certain medications, and pollen. When these antigens come in contact with antibodies in the body fluids or lymphocytes, an immune response is normally initiated. This response provides the body with a number of mechanisms to neutralize the effects of the antigen. On occasion, the immune response itself may cause allergic reactions (hypersensitivity).

Endogenous antigens (d) are those that arise within the body. Here we illustrate tissue antigens from the thyroid gland that interact with antibodies (e) in the thyroid tissues, inducing an immune response and tissue injury (f). Normally, thyroid gland tissue would be recognized as "self," and antibodies would not be produced to react with it. In autoimmune disease of the thyroid, the immune system fails to recognize the thyroid tissue as "self," and antibodies interact with the tissue as antigens. The resulting immune responses induce inflammation and other deleterious changes in the thyroid.

ANTIGENS

**BACTERIAL
ANTIGENS** (a)
   FLAGELLUM a¹
   CAPSULE a²
   CELL WALL a³
   CELL MEMBRANE a⁴
   CYTOPLASMIC ORGANELLE a⁵
   EXOTOXIN a⁶

ANTIGEN MOLECULE b
   ANTIGENIC
   DETERMINANT b¹
EXOGENOUS ANTIGEN c
ENDOGENOUS ANTIGEN d
   ANTIBODY e
   TISSUE INJURY f

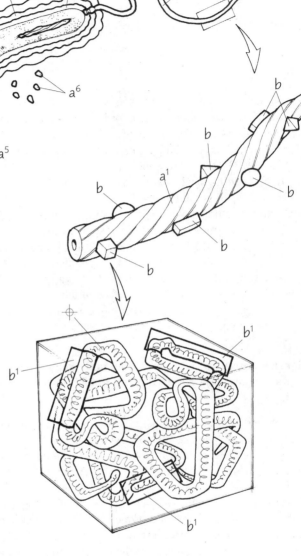

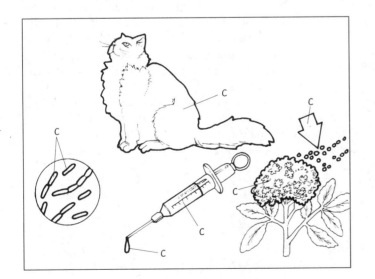

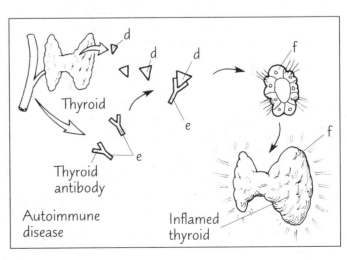

Thyroid

Thyroid antibody

Autoimmune disease

Inflamed thyroid

CELL-MEDIATED IMMUNITY

The character of the immune response to the presence of antigen is dependent upon the kind of lymphocyte involved. There are two kinds of lymphocytes that are activated by antigen: B lymphocytes and T lymphocytes (also known as B and T cells). Here we are concerned with T lymphocyte activation, resulting in what is called cell-mediated immunity.

Cell-mediated immunity involves an immune response in which antigen interacts with T lymphocytes. Antibodies are not a part of cell-mediated immune responses; however, B lymphocyte activation and subsequent antibody production may follow T lymphocyte activation. T lymphocytes do not recognize any antigen that is not protein, and then only recognize it if it is attached to the surface of other cells, such as phagocytes. T cells do not recognize soluble proteins because they cannot bind to them.

Color the structures and related titles in the sequence from above to below. Use colors that do not obscure cellular detail.

Cell-mediated immunity begins with the presentation of antigen by a phagocyte to a T lymphocyte; specifically, a helper T lymphocyte. The antigen may be any one of a number of molecular parts of a substance or microorganism. In the case shown here, the antigen is a breakdown product of phagocytosis (b¹) of a bacterium (a). Once taken up by the phagocyte (b), the bacterium is lysed by enzymes from lysosomes. The resultant fragments are discharged from the cell, and certain parts of the bacterium remain attached to the cell surface as antigens (c).

The antigen must be a protein and attached to a cell for the T cell to recognize the antigenic determinant. The antigen-bearing phagocyte is known as an antigen-presenting cell (or APC). When the antigen-presenting cell encounters a helper T lymphocyte (d; T_H cell) in the blood or intercellular tissues, the highly specific receptor site (d¹) of the T lymphocyte interacts with the corresponding antigenic determinant (c¹) of the antigen. This interaction sensitizes the helper T cell and activates it (d²). These activated helper T lymphocytes begin to proliferate rapidly.

They also secrete cytokines that induce the proliferation of cytolytic T cells. Cytokines (f; also called lymphokines) are activating substances and are produced in very small concentrations by the T_H cell. Cytokines enhance the inflammatory response (e), and activate neighboring B lymphocytes (h), inducing humoral immunity (next plate). Cytokines from helper T cells also activate phagocytes (i), facilitating phagocytosis of microorganisms and other non-self substances.

Some helper T cells differentiate into memory cells (g), memorizing the original T cell-antigen interaction. These memory cells live long lives and will rapidly activate in a second encounter with the original antigen. In this way, long term immunity to disease develops.

Cytolytic T cells (j; called CTLs or cytotoxic cells) proliferate rapidly. These cells directly engage cells expressing certain antigens, and lyse (destroy) them. The CTLs are specifically sensitive to cells containing viruses and to cancer cells. The CTLs also differentiate into memory cells; like those from T_H cells, these cells retain memory of the structure of the antigen initially encountered.

All of the discussed mechanisms target the offending microorganism or foreign particle and its antigen(s). In one or more ways, the antigens are contained and destroyed to the extent they do not represent a threat to the host organism.

CELL-MEDIATED IMMUNITY

BACTERIUM a
PHAGOCYTE b
PHAGOCYTIC PROCESS b¹
ANTIGEN c
ANTIGENIC DETERMINANT c¹

HELPER T
LYMPHOCYTE d

RECEPTOR SITE d¹

ACTIVATED
HELPER
T LYMPHOCYTE d²

INFLAMMATORY
RESPONSE e

CYTOKINE f
ACTIVATED
B CELL h
ACTIVATED
PHAGOCYTE i

CYTOLYTIC (CTL)
T LYMPHOCYTE j

MEMORY
CELL g

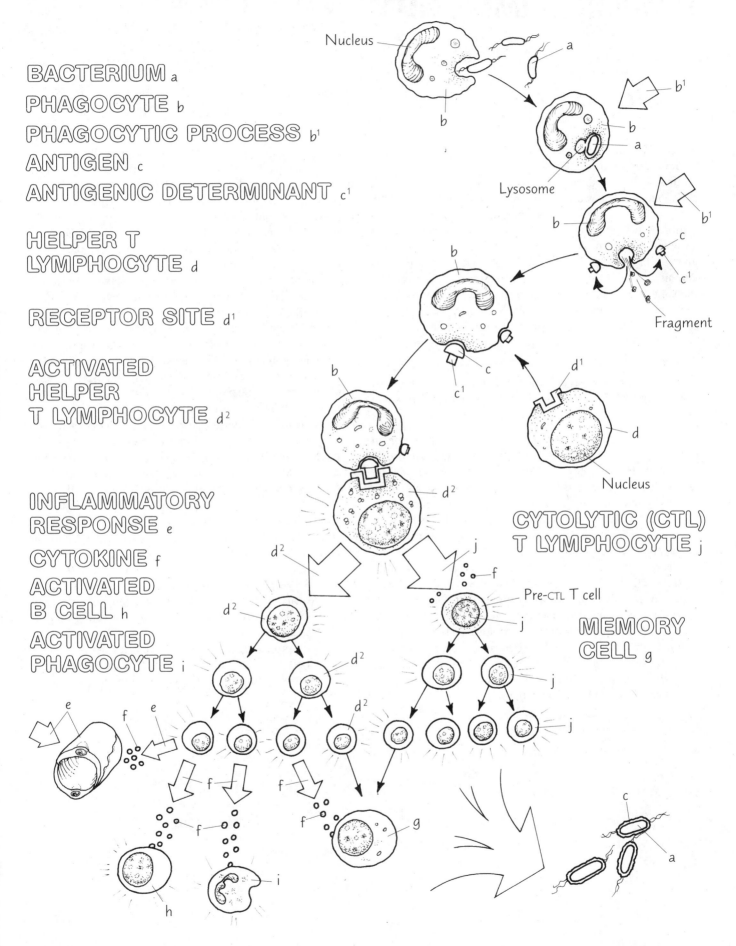

58
HUMORAL IMMUNITY

Humoral immunity is concerned with the interactions between antigens and antibodies that induce immune responses. Antibodies are produced by plasma cells differentiated from B lymphocytes. They are found in body fluids, primarily blood, and they are located on the membrane surface of B cells. B lymphocytes (and antibodies) can react with a variety of antigens, not just proteins, as is the case in cellular immunity and T lymphocytes. Further, B lymphocytes and antibodies can interact with soluble antigens (not attached to cells). It is the soluble character of antibodies that gives meaning to the term "humoral immunity" (humoral referring to fluid). As you learned in the previous plate, humoral immunity can be activated by helper T cells during cell-mediated immunity.

Color the titles and structures (a) through (g), starting at the top right of the plate. Consider using shades of the same color or color group for (b), (c), (d), and (e).

When microorganisms (such as bacteria; a) come in contact with B lymphocytes (b), an immune response is generated. The receptor sites (c^1) of a B cell are located on antibodies (c) on the surface of the B cell. One of the antigenic determinants (a^2) on the antigens (a^1) of the bacteria is contoured so that it is the precise reciprocal of the structure of the receptor site on the antibody molecule. This fit permits a contact that activates the B lymphocyte.

The initial feature of the immune response is activation of the B cell (b^1). The cell enlarges, and the cytoplasm is seen to become heavily populated with ribosomes and other protein-synthesizing organelles. The activated B cells proliferate rapidly. A number of these activated B lymphocytes differentiate into plasma cells (d). Plasma cells, and some B lymphocytes, produce antibodies (c) and release them into the body fluids, primarily blood. Several thousands of them are produced per second in the human body but only one kind of specific antibody is produced by each plasma cell, and, where applicable, each B cell. The Y shape of the antibody reflects the actual configuration of the molecule. Like antigen-T cell and antigen-B cell interactions, the antigen-antibody connection is made possible due to the structural reciprocity of the combining elements. Structural reciprocity refers to the precise fit of the antigenic determinant to the complementary structure of the receptor. It is this structural reciprocity that reflects the definition of specific resistance (Plate 54; specificity). The antibody facilitates phagocytosis (f) of the antigen, allowing the body to rid itself of microorganisms and eliminate infection. As the microorganisms are gradually eliminated, the activity of the plasma cells diminishes and the antigen-antibody response is concluded. Evidence suggests that a suppressor type T cell (not shown) may be responsible for inhibiting immune responses.

The proliferation of activated B cells includes development of memory cells (e). Like memory cells born from T cells, these memory cells retain a memory of the structural character of the antigen initially encountered. They have long lives and enable an individual to respond to the same antigen some time later. These subsequent responses are usually faster and greater than the initial responses, and the microorganisms generally do not have an opportunity to establish a foothold in the body the second time around. Indeed, the word immune means to "be free from."

In the case of protein antigens, the antigen is processed by the B lymphocyte and becomes attached to certain molecules on the surface of the B cell. This antigen is presented directly to the helper T lymphocytes (g), activating them to produce cytokines (not shown) that enhance the proliferation of antibody-producing plasma cells. Recent evidence suggests that helper T lymphocytes can stimulate antigen-carrying B lymphocytes in more subtle ways than the formation of cytokines.

HUMORAL IMMUNITY

BACTERIUM a

ANTIGEN a¹

ANTIGENIC DETERMINANT a²

B LYMPHOCYTE b

ANTIBODY c

RECEPTOR SITE c¹

ACTIVATED B CELL b¹

PLASMA CELL d

MEMORY CELL e

PHAGOCYTE f

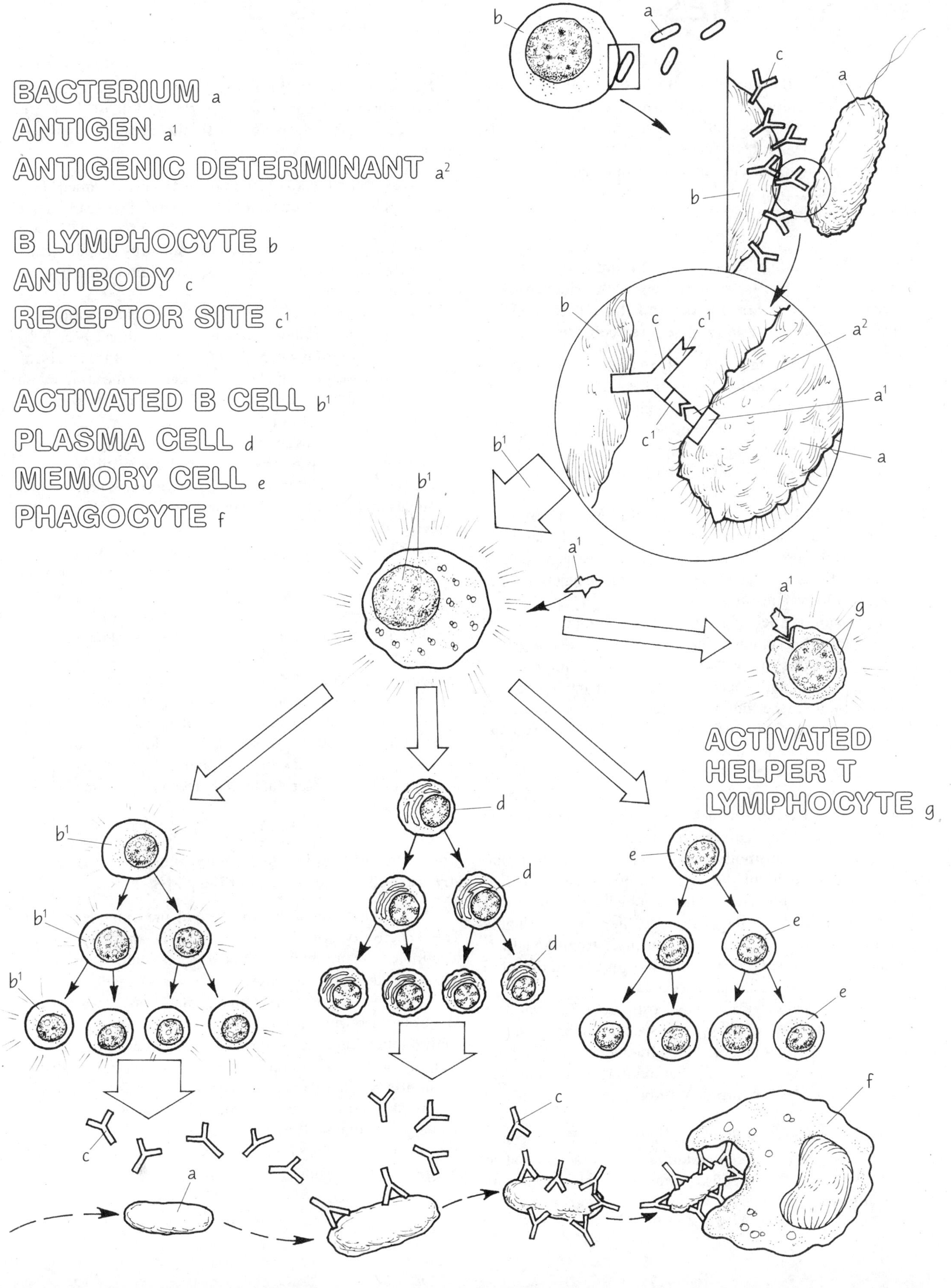

ACTIVATED
HELPER T
LYMPHOCYTE g

59
ANTIBODIES

Antibodies are the key elements in humoral (antibody-mediated) immunity. They are formed in B lymphocytes (B cells) and plasma cells, and are found in and on B cells, phagocytes, and in the body fluids, primarily blood. Antibodies pour into the circulation after stimulation of B cells by antigens. At any one time, approximately 17% of the protein in the blood consists of antibody molecules.

Color the subheading Structure and the antibody molecule. The subscripts are related to the terms they identify. Use contrasting light colors for (h) and (l), and contrasting darker colors for (v) and (c). Use a red color for (s).

Antibodies are composed solely of protein. The protein is organized into four units. Each unit contains amino acids bonded to one another by peptide bonds. Two of these units are heavy chains (h) with approximately 400 amino acids each. The other two units are light chains (l), each with about 200 amino acids. The heavy chains are longer than the light chains. The heavy chains of antibodies are similar within each type (see Antibody Types below); they function in binding complement, and help determine the activity of the antibody in immune responses. The light chains, unlike the heavy chains, do not significantly influence antibody immune function. Each set of four chains constitutes a monomer.

The chains are held together at certain points by bonds of sulfur (s; disulfide bonds). The "arm" portion of the heavy chain is connected to the "stem" portion by a hinge region (h¹). This is a region of increased molecular motion, giving some flexibility to the antibody shape. The antibody molecule flares at the hinge region to form a Y shape. Those parts of the heavy chains below the hinge regions (stem of the "Y") tend to crystallize when separated from the arms of the "Y" by proteolytic enzymes. For this reason, the stem of the antibody molecule is called Fc (fragment, crystalline). The Fc region activates the antibody molecule in an immune response.

Within each antibody molecule, there are terminal domains at the ends of the light and heavy chains called variable regions (v). These are where antibody molecules differ structurally from one another. These variable regions are highly specific and react with the antigen that stimulated their production. This specificity arises from the amino acid composition and sequence in these areas.

The combination of the outermost parts of the variable regions of both light and heavy chains are the antigen binding sites (called Fab, for antigen-binding fragment). The diversity of amino acid sequences in the Fab regions of the antibody make possible the binding of millions, if not billions, of different antigens.

At the lower parts of the "arms" of the antibody molecule, as well as the "stem," the configuration of amino acids is called the constant region (c). The constant region contains complement-binding sites (c¹).

Color the subheading Antibody Types, the antibodies, and related abbreviated identifiers in the middle of the plate. The subscripts are assigned to the type of antibody illustrated.

Five different types of antibodies exist in the human body. Each is called an immunoglobulin (abbreviated Ig), a name based on the fact that each belongs to a class of proteins known as globulins, and that each of these globulins is specifically concerned with immune responses. Each immunoglobulin is identified by a letter.

Immunoglobulin M (IgM) is the largest of the antibodies, consisting of five antibody subunits. Each subunit is called a monomer, thus IgM is a pentamer (penta, five). IgM (m) is the first antibody to appear in the circulation after the stimulation of B cells by antigen. Roughly 5 to 10% of the antibodies in the circulation are IgM molecules.

IgG (g) constitutes about 80% of the circulating antibodies, and is usually what is referred to by the term "antibody." This antibody is often called gamma globulin (g, gamma). IgG continues and completes the neutralization of antigen begun by IgM.

Antibodies IgE (e) and IgD (d) are monomers like IgG but exist in very low concentrations. IgE is found in persons with allergies and one who is experiencing allergic reactions (exaggerated sensitivity to antigen). The function of IgD, barely perceptible in the blood, is not well established but it is believed to be the antibody on the surface of the B cell where the antigenic stimulation takes place. IgA (a) is present in both blood and body secretions. It provides specific resistance to microorganisms in the tears, saliva, and in the gastrointestinal and respiratory tracts. IgA is also secreted in breast milk and is passed onto the suckling child, enhancing resistance to infection. Unlike other antibodies, IgA is composed of two monomers (dimer).

Color the subheading Specificity and related title and structures in the lower part of the plate.

Specificity is a cardinal feature of the immune system. This specificity is reflected very well in antibody-antigen reactions. Here are shown a number of antigenic determinants (b) on the surface of an antigen. Each determinant has its own molecular configuration. Only a matching (structurally reciprocal) configuration can attach to it. Such matching configurations are established on the antigen binding sites (Fab) of the arms of the antibody molecule (in this case, using the antibody IgG). When the connection is made between reciprocally structured sites, the first step of the immune response is initiated. Here we show two IgG antibodies, each with their own unique binding sites, finding complementary antigenic determinants.

ANTIBODIES

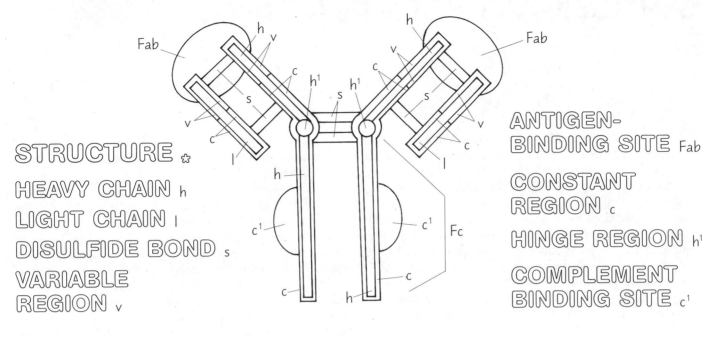

STRUCTURE ✿

HEAVY CHAIN h
LIGHT CHAIN l
DISULFIDE BOND s
VARIABLE
REGION v

ANTIGEN-
BINDING SITE Fab

CONSTANT
REGION c

HINGE REGION h[1]

COMPLEMENT
BINDING SITE c[1]

ANTIBODY TYPES ✿

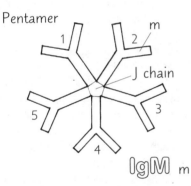

Pentamer

J chain

IgM m

Monomers

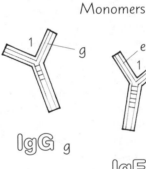

IgG g

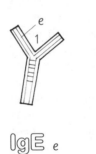

IgE e

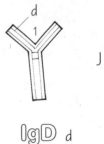

IgD d

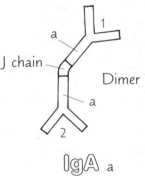

J chain

Dimer

IgA a

SPECIFICITY ✿

ANTIGENIC DETERMINANT b

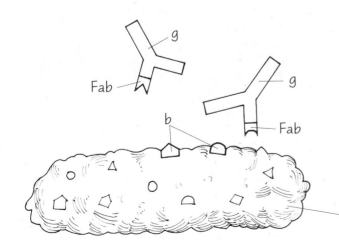

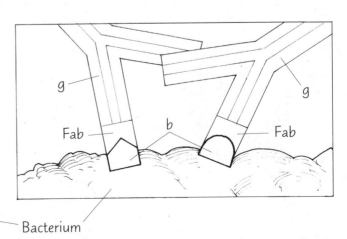

Bacterium

You will recall that antigen-antibody reactions are a part of antibody-mediated immunity (Plate 58). Antibodies are secreted by plasma cells and each consists of a protein of four units arranged in a Y shape (recall Plate 59). When an antibody interacts with an antigen at complementary structural sites, specific resistance takes place. This is where the immune function combines with nonspecific resistance such as phagocytosis to bring about resistance to disease. Understanding how these interactions take place gives us a glimpse of resistance to disease. A number of different reactions or events may occur as a consequence of antigen-antibody reaction. Some specific reactions are illustrated here.

Color the types of antigen-antibody reactions shown and the related titles. Use contrasting colors for (c) and (g).

One way the body develops specific resistance to viruses is by forming antibodies that prevent the entry of viruses into host cells. A virus (a) has on its surface a number of antigens with antigenic determinants (a^1) that match receptors sites (b^1) on host cells (b). Should these viruses make contact with the complementary receptor sites on these cells, they would invade the cells and make changes that would lead to destruction of the host cells. Soluble antibodies (c; IgG in this case) in the blood may already have antigen binding sites (c^1) complementary to the antigenic determinants of the virus. They can be formed following the initial encounter of antigen with B cells and the subsequent formation of plasma cells. These antibodies then bind to the viruses and prevent the virus from entering the cell. Viruses so affected remain in the fluid outside the cell and are eventually engulfed by phagocytes. Bacterial or toxin neutralization occurs similarly (not shown). Antibodies that interact with toxins are called antitoxins.

When antibodies (c) are allowed to react with antigens on bacterial surfaces (e), the bacteria tend to congregate in clumps, a phenomenon called agglutination (d). These clumps are vulnerable to phagocytosis, presumably because of the availability of multiple sites for antibody-phagocyte interaction. Also, the large clumps of bacteria attract phagocytes because phagocytes react with large particles better than with smaller particles. Agglutination tests are used routinely for performing diagnostic tests for bacteria. For example, to detect *Salmonella* species, a sample of unknown bacteria is mixed with anti-*Salmonella* antibodies. If clumping takes place, the bacteria are identified as a *Salmonella* species. If no clumping takes place, the bacteria are something other than *Salmonella*.

If antigens are soluble and not attached to cell surfaces, they react with antibodies in a manner that forms enlarged antigen-antibody complexes. A cross-linking occurs between antigen and antibody molecules when one antibody molecule links to two or more antigen molecules at the same time. If there is sufficient antibody present, usually in the form of immunoglobulin M (IgM), a pentamer, the antibody molecule (c^2) will fill up its multiple antigen binding sites with soluble antigen (g). This complex is much larger and heavier than the surrounding medium, resulting in its precipitation (f). These precipitated complexes are vulnerable to enhanced phagocytosis. Cross-linking and precipitation of antigen-antibody molecules is also the basis of laboratory precipitation tests.

In enhanced phagocytosis, the variable regions of the antibody molecule bind with antigen. The constant regions of the antibody molecule at the bottom of the Y shaped molecule attach to receptor sites on the phagocyte (h^1). This combination permits rapid phagocytosis (h) of the antigen. This enhanced phagocytosis is a called opsonization (Plate 52).

ANTIGEN-ANTIBODY REACTIONS

VIRAL NEUTRALIZATION ✱

VIRUS a

ANTIGENIC DETERMINANT a^1

CELL b

RECEPTOR SITE b^1

IgG c

ANTIGEN BINDING SITE c^1

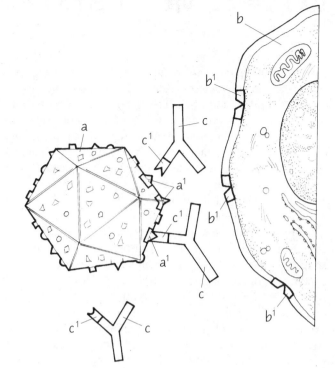

AGGLUTINATION d

BACTERIUM e

PRECIPITATION f

SOLUBLE ANTIGEN g

IgM c^2

ENHANCED PHAGOCYTOSIS h

PHAGOCYTE h^1

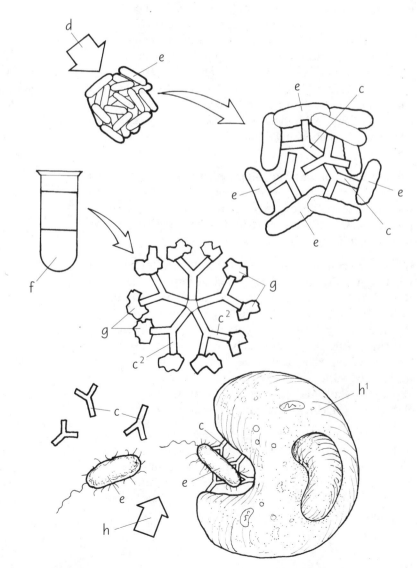

COMPLEMENT SYSTEM

The complement system consists of a number of proteins in the plasma of blood. They are functionally linked in a process called the "complement cascade." The complement cascade causes: (1) disruption of the cell membranes of bacteria by osmotic lysis, (2) induction of an inflammatory response following fragmentation of certain complement proteins, and (3) enhancement of phagocytosis by opsonization of bacteria (recall Plate 52). The complement system "complements" the function of the immune system.

The complement system employs two pathways to achieve its ends: classical and alternate.

The classical pathway begins with activation of a complement protein (C1) bound to the site of an antibody-antigen reaction on the surface membrane of a microorganism. Binding of complement to an antibody is called "complement fixation." As a result of C1 activation, a series of linked reactions ("cascade") occurs among complement proteins. These reactions, consisting of the splitting and binding of complement proteins, involve certain complement proteins functioning as proteolytic enzymes. The pathway ends with the formation of a "membrane attack complex" that perforates the cell membrane, destroying the microorganism (cell).

The alternate pathway is associated with non-complement protein factors and the polysaccharide layer of the surface membrane of a microorganism in the *absence* of an antibody-antigen reaction. The alternate pathway increases the enzymatic breakdown of C3 that sets in motion a series of processes shared with the classical pathway resulting in cell membrane disruption and death of the microorganism.

Choose four contrasting colors, each with three distinct shades, for the complement proteins C2, C3, C4, C5, and their proteolytic products. You will need five other colors for C1, C6, C7, C8, and C9. Refer to the Using Color section if you need to create more shades.

Color the structures and titles (d), (d¹), and (e) at upper right. Color the subheading Classical Pathway gray and the related arrows. Color the complement proteins in numerical order, following the path of the arrows. Note that the titles of the complement proteins are to be colored, but these titles are without subscripts as they are linked directly to the related complement protein. The arrows designated with a bullet (•) are to be colored black; they represent the participation of the alternate pathway. Subscripts (a), (b), and (c) are not used.

In the classical pathway, the complement protein C1 is activated by IgG or IgM binding to specific antigens (antibody-antigen reaction) on the surface of the cell membrane (d) of the bacterium. Prior to activation, the proteins of the complement system are in solution and unattached. To induce activation, a C1 molecule must be bound to at least two constant regions of an antibody, such as the pentamer IgM (not shown) or two molecules of IgG (e). The antigen-binding sites (Fab) of the variable regions of the antibodies are attached to the antigenic determinants (d¹) of the bacterial cell membrane.

Three soluble complement proteins are activated by an antibody-antigen reaction: C1, C2, and C4. C1 is a large molecule, and enzymatically (z) cleaves C2 to form C2a and C2b. C1 also cleaves C4 to form C4a and C4b. C2b and C4a remain in solution, but C2a and C4b bind to form the enzyme C3 convertase (C2aC4b). The formation of C3 convertase occurs in both alternate and classical complement cascade pathways. The action of this enzyme, to split C3 into C3a and C3b, is critical to the formation of the membrane attack complex and the opsonization of the bacterium.

C3a joins complement protein C5a in accelerating and facilitating the inflammatory response. This includes inducing vasodilatation and the associated movement of white blood cells out of the dilated blood vessels and into the connective tissues (f; diapedesis), and providing direction to inflammatory cells migrating toward masses of microorganisms.

Complement protein C3b functions in opsonization (g) by attaching to the bacterial cell in a manner described as "coating" or "buttering" the antigen-carrying organism. Receptors for the C3b protein on the phagocyte's surface assure capture of the C3b protein, enhancing phagocytosis of the bacterium.

Complement protein C3b also joins combined proteins C2aC4b to form C2aC4bC3b or C5 convertase. This enzyme splits C5 into C5a and C5b proteins.

Protein C5a joins with C3a to stimulate the inflammatory response. C5b becomes attached to the bacterial cell membrane. Subsequently, and in sequence, complement proteins C6, C7, C8, and C9 bind with C5b on the bacterial cell membrane. The combination of these proteins forms the structure of the membrane attack complex (h).

The membrane attack complex has a cytolytic effect (h¹) on the bacterium by producing pores in the membrane, permitting the escape of small intracellular molecules and water. This is effective in destabilizing osmotic equilibrium, killing the microbe.

The complement cascade would cause significant damage to host tissues (tissue digestion, inflammation, immune complexes) if it were not strongly regulated. This regulation (not shown) occurs in the form of soluble proteins functioning to accelerate, decelerate, inhibit, or restrict the complement cascade. Activation of the cascade only occurs in the presence of antibody-antigen reactions taking place on the cell membrane of the microorganism. The alternate pathway (functioning in the absence of an antibody-antigen reaction) is subject to the same regulators and is additionally unable to enter the cascade unless C3b is attached to the microbial cell membrane and certain biochemical relationships exist.

COMPLEMENT SYSTEM

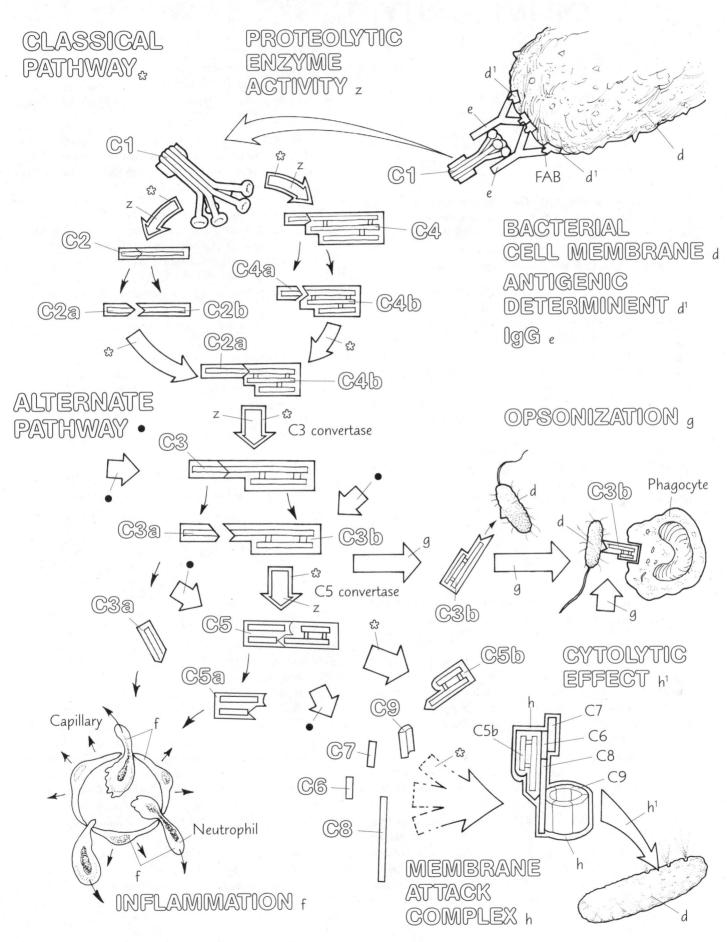

CLASSICAL PATHWAY ✳

PROTEOLYTIC ENZYME ACTIVITY z

C1

C2

C2a — C2b

C2a

C4

C4a

C4b

C4b

C3 convertase

ALTERNATE PATHWAY ●

C3

C3a — C3b

C3a

C5 convertase

C5

C5a

C7

C6

C8

Capillary

Neutrophil

INFLAMMATION f

BACTERIAL CELL MEMBRANE d

ANTIGENIC DETERMINENT d¹

IgG e

FAB

OPSONIZATION g

C3b

Phagocyte

C3b

CYTOLYTIC EFFECT h¹

C5b

C9

C5b

C7

C6

C8

C9

MEMBRANE ATTACK COMPLEX h

62
TYPES OF IMMUNITY

Immunity to disease may be innate (inborn) or acquired. Innate immunity begins with embryonic development and is derived from genetic and other factors in the body. Acquired immunity begins with the first exposure to antigens and the subsequent formation of antigen-specific antibodies. Acquired immunity may occur during fetal development or later throughout life. Here we illustrate four types of acquired immunity.

Select four contrasting but light colors for each of the four vignettes shown (a), (c), (d), and (e). Color the title, arrows, and structures (a) associated with naturally acquired active immunity. Select a color for (b) that will be used in three of the four vignettes. Read below before going on.

Naturally acquired active immunity (a) usually follows an episode of microorganism-induced illness. When pathogenic bacteria (b), for example, enter the body (station 1), they come in contact with B lymphocytes (station 2) which become activated and differentiate into plasma cells (station 3). The plasma cells produce antibodies with receptor sites structured to match the antigenic determinants on the surface of the bacteria (station 4). The antibodies bind to the antigens and lead them to phagocytosis (station 5) and destruction. The healing processes predominate and recovery to health ensues.

Memory cells remain after the infectious event to resist a second encounter with the same bacterium; thus, we rarely suffer two episodes of the same disease. The immunity is natural because it develops in the natural course of events. It is active because the body actively produces antibodies in response to the bacterial attack. Normally, this type of immunity, following each exposure to a different micro-organism, remains for a person's lifetime.

Color the title, arrows, and structures (c) associated with artificially acquired active immunity.

Artificially acquired active immunity (c) develops in an individual following an injection of a vaccine or toxoid. A vaccine may consist of weakened whole viruses or bacteria (first generation vaccine), fragments of microorganisms (second generation vaccine), or genetically engineered antigens from microorganisms (third generation vaccine). Toxoids are chemically altered toxins from microorganisms, and are used to immunize against such diseases as tetanus and diphtheria. When a toxoid or vaccine is injected into the body (station 1), B lymphocytes are activated with contact (station 2), and plasma cells secrete antibodies specific to the injected vaccine/toxoid (station 3). Memory cells remain in the body for long periods to provide protection (from activated B lymphocytes and so forth) when challenged in the future by invading related microorganisms (b^1; station 4). Protection against measles, mumps, rubella, polio, hepatitis B, pertussis, and other diseases is rendered this way. The immunity is considered artificial since it relies on vaccination and was not developed as a natural

function of acquired immunity. It is active because the body is producing its own antibodies. This type of immunity usually lasts for many years. In some instances, it will last a lifetime.

Color the titles, related arrows, and structures (d) associated with naturally acquired passive immunity. Use light shades for (d^1) and (d^2). Combine these colors in the placental barrier.

Naturally acquired passive immunity (d) is achieved during fetal development. During this time, maternal antibodies pass from the mother's circulation (d^1; station 1) through the placenta (d^1 and d^2; station 2) to the fetal circulation (d^2; station 3). The antibodies remain in the fetus (station 4) for a period of three to six months after birth, protecting the child from a variety of infectious diseases. Although the child's immune system is active and developing, it is not always adequate to meet the environmental challenges. This type of immunity is natural, developing between mother and child. It is passive because the antibodies are transferred to the child from the mother; the child did not actively produce the antibodies. Since the immunity conferred is passive, the immune system of the child does not actively form activated B lymphocytes and memory cells; the immunity is, therefore, short-lived.

Color the title, related arrows, and structures (b) and (e) associated with artificially acquired passive immunity.

Artificially acquired passive immunity (e) is conferred upon a person by antibodies received from another person or animal. The antibodies may be obtained from a person who has recovered (station 1) from the same disease the recipient (acute patient) is experiencing (station 2). In some cases, the antigens or microorganisms causing the disease may be injected (station 3) into an animal, such as a horse. In this manner, antibodies to the antigen are generated (station 4). In effect, the horse is a type of factory for producing antibodies. The antibodies are then separated from the blood of the horse, purified, concentrated, and injected into the recipient (station 5). This type of immunity is artificial as it does not occur naturally in the body of the person gaining the immunity. It is passive because the antibodies are not generated in the body of the person to be immunized, but are formed and received from another source. Such immunity is short-lived, lasting for only a few days or weeks. This is so because B lymphocytes are not activated and no memory cells are formed in the recipient as a result of the injection. In addition, the antibody preparation may pose a risk because it contains proteins to which the recipient may be allergic. Nevertheless, the antibody preparation is often used when a disease is due to a toxin or where a serious condition exists and no antibiotic is available. Certain viral diseases, such as hepatitis A, often respond to such therapy.

TYPES OF IMMUNITY

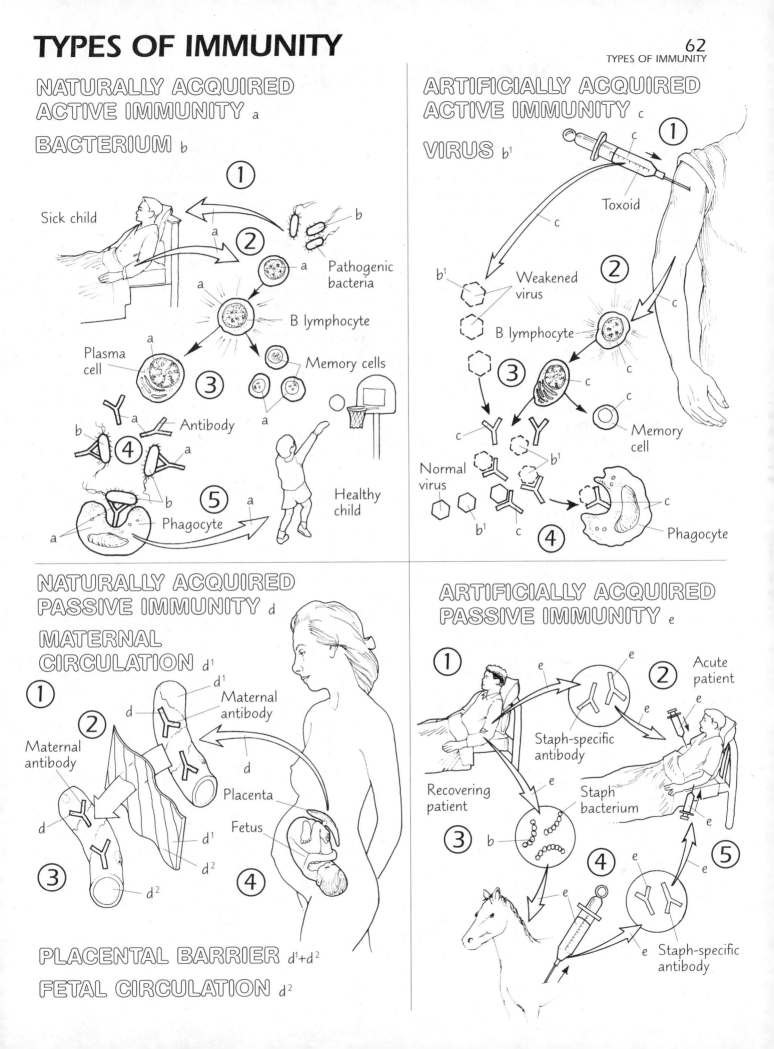

NATURALLY ACQUIRED ACTIVE IMMUNITY a
BACTERIUM b

①

Sick child

a

b

②

a

Pathogenic bacteria

a

B lymphocyte

a

Plasma cell

③

a

Memory cells

a

Antibody

b

④

a

b

a

Phagocyte

⑤

a

Healthy child

ARTIFICIALLY ACQUIRED ACTIVE IMMUNITY c
VIRUS b¹

①

Toxoid

c

c

b¹

Weakened virus

②

c

B lymphocyte

③

c

c

c

Memory cell

c

Normal virus

b¹

b¹

c

④

c

b¹

Phagocyte

NATURALLY ACQUIRED PASSIVE IMMUNITY d
MATERNAL CIRCULATION d¹

①

d¹

d

Maternal antibody

②

Maternal antibody

d

d¹

d

Placenta

Fetus

d

d²

③

d²

④

PLACENTAL BARRIER d¹+d²
FETAL CIRCULATION d²

ARTIFICIALLY ACQUIRED PASSIVE IMMUNITY e

①

e

e

② Acute patient

e

e

Staph-specific antibody

Recovering patient

e

Staph bacterium

③ b

④

e

e

⑤

e

e

Staph-specific antibody

63
MONOCLONAL ANTIBODIES

Having commercially-available antibodies to react with antigens causing disease (infection and cancer) is a desirable concept. The process of producing such antibodies became a reality with the findings of Georges J. F. Kohler of West Germany and Cesar Milstein of Argentina in 1975. These investigators shared the Nobel Prize in Physiology and Medicine in 1984 for their technique in which cells hybridize by fusion and produce colonies (clones) of antibody-producing cells. The antibodies formed by these cells are called monoclonal antibodies.

Look over the plate before coloring and note that the technique is broken down into eight stations. Color the subheading Production Technique gray. Color the parts and related titles associated with each station. Use light colors where necessary to prevent obscuring detail. The spleen is reddish-brown in mice. Use shades of one color for (c), (c^1), and (c^2), and shades of another color for (g), (g^1), and (g^2).

The manufacturing process begins with the isolation and purification of antigen (a) and the injection of the antigen into a healthy mouse (station 1). The antigen stimulates the immune system of the mouse, and the process leads to activating B cells (c), forming plasma cells (c^1), and producing antibody (c^2; station 2). Some of these plasma cells are extracted from the spleens of injected mice (station 3) and placed in a culture dish (d). Unfortunately, these activated plasma cells can remain alive in the culture dish only for a short time.

The value of the activated plasma cells is their ability to produce antibodies. But the cells must be fused with longer-living "immortal" cells whose progeny will survive indefinitely and produce the antibodies. Such immortal cells are cancer cells found in lymphoid tissue tumors called myelomas (-oma, tumor). These myeloma cells are generated from cell lines that have transformed significantly from the normal lymphoid cells. Myeloma cells are extracted from the spleens of mice harboring myelomas (station 4) and mixed with the culture of plasma cells, forming a mixed culture (f; station 5).

In the mixed culture dish, many of the myeloma cells fuse with the plasma cells, forming new cells (hybrids). These cells are called hybridomas (g^1; station 6). The hybridomas proliferate quickly at the expense of the unfused cells, which die in the culture medium. The hybridomas inherit the "immortality" of the myeloma cells, and they retain the antibody-producing capacity of the previously-cultured plasma cells. Masses of these identical hybridomas constitute a clone.

As the clone proliferates, it produces only the antibody (station 7) programmed by the antigen injected earlier (recall stations 1 and 2). Since these antibodies are derived from one clone of cells, they are called monoclonal antibodies (g^2). After a time in culture, the hybridoma cells secrete a large quantity of monoclonal antibodies, more pure and uniform than those antibodies isolated from the blood. This antibody production is in sharp contrast with an individual's normal contingent of plasma cells that produce a great variety of antibodies. Further, in the body, more than one kind of plasma cell is available for responding to different antigens.

Monoclonal antibodies can be used a number of different ways. They can be employed to seek out specific antigens, in this case, Ag^1 (h), in a mass of multiple antigens on the surface of a microorganism (station 8). The process is called target antigen identification (h). The target antigens can then be isolated for use in a vaccine. Monoclonal antibodies can also be used in diagnostic tests such as that used for gonorrhea (see ahead to Plate 67). Monoclonal antibodies can be produced against tumor cells in culture and then used to attack those cells in the body. In some cases, these antibodies can be used to "cleanse" bone marrow prior to transplant. Research scientists foresee the day when medications can be chemically attached to monoclonal antibodies and injected into the body. When the antibodies find their target antigens, the medication would be delivered to that site. Clearly, monoclonal antibodies represent one of the most elegant expressions of modern biotechnology.

MONOCLONAL ANTIBODIES

PRODUCTION
TECHNIQUE *

ANTIGEN a

SPLEEN b

ACTIVATED B
LYMPHOCYTE c

PLASMA CELL c¹

ANTIBODY c²

PLASMA CELL
CULTURE d

MYELOMA CELL e

MIXED
CULTURE f

HYBRID
CULTURE g

HYBRIDOMA
CELL g¹

MONOCLONAL
ANTIBODY g²

TARGET ANTIGEN
IDENTIFICATION h

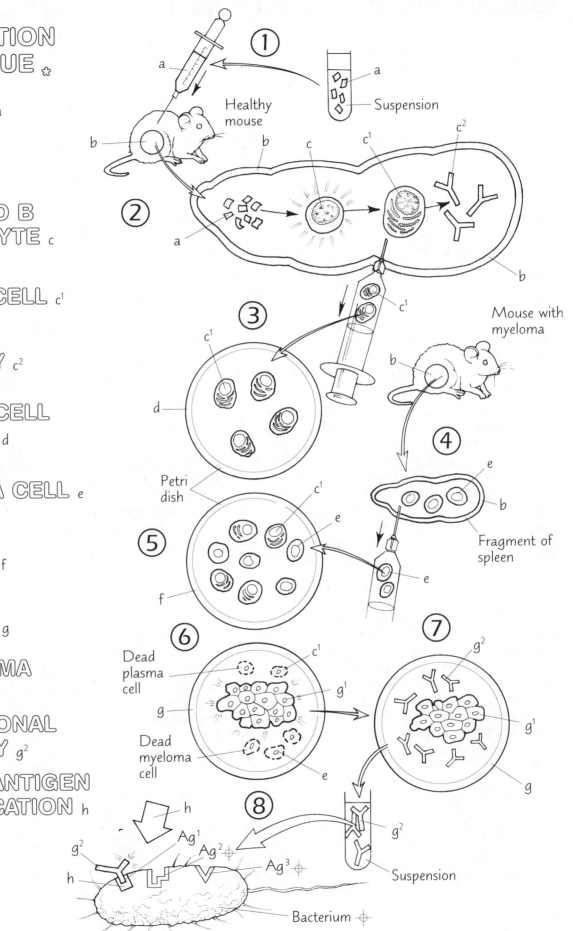

Healthy mouse

Suspension

Mouse with myeloma

Petri dish

Fragment of spleen

Dead plasma cell

Dead myeloma cell

Suspension

Bacterium ⊕

Ag¹ Ag² ⊕ Ag³ ⊕

COMPLEMENT FIXATION TEST

The reaction between antigens, antibodies, and complement forms the basis for an important laboratory procedure called the complement fixation test (recall Plate 61). This test is used to detect antibodies for a particular disease organism in the blood of a patient. If the antibodies are present, then there is a strong possibility that the suspected organism is also present. Here we examine the basis for this test. For this plate, we are testing the serum of two patients suspected of having syphilis. The complement fixation test will show that one of them is infected with syphilis spirochetes and one is not. This test, now rarely used for syphilis (Wasserman test), is used in the laboratory diagnosis of diseases caused by viruses and fungi.

Set aside shades of red for (e) and (e¹), and a neutral color for (d). Color the structures and titles associated with stage 1. Use contrasting colors for (a), (b), and (c). Work the samples of both patients (1 and 2). Wait to color stage 2 for now.

Prior to beginning the complement fixation test, a sample of each patient's blood is taken. Serum is obtained by allowing the blood to clot for about 10 minutes. This removes the clotting proteins from the plasma. The clotted blood is then subjected to centrifugation, and the fluid on top of the packed clotted blood is removed. This fluid is serum. It is then heated to eliminate any complement components. The heating must be carefully controlled with respect to temperature and time (56° C, 30 minutes) to prevent inactivating the antibody. The patient having specific antibodies to the suspected microorganism will carry them (a) in the serum. The patient that is free of the disease will not have syphilis-specific antibodies in the serum.

A sample of commercially-prepared antigen is added to the serum of each patient. In the case here, it is syphilis antigen (b). In general, the antigen selected will be the one suspected to be associated with the disease investigated. A quantity of commercially-prepared complement (c) is combined with the serum and antigen. The tube containing the three elements is incubated (d) for 90 minutes at 37° C. In the case of the patient infected with syphilis, the incubation period permits the antibody to bind the antigen and the antibody to bind the complement. (The illustration shows the antibody-antigen

relationship conceptually. Actually, the antigen binds to the Fab site at the end of the "arms" of the antibody.)

The antibody-antigen complexes are fixed by complement in such numbers that they can be visualized in stage 2 of the procedure. In the case of the antibody-free mixture of patient 2, incubation (d) fails to fix the complement because there is no specific antibody to which the complement can attach. The combination is said to be "non-fixed." The non-fixed state, like the fixed state, cannot be visualized without taking the mixture through stage 2.

Color the structures and titles associated with stage 2. Use bright red for (e) and a contrasting color for (f). Color the solution labeled (e¹) a reddish color but slightly lighter than (e). Color over the structures (c), (d), and (f) with the color for (e¹) but do not obscure the detail of these structures.

The first test tube mixture is now combined with sheep red blood corpuscles (e; RBCs) and hemolysin (f), an antibody that will bind to and dissolve (hemolyze) the sheep RBCs. The second mixture is treated the same way. Both mixtures are incubated at 37° C for about 2 hours.

In the case of the "fixed" mixture, the complement is not available to bind with the hemolysin-RBC complex because it is already bound to the syphilis antibody-antigen complex. Hemolysin cannot hemolyze the RBCs without complement. When the tube is centrifuged, the RBCs and antibodies will pack at the bottom of the tube and the fluid above will be clear. This result indicates that the complement fixation test is positive. The conclusion is that syphilis antibodies were present in the serum of patient 1.

In the case of the "non-fixed mixture" (patient 2), complement is available to bind with the hemolysin-RBC complex. This binding permits the antibody to lyse the RBCs (hemolysis) and disrupt their membranes and release the hemoglobin. When the mixture is centrifuged, there will be no packing of RBCs at the bottom of the tube; instead, the hemoglobin will diffuse throughout the mixture, reddening it. This reddening of the fluid indicates that the complement fixation test is negative (e¹). The conclusion is that syphilis antibodies were not present in the patient's blood, and that he/she probably does not have syphilis.

COMPLEMENT FIXATION TEST

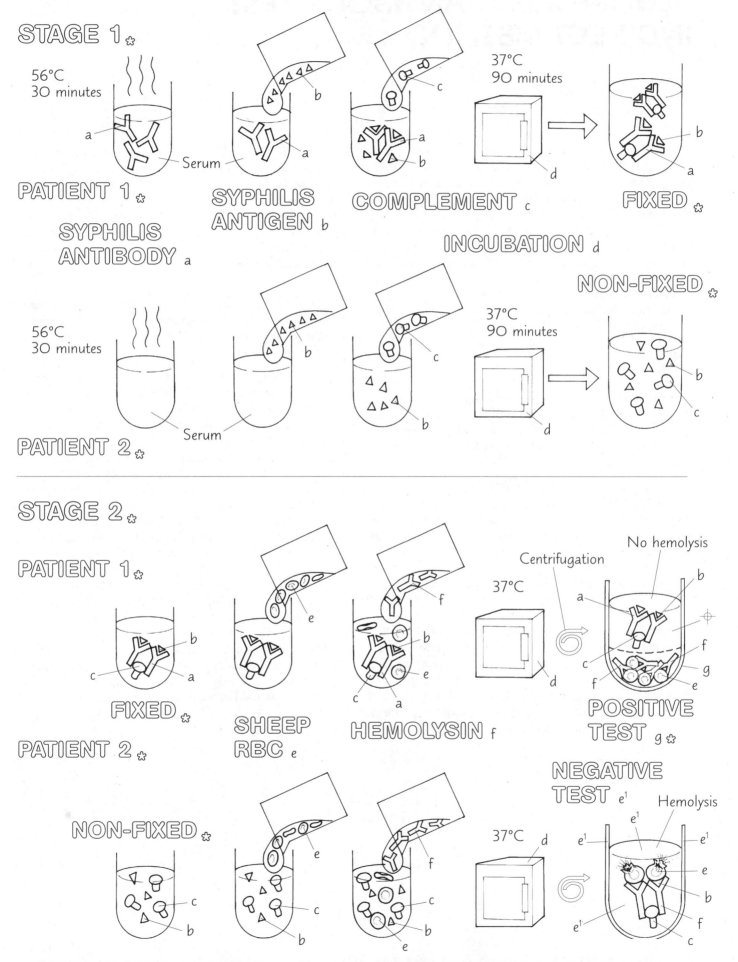

STAGE 1 ✿

56°C 30 minutes — a

PATIENT 1 ✿

SYPHILIS ANTIBODY a

Serum

SYPHILIS ANTIGEN b

COMPLEMENT c — a, b

37°C 90 minutes — d

INCUBATION d

FIXED ✿ — b, a

PATIENT 2 ✿

56°C 30 minutes

Serum — b

c — b

37°C 90 minutes — d

NON-FIXED ✿ — b, c

STAGE 2 ✿

PATIENT 1 ✿ — b, c, a

FIXED ✿

SHEEP RBC e — e

HEMOLYSIN f — f, b, e, c, a

37°C — d

Centrifugation — a, c, f

No hemolysis — b, f, g, e

POSITIVE TEST g ✿

PATIENT 2 ✿ — c, b

NON-FIXED ✿

e — c, b

f — c, b, e

37°C — d

Hemolysis — e¹, e¹, e¹, e, b, f, c

NEGATIVE TEST e¹

FLUORESCENT ANTIBODY TEST: INDIRECT METHOD

The fluorescent antibody test is a laboratory procedure used for the diagnosis of bacterial diseases. The test may employ the direct method (identifying the bacterium directly in the patient's serum on a slide) or the indirect method shown here. The indirect method of the fluorescent antibody technique involves the combining of antibody with a fluorescent dye in serum containing the bacterium, The object of the test is to detect specific antibodies in the blood of a person suspected to have a certain bacterial disease. In this plate, we are seeking to determine the presence of the antibody to the spiral bacterium *Treponema pallidum*, the spirochete responsible for syphilis. When used to test for *Treponema*, the fluorescent antibody test is called the FTA or Fluorescent Treponemal Antibody test.

Color the titles and structures (a) through (f) beginning at the top of the plate and working down. Wait to color the two lower circles for now. Select contrasting colors for (a) and (d); (d) and (d¹) should be orange or bright green. Use a yellow color for (b).

To begin the test, a sample of the commercially-available syphilis organism *Treponema pallidum* (a) is placed on a glass slide. In the microscopic (circled) view, the spiral character of the bacterium can be seen. The organism is allowed to dry on the slide.

A sample of the patient's serum (b), that is, the yellow fluid remaining after the blood has clotted, is added to the slide. If the patient has syphilis, the serum will contain antibodies (c) to the organism (patient 2). If the patient does not have syphilis, there will be no antibodies (patient 1). The serum antibodies, if present, will bind the spirochete antigens.

A third component consisting of antibodies to which a fluorescent dye is chemically attached is added to the slide. These antibodies are prepared by injecting a laboratory animal with human antibodies. The animal's immune system interprets the antibodies as antigens, and it responds by producing antibodies to these antigens. Thus, antihuman antibodies are obtained. These antibodies will react with any human antibodies. Molecules of a fluorescent dye are then chemically attached to the antihuman antibodies to produce fluorescent-tagged antihuman antibodies (d). These antibodies are added to the slide; after a few moments, the slide is washed, and unattached antibodies are swept away. This washing prevents non-specific fluorescence.

Color the lower two circles: the one at left (e), representing the view seen under the fluorescence microscope after the wash, is to be colored black. Color the view at right. Use a bright color for (f).

To interpret the test, each slide is placed under a special microscope that directs ultraviolet light onto the specimen. Ultraviolet light will cause the fluorescent dye to glow when the dye has accumulated on a surface. In the case where no syphilis antibodies exist, there were no antibodies to which the fluorescent dye could attach, and therefore the fluorescent antibodies would wash away. Without fluorescence, the view is black to the observer. The conclusion is that this patient does not have (is negative for) syphilis (e).

In the case at right, the fluorescent dye is attached to the antihuman antibodies (d) that are attached to the human syphilis antibodies (c). When the fluorescence microscope is used to study the slide, the ultraviolet light will cause the fluorescent dye to glow an orange or bright green color when the dye has accumulated on the surface. The conclusion is that syphilis antibodies were present in the patient's serum, indicating that the patient is positive for syphilis (f).

Fluorescent antibody tests are adaptable to a broad variety of antigens and microorganisms, and are widely used in diagnostic laboratories. By varying the commercially-prepared antigen, one can test for numerous different antibodies.

FLUORESCENT ANTIBODY TEST: INDIRECT METHOD

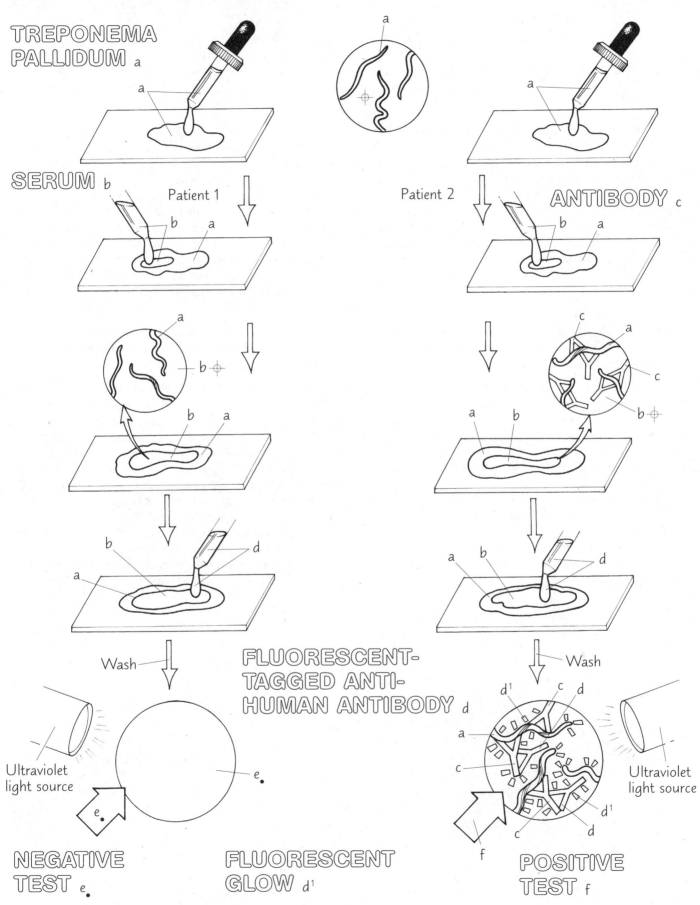

TREPONEMA PALLIDUM a

Patient 1

Patient 2

SERUM b

ANTIBODY c

Wash

FLUORESCENT-
TAGGED ANTI-
HUMAN ANTIBODY d

Wash

Ultraviolet
light source

Ultraviolet
light source

NEGATIVE TEST e

FLUORESCENT GLOW d¹

POSITIVE TEST f

RADIOIMMUNOSORBENT ASSAY (RIA)

Immunosorbent tests are diagnostic procedures in which antigens or antibodies are attached to a solid surface, such as polystyrene beads. A sample of the patient's serum is added to see whether the serum contains substances that will react with the surface of the beads. If specific substances are present, they will combine with (or immunosorb to) that solid surface. Antigens or antibodies in a patient can thus be detected. Here, the radioimmunosorbent assay (RIA) uses radioactive substances to determine whether antibodies have combined with antigens on the surface of a particle.

Color the titles and related materials (a) through (f) associated with patient 1 and patient 2. Wait to color station 4 for now. Use contrasting colors for (a) and (b), (c) and (d), and (e) and (f).

The RIA test uses antigens from a microorganism bonded to solid particles. In our case here, the particles are commercially-available polystyrene beads (a) of microscopic size bonded to antigens (b) from hepatitis viruses. To perform the test, samples of serum (c) are taken from two patients; the sample on the left (patient 1) is without antibodies to hepatitis antigen; the sample on the right (patient 2) contains hepatitis antibodies (d). These serum samples are added to the hepatitis antigen-coated polystyrene beads (station 1). In the sample at left (station 2), there will be no antigen-antibody reaction because there is no antibody present specific for the hepatitis antigen. In the sample at right (station 2), the antibodies for the antigen are present, and they bind on the surface of the beads. The reaction cannot be easily observed, however. Therefore, an antihuman antibody against the hepatitis antibody (e; recall Plate 65), combined with a radioactive substance is added to the mixture (station 3) so the reaction may be observed.

In the serum where there are no hepatitis antibodies attached to antigen-coated beads (left side, station 3), there is no reaction with the radioactive antihuman antibodies. In the serum containing hepatitis antibodies attached to antigen-coated beads, the radioactive antihuman antibodies react with the hepatitis antibodies (right side, station 3).

Color the titles and related radioactivity detector (g) and radioactive signals (f^1) at station 4. Use a light neutral color for (g).

After the two samples are washed to eliminate any unattached radioactive antibodies, their radioactivity is measured by a detector (g). The serum on the right, containing the radioactive beads, will register a level of radioactivity (f^1; station 4) proportional to the amount of antibody in the patient's serum. This means the patient has antibodies to the hepatitis virus. Different hepatitis antibodies indicate to which hepatitis virus the patient was exposed. The antibodies also suggest how recent the exposure occurred (recall Plate 59).

The serum containing no hepatitis antibodies (left side, station 4) will not generate any radioactivity because the radioactive antihuman antibodies have no hepatitis antibodies with which to bind and will be removed in the wash step.

RADIOIMMUNOSORBENT ASSAY (RIA)

BEAD a

ANTIGEN b

SERUM c

HEPATITUS ANTIBODY d

ANTIHUMAN ANTIBODY e

RADIOACTIVE SUBSTANCE f

RADIOACTIVITY DETECTOR g
SIGNAL f¹

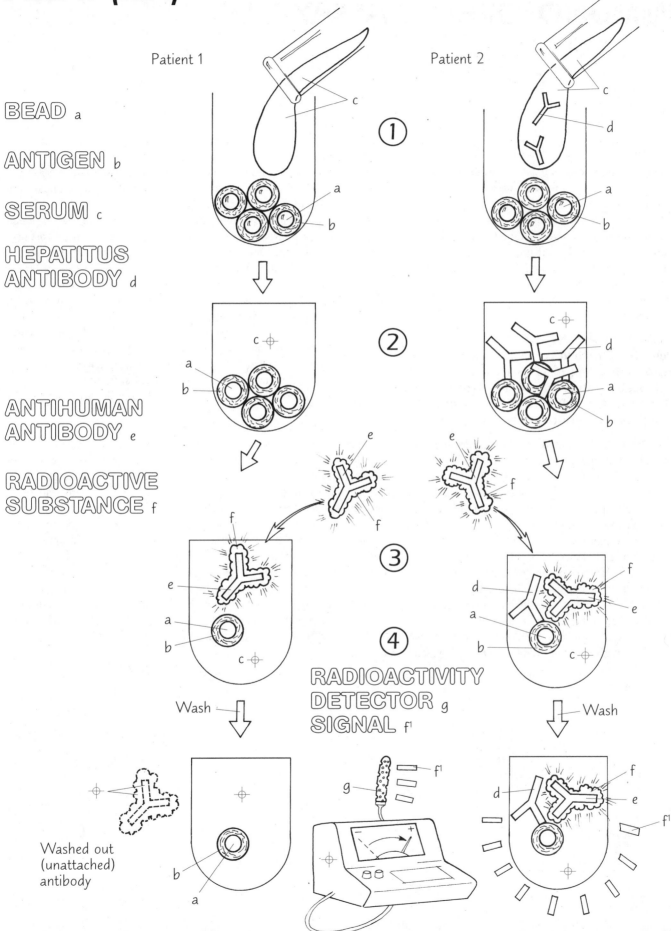

Patient 1

Patient 2

Washed out (unattached) antibody

Wash

67
ENZYME-LINKED IMMUNOSORBENT ASSAY (ELISA)

Immunosorbent tests and assays are diagnostic procedures in which antigens or antibodies are attached to a solid surface. Where the radioimmunoassay (RIA) uses radioactive substances (Plate 66), another immunosorbent test uses an enzyme reaction to detect antibodies or antigens. This test is called the enzyme-linked immunosorbent assay (ELISA; or enzyme immunoassay or EIA). The ELISA test used for detecting antigens is called the direct ELISA; the test for antibodies is called the indirect ELISA. The direct ELISA is illustrated here.

Color the titles and related structures (a) through (e¹), beginning at the top of the plate and working down (stations 1 through 4). Reactions to the presence of antigen are illustrated at the left; lack of reaction due to the absence of antigen is illustrated at right.

The assay illustrated here is to determine the presence or absence of antigens associated with the gonorrhea organism *Neisseria gonorrhoeae*. This test uses commercially-available polystyrene beads (a) bonded to gonorrhea antibodies (b). Swab samples are taken from the reproductive tract of patient 1, believed to be infected with gonorrhea, and from patient 2. The sample-laden swabs are placed in the containers of beads bonded to gonorrhea antibodies (station 1). If gonorrhea antigens (c) are present, they will react with (immunosorb to) antibodies on the surface of the beads.

A sample of enzyme-linked gonorrhea antibodies (d) is added to each of the containers (station 2). These are antibodies which have a certain enzyme bound to them. These antibodies will react with the gonorrhea antigens, if present (station 3, left side). If no gonorrhea antigen is present (station 3, right side), the enzyme-linked antibodies will be flushed out of the container when the samples are washed.

The specific substrate on which the enzyme acts is added to the mixtures. This is called an enzyme-specific substance (e). To encourage the reaction between the enzyme and its substrate, the mixtures are incubated. During this time, the enzyme will react with its substrate (station 4, left), causing a color change which can be detected and quantified by colorimeter or spectrophotometer. Sometimes the color change can be seen without visual aids. The color change indicates that an enzyme reaction has taken place. The reaction could not have occurred if there had not been gonorrhea antigen to which the enzyme-linked gonorrhea antibody had attached. Therefore, the color change is evidence that patient 1 has gonorrhea antigens in his/her body and, therefore, is infected with *Neisseria gonorrhoeae*.

In the container without the gonorrhea antigen, the free enzyme-linked antibody will be removed from the container by washing (station 3). This is so because there is no gonorrhea antigen to which it can attach. Later, when the substrate (e) is added, no reaction will occur because there is no enzyme present in the container. Thus, the lack of color change is evidence that patient 2 does not have gonorrhea antigens in his/her body and, therefore, is not infected with *Neisseria gonorrhoeae*.

This ELISA procedure can be adapted to a variety of tests simply by varying the antigen or antibody on the bead's surface. Thus, any of a multitude of infectious diseases can be detected by these immunological procedures.

ENZYME-LINKED IMMUNOSORBENT ASSAY (ELISA)

BEAD a

GONORRHEA
ANTIBODY b

GONORRHEA
ANTIGEN c

ENZYME-LINKED
GONORRHEA
ANTIBODY d

ENZYME-SPECIFIC
SUBSTANCE e

COLOR CHANGE e¹

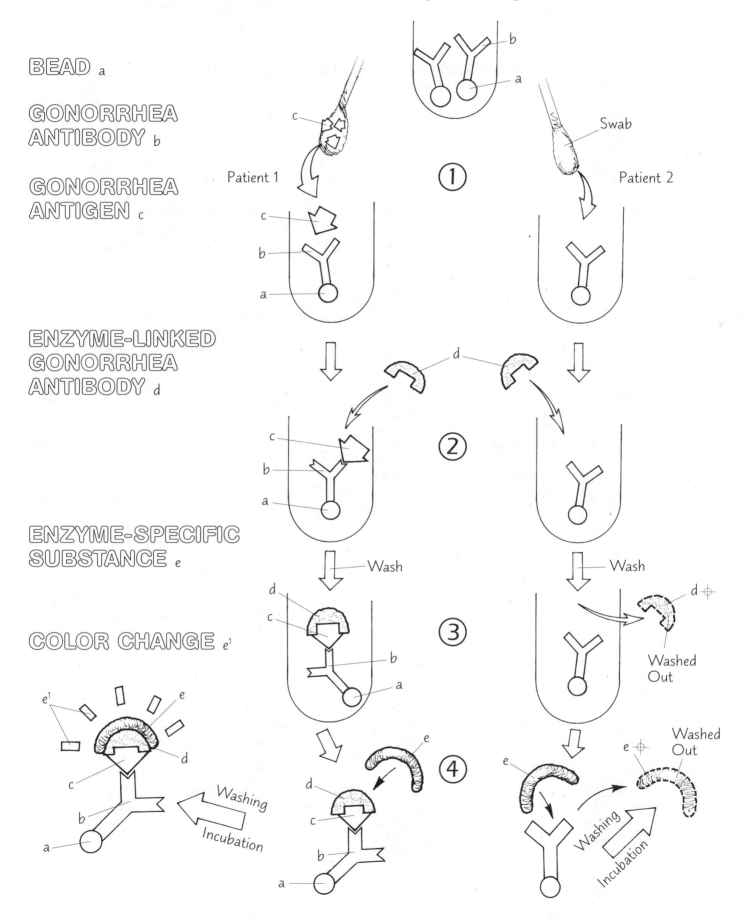

68
NEUTRALIZATION TEST

The diagnostic laboratory uses procedures called neutralization tests to detect antibodies produced by the patient against toxins and viruses. These antibodies are called antitoxins or virus neutralizing antibodies. A toxin consists of molecules that cause physiological damage to cells and organisms. Endotoxins are toxins that are bound to the cell membranes or walls of bacteria, and which are released during bacterial lysis or disintegration. Toxins that are produced as metabolic products of bacteria and then released to the environment are called exotoxins. Exotoxins, and not endotoxins, are proteins that can be neutralized by specific antibodies (antitoxins) in neutralization tests.

Neutralization tests tend to be tedious because the neutralization reaction usually requires live animals to determine whether the reaction has taken place. In this plate, we present two types of neutralization tests performed in the diagnostic laboratory. Since their object is to detect antibodies, both are immunological procedures

Color the subheading Normal Toxic Reaction gray, and the arrow labeled (*¹). Starting at upper left, color the titles and related structures (a) through (c), following the directional arrows. Use a light color for (a) and a lighter shade for (a¹). Color the subheading Toxin Neutralization Reaction/Test and arrow labeled (*²) gray, and the related titles and illustrations. Use a light color for (d); use the same color for (c¹) as you did for (c).

The toxin neutralization test seeks to determine whether a patient has antibodies (antitoxins) that can neutralize toxin molecules. In a normal toxic reaction, a healthy mouse (b) with normal cells (a) can be injected with toxin molecules (c). The toxin molecules will act on the cells of the mouse, causing cell destruction (a¹) within a few hours. This pathogenic effect will usually cause the death of the mouse (b ⊕). One toxin that can cause such an effect is the botulism exotoxin (botulin, or botulinum toxin) produced by the anaerobic, Gram-positive bacterium *Clostridium botulinum.*

In a typical clinical case, a patient presents at the hospital emergency department with slurred speech, blurred vision, muscular weakness, and other signs of neurological impairment. If botulism is suspected, a sample of blood is taken from the patient and sent to the clinical microbiology laboratory for a toxin neutralization test. If the patient was exposed to botulinum toxin, the immune system would have formed antibodies to the toxin (e; antitoxin). These antitoxins will be in the blood taken from the patient.

In the laboratory, serum (d) is separated from the blood sample. An amount of botulinum toxin (c¹) is added to the serum containing antitoxin. Following a short period of incubation, this toxin/antitoxin mixture (d¹) is injected into a laboratory rodent (in this case, a mouse). The mouse remains healthy. A biopsy of the mouse's tissue would reveal that the cells (a) are unaffected by the toxin. This is so because the toxin had reacted with the antitoxin during the incubation period before injection into the mouse, and the toxin was neutralized and rendered harmless. Therefore, the test is positive for the presence of botulinum antitoxin, and the presence of the toxin can be inferred. The physician would be advised of the results and the patient treated accordingly. If the mouse died, it would indicate that the patient was not exposed to botulinum toxin and did not form antibodies to it.

Color the subheading Normal Reaction to Virus gray, as well as the arrow labeled (*³). Color the title (f) and related structures (a), (a¹), and (f) working from left to right. Use a light color for (f). Then color the subheading Viral Neutralization Reaction/Test gray, and color the titles and related structures (a), (d), (d²), (e¹), and (f¹). Use the same color for (f¹) as you did for (f).

The object of the viral neutralization test is to determine whether antibodies for a specific virus are in the patient's serum. Normally, if viruses (f) are added to a culture of normal cells (a), the viruses will infect the host cells. They will multiply and destroy the host cells (a¹).

A positive reaction to this test is demonstrated in the following clinical example. A patient experiences abdominal pain, enlarged liver (hepatomegaly), dark urine, and skin yellowing (jaundice); the evaluating physician suspects the viral infection hepatitis A. A blood sample is taken from the patient and sent to the microbiology lab. If the patient has hepatitis A, the blood serum (d) will contain hepatitis antibodies (e¹). The serum is mixed with hepatitis virus (f¹), and the virus/antibody mixture (d²) is added to a culture of normal cells (a). The viruses are neutralized by the antibodies, and are therefore unable to penetrate the cells, multiply, and destroy them. After a period of incubation, the cultured cells appear normal. The conclusion is drawn that the serum contained hepatitis antibodies, and therefore, is presumed positive for the presence of hepatitis A virus.

Neutralization tests are valuable diagnostic tools but the materials needed for the tests are substantial. For this reason, the neutralization tests are normally available only in sophisticated laboratories. Neutralization tests are also lengthy, and in some laboratories have been replaced by immunoassays.

NEUTRALIZATION TEST

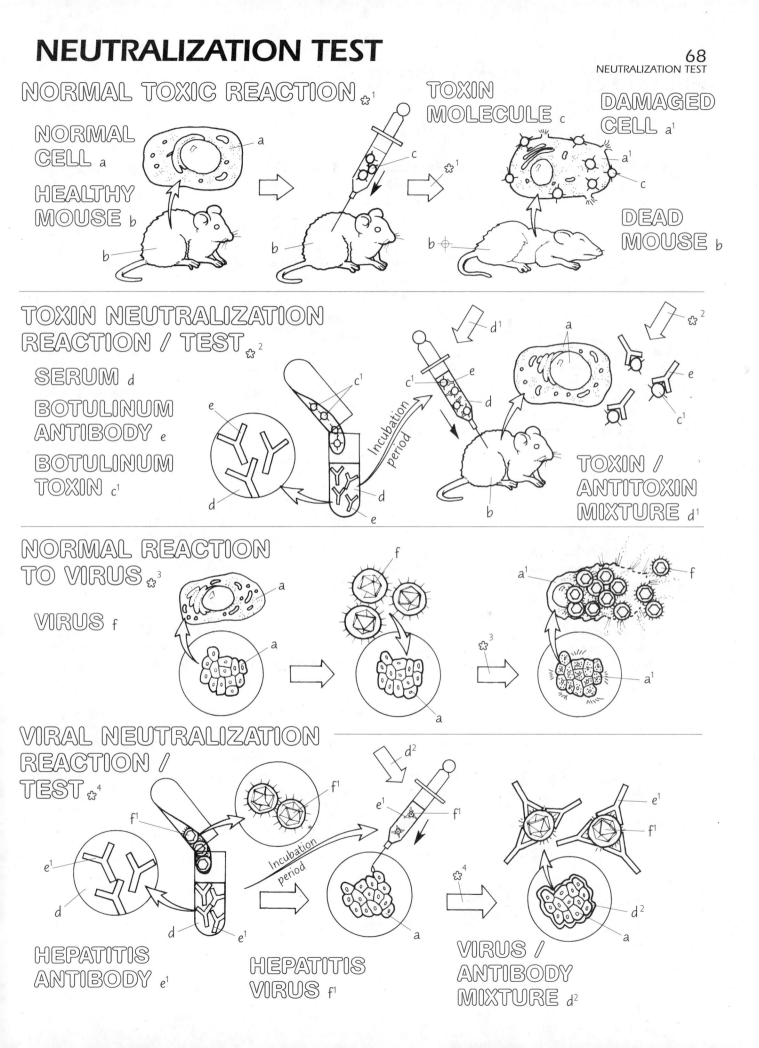

NORMAL TOXIC REACTION ✿1

NORMAL CELL a
HEALTHY MOUSE b

TOXIN MOLECULE c

DAMAGED CELL a1

DEAD MOUSE b

TOXIN NEUTRALIZATION REACTION / TEST ✿2

SERUM d
BOTULINUM ANTIBODY e
BOTULINUM TOXIN c1

Incubation period

TOXIN / ANTITOXIN MIXTURE d1

NORMAL REACTION TO VIRUS ✿3

VIRUS f

VIRAL NEUTRALIZATION REACTION / TEST ✿4

Incubation period

HEPATITIS ANTIBODY e1

HEPATITIS VIRUS f1

VIRUS / ANTIBODY MIXTURE d2

69
AGGLUTINATION TEST

The agglutination test is one in which cells or particles clump together, or agglutinate, when combined with specific antibodies. The reaction forms the basis for a number of diagnostic procedures used in the immunology or microbiology laboratory. One object of the tests is to determine whether the patient's serum contains antibodies specific for a particular antigen. If the test is positive for such antibodies, then the tested person may be harboring the suspected microorganism containing the antigen. This test can also be used to test for the antigen itself.

Color the subheading Slide Test, the titles (a) through (c), and the related structures in the Petri dish and on the glass slides at the upper left part of the plate. Use a light yellow color for (a), and a bright color for (c). Then color the subheading Tube Test and the related titles and structures on the upper right part of the plate. Use a light, neutral color for (d).

The agglutination test may be accomplished with glass slides (slide test) or test tubes (tube test). In the slide test, a sample of the patient's serum (a) is mixed with cells of the suspected microorganism (b), in this case *Salmonella.* If the patient is infected with *Salmonella,* then there will be antibodies in the serum that are specific against that bacterium. While the mixture is permitted to stand for a few minutes, the antibodies (not shown) will react with the antigens (not shown) on the surface of the bacteria, causing a clumping or agglutination (b¹) of the bacterial cells. This clumping of cells is a (c) diagnostic indication of salmonellosis, a foodborne intestinal disease. Specifically, the test shows that antibodies to the *Salmonella* microorganisms are present; the actual microorganisms may be present or they may be absent but were present in the past. If there were no such microorganisms in the patient's serum, and never had been, there would be no antibodies, and no clumping would take place (not shown).

The tube agglutination test works on the same principle as the slide agglutination test. A sample of the patient's serum (a) is placed in a test tube, and a quantity of suspected antigen (b; *Salmonella*) is added. The mixture is incubated at 37° C for several hours and observed. As before, if the antibody to the bacterium is present in the serum, it will mix with the antigen on the surface of the microorganism, causing the micro-

organism to agglutinate. Agglutination is a positive sign (c) of the current or past presence of antibodies to *Salmonella*, and indirectly, of salmonellosis. The lack of agglutination (b) constitutes a negative test (d).

Certain bacteria lend themselves to slide tests, and others to tube tests. The method selected for laboratory diagnosis will depend upon which antibodies in the serum are suspect.

Color the subheading Hemagglutination Test and the titles and structures in the lower half of the plate. Use a medium red for (e) and (e¹). Then color the subheading Hemagglutination Inhibition Test, and the related titles and structures. Use a color for (g) that contrasts sharply with (f).

Hemagglutination refers to the binding reaction between substances, including certain microorganisms, and red blood corpuscles (RBCs). It is not the kind of agglutination reaction associated with antibodies. When the reaction takes place, the RBCs clump together (agglutinate) with the virus only. In the case illustrated, a patient's blood is being tested for the presence of the measles virus (f). When the RBCs of that blood sample are mixed with a quantity of the virus, the RBCs clump together (hemagglutination). This reaction can be seen with the unaided eye. Viral hemagglutination is not an antibody-antigen reaction because antibodies are not involved. However, if a person has *antibodies* against the measles virus, the hemagglutination inhibition test can reveal it by preventing hemagglutination with antibody.

Consider the case of a person with clinical signs of measles. This person would be expected to have antibodies against the measles virus in their serum. A sample of serum (a) containing antibodies to the measles virus (g) is mixed on a slide with a quantity of measles viruses (f). The antibodies bind the viruses. When a sample of RBCs is added to the mixture, the antibody-bound viruses cannot react with the RBCs. Thus, no hemagglutination occurs.

It can be concluded that the serum contained antibodies to the measles virus. The person either has the measles virus present at the time of the test or had it earlier. If there had been no antibodies to the virus, then the added virus would have been free to agglutinate the RBCs, and hemagglutination would have been observed.

AGGLUTINATION TEST

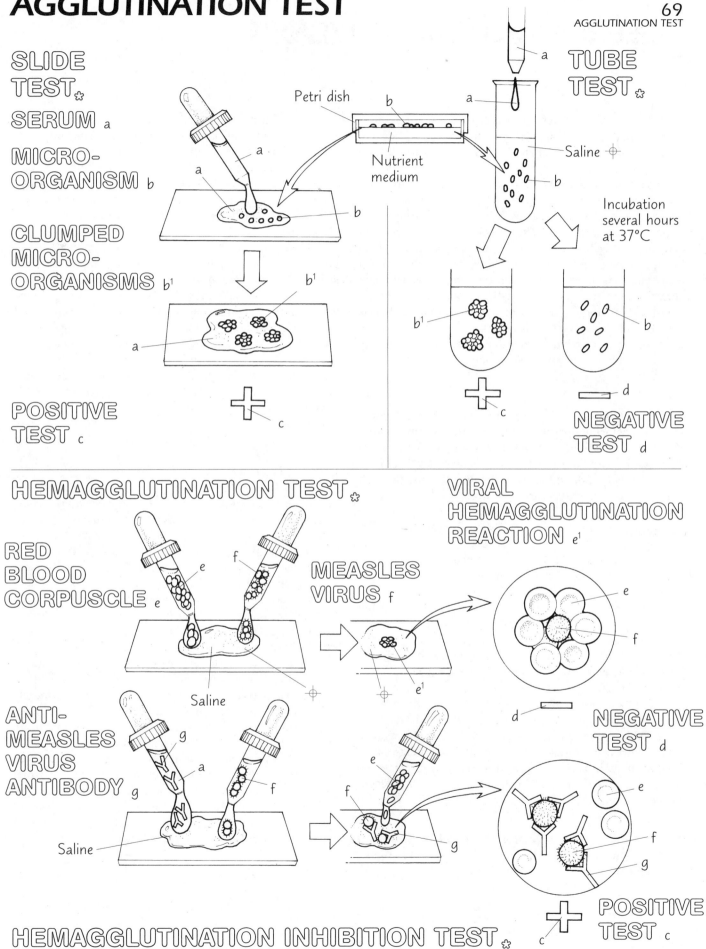

SLIDE TEST ✿

SERUM a

MICRO-ORGANISM b

CLUMPED MICRO-ORGANISMS b¹

POSITIVE TEST c

TUBE TEST ✿

Petri dish

Nutrient medium

Saline

Incubation several hours at 37°C

NEGATIVE TEST d

HEMAGGLUTINATION TEST ✿

RED BLOOD CORPUSCLE e

MEASLES VIRUS f

ANTI-MEASLES VIRUS ANTIBODY g

Saline

VIRAL HEMAGGLUTINATION REACTION e¹

NEGATIVE TEST d

POSITIVE TEST c

HEMAGGLUTINATION INHIBITION TEST ✿

ALLERGY AND ANAPHYLAXIS

The body's immune system gives protection against disease-causing organisms and other antigens that enter the body. On occasion, however, the immune system may malfunction and give rise to immune disorders. These disorders are referred to as hypersensitivities (hyper-, excessive) because they are characterized by excessive degrees of sensitivity to certain antigens. Allergies are types of immediate hypersensitivity (type I). In an allergy, certain antibodies react with certain antigens called allergens and induce multiple reactions. These include dilation of small blood vessels (vasodilatation) and contraction of smooth muscle in certain organs (e.g. bronchial constriction, increased gastrointestinal contractions) in local areas. A severe form of immediate hypersensitivity (type I) is called anaphylaxis (ana-, without protection or prophylaxis). Here the hypersensitivity is life-threatening, often resulting in asphyxiation from spasmodic contractions of bronchial (airway) smooth muscle.

The illustrations are arranged into eight stations. Begin at station 1, coloring the title (a) and the shapes representing the allergens a bright color. Titles and structures (b) through (b²) can be colored with a single color. Be sure to select colors that do not obscure detail. Continue through station 4 and titles/structures (b) through (e¹).

Antigens that cause allergies are called allergens. Antigens, you will recall, are substances that provoke immune responses. Allergens may come from plants, foods, molds, animal proteins, or insect venoms. Here we show an allergen (a) from an insect (station 1). Allergens are molecules, and they can enter the body (station 2) through ingestion by mouth (b), inhalation (b¹) into the respiratory tract, or a break (abrasion, laceration, or puncture) of the skin (b²). In the case illustrated here, the insect breaks the skin and its antigenic insect venom enters the body through that small laceration or puncture. Once in the vascular system, the allergens are engulfed by phagocytes (c). These phagocytes will carry the allergen (antigen) to the B lymphocytes (station 3), which are activated in response (d). The activated B cells differentiate into plasma cells (e) and secrete specific antibodies called immunoglobulin E (e¹; designated IgE). The IgE enters the body fluids, including the blood, and circulates throughout the body (station 4).

Color the titles and structures at stations 5 through 8 beginning with (f). Use shades of the same color for (i) through (i²).

The cellular basis for allergic and anaphylactic hypersensitivity is a reaction between antigens and IgE on the surface of basophils and mast cells. The IgE attaches to the cell membranes (station 5). The attachment of the IgE to the cell "sensitizes" the mast cell and basophil. This sensitization may last for weeks, months, or years. Subsequent contact with allergens following long periods of non-contact may incite an allergic reaction.

Mast cells (f) are found in the loose connective tissues in the walls of organs, and in the skin and in the connective tissues deep to the skin (station 6). They are large cells filled with granules. Basophils (f) are white blood cells that form in the bone marrow. They make up only a small part of the white blood cell population (less than 1%). These cells contain great numbers of granules as well; the granules are so numerous that they obscure the nucleus of the cell.

When the allergen makes contact (g) with the sensitized mast cells and basophils, a biochemical signal triggers the release of the granules (h) from these cells (station 7). As the granules accumulate in local and distant tissues, they release one or more very active substances called mediators (h). These mediators include histamine, serotonin, bradykinin, leukotrienes, and certain cytokines (not shown).

The mediators bring on the reactions seen clinically as the allergic or anaphylactic response. These reactions include vasodilatation, bronchial constriction, and contraction of the smooth muscles in the gastrointestinal tract (station 8). Vasodilatation (i) permits the blood fluid (plasma) to fill the local connective tissues. Dilated vessels lead to the movement of fluids out of the vessels, causing swelling. This swelling (edema) may be local or more generalized (global). Marked swelling in the walls of the larynx can bring on airway obstruction and require intubation or tracheostomy to maintain life. Contraction of smooth muscle in the airways (bronchi, bronchioles) of the lower respiratory tract (i¹) can also induce breathing difficulty where inhalation of air is difficult or labored (asthma). Finally, contraction of the smooth muscle of the stomach and intestines (i²) cause sensations of cramping.

Anaphylaxis is remarkable for its life-threatening potential. In addition to the respiratory threat, there is the danger of system-wide vasodilatation. The generalized loss of smooth muscle tone in the vessels throughout the vascular system can precipitate a pronounced drop in blood pressure and the onset of shock (anaphylactic shock). Rapid resuscitation with epinephrine or a similar shock-reversing substance may be required to prevent death.

ALLERGY AND ANAPHYLAXIS

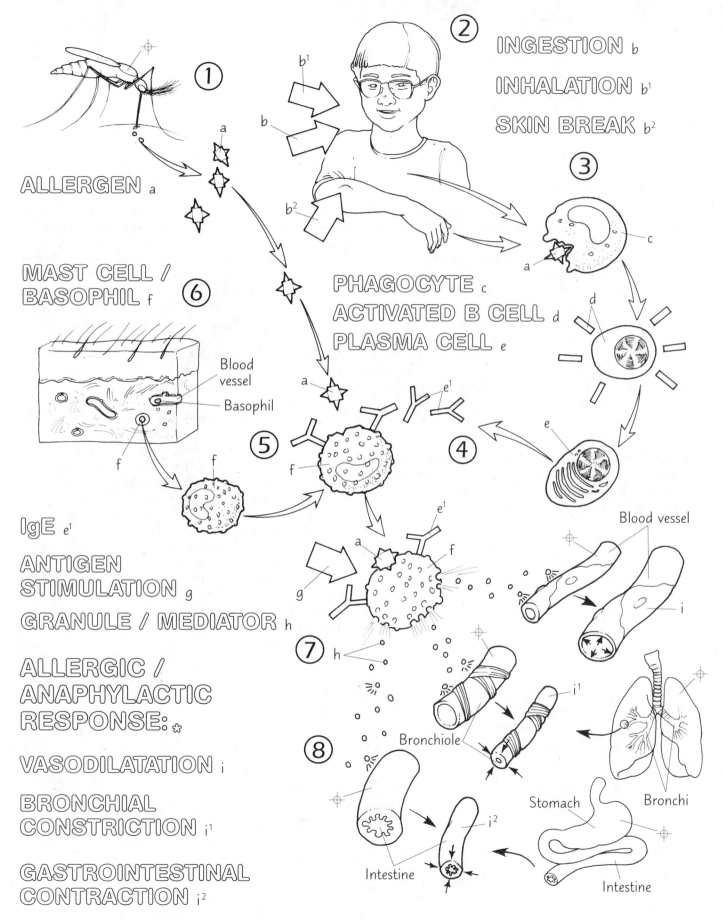

① ALLERGEN a

② INGESTION b
INHALATION b¹
SKIN BREAK b²

③

PHAGOCYTE c
ACTIVATED B CELL d
PLASMA CELL e

④

⑤

⑥ MAST CELL /
BASOPHIL f

Blood vessel
Basophil

IgE e¹

ANTIGEN STIMULATION g

GRANULE / MEDIATOR h

ALLERGIC / ANAPHYLACTIC RESPONSE: ✿

VASODILATATION i

BRONCHIAL CONSTRICTION i¹

GASTROINTESTINAL CONTRACTION i²

⑦

⑧

Blood vessel

Bronchiole

Stomach

Bronchi

Intestine

Intestine

THROMBOCYTOPENIA

Thrombocytopenia is a condition characterized by reduced numbers of thrombocytes (-penia , few). It is a form of type II (cytotoxic) hypersensitivity. Thrombocytes are also known as platelets. They function in the early stages of the blood clotting mechanism. Significant reductions of thrombocytes in the blood (less than 200/mm^3; normal is 250 to 400/mm^3) reduce the ability of the blood to clot in the event of hemorrhage. Thrombocytes are not true cells, but are fragments from larger cells (megakaryocytes) in the bone marrow. For immune purposes, they are treated as cells, and the condition of thrombocytopenia is an example of cytotoxic (cyto-, cell) hypersensitivity. Thrombocytopenia is an example of an autoimmune disease because the person produces antibodies that interact with and destroy the person's own body cells.

Consider using colors from the previous plate for the same structures on this plate. Color the titles and structures (a) through (h), starting at the top of the plate and working down. Do not use colors that obscure detail.

Certain medications, such as aspirin (acetylsalicylic acid) and some antibiotics, can induce thrombocytopenia. In sensitive individuals these medications are treated as antigens by the immune system.

Upon entering the body, these medications (a) are engulfed by phagocytes (b) and in a sensitive individual, an immune response is initiated. Scientists are unsure why one person is sensitive while others are not, but the basis may be in the genetic constitution of the person. In the sensitive person, the response is characterized by activation of B lymphocytes (c). The B lymphocytes differentiate into plasma cells (d), and they secrete antibodies. The antibodies are specifically structured to interact with the medication acting as an antigen (d^1; anti-medication antibody). These antibodies are usually types of immunoglobulin G (IgG).

In sensitive individuals, the IgG molecules bind to the medication molecules which have become bound to the surface of the thrombocytes (e). The medication molecules are, in effect, antigens on the thrombocytes' surface. A cell surface antigen-antibody reaction (f) ensues, activating a group of plasma proteins called complement (g; recall Plate 61), and a complement cascade (g^1) is initiated. The products of the complement cascade (membrane attack complex) fix to the surface of the thrombocytes. They chemically and physically alter the surface structure of the thrombocytes, permitting cytoplasm to leak to the exterior, and causing destruction of the thrombocytes (h).

The cytotoxic hypersensitivity effect of these medications is usually short-lived, and it subsides as the medication is withdrawn. In some cases, several exposures to the medication may be necessary before the condition of thrombocytopenia develops. When the medication binds to red blood corpuscles, and an antigen-antibody reaction occurs, the condition is called hemolytic anemia.

THROMBOCYTOPENIA

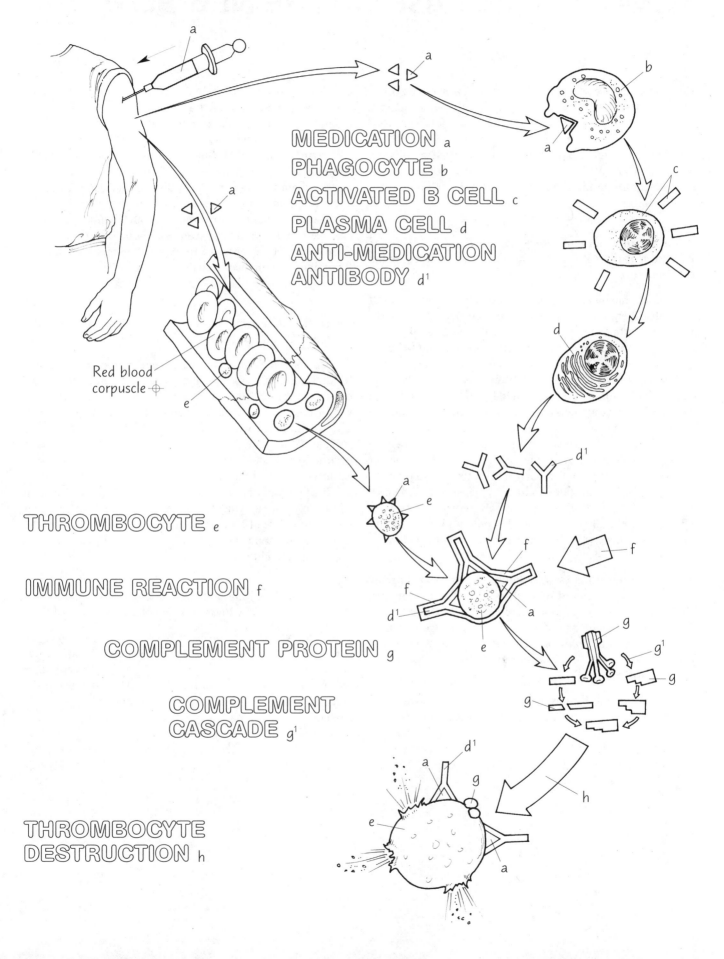

MEDICATION a
PHAGOCYTE b
ACTIVATED B CELL c
PLASMA CELL d
ANTI-MEDICATION
ANTIBODY d¹

Red blood
corpuscle ⊕

THROMBOCYTE e

IMMUNE REACTION f

COMPLEMENT PROTEIN g

COMPLEMENT
CASCADE g¹

THROMBOCYTE
DESTRUCTION h

HEMOLYTIC DISEASE OF THE NEWBORN

Hemolytic disease of the newborn (also known as erythro-blastosis fetalis) is a type II cytotoxic hypersensitivity occurring in the near-term fetus and newborn. It is caused by the mixing of fetal Rh-positive red blood corpuscles (RBCs) with anti-Rh antibodies produced in the maternal circulation. The condition involves destruction or lysis of red blood corpuscles, hence the term hemolysis (hem-, blood; -lysis, destruction).

Color the titles and structures (a) through (b^1) at upper right. Use a red color for (a) and a contrasting color for (b). Do not color (b^2).

The Rh factor was first discovered in Rhesus monkeys. It was subsequently found in humans as well. The Rh factor (b) is an antigen that exists on the surface of RBCs (a) of certain people, but not all. When it exists, the blood is said to be Rh positive (b^1); on those in which it does not exist, the blood is said to be Rh negative (b^2 ϕ). There are significantly more people in the world with Rh positive blood than Rh negative. The Rh factor is an antigen because it induces an immune response in the blood of those who are Rh negative.

If the parents of a newborn are both Rh positive or both Rh negative, there is no problem for the newborn. If the mother is Rh positive and the father Rh negative, there is no problem for the newborn. If the father is Rh positive and the mother Rh negative (as illustrated), there is a 75% probability that the newborn will be Rh positive; and that presents the possibility of hemolytic disease in the fetus of the second pregnancy.

Color the subheading 1st Pregnancy and the related arrow (*^{1}). Color the placenta (c) and its title. Then color (a^1), (b), (d), (d^1), and (a^2) in that order. Use a light color for (d). Use the same color of red for (a) and (a^1); use a lighter shade of the same color for (a^2). The title for (a^2) is listed under 2nd Pregnancy: Untreated. Color the "positive" signs on the fetus.

When the first Rh positive fetus (b^1) develops within an Rh negative woman (*^{1}), there is little effect on either of them because the fetal RBCs do not normally cross the placenta into the maternal circulation. At birth, however, some of the fetal Rh positive RBCs (a^1) may cross a torn placenta (c), before the placenta is discharged. In this way, fetal Rh positive cells enter the mother's circulation. The Rh antigen, previously unknown to the woman's blood, will induce an immune response (d) in her body. B lymphocytes will be activated and the plasma cells will form anti-Rh antibodies (d^1). These antibodies will normally attack the Rh positive RBCs and destroy them (a^2; by hemolysis; discussed ahead). Since the baby is already born, these antibodies cannot reach the newborn's circulation. However, the mother's immune system has been sensitized to the event (immunological memory). Should her body experience Rh positive antigens again, anti-Rh antibodies will react with those antigens. These antibodies *can* cross an intact placenta (placental barrier).

Color the subheading 2nd Pregnancy: Untreated, and the related titles and structures.

Consider the case where an Rh negative woman with anti-Rh antibodies becomes pregnant a second time with an Rh positive fetus (*^{2}), the anti-Rh antibodies in her circulation will diffuse across the placenta and enter the fetal circulation. These antibodies will interact with the Rh antigen on the surface of the fetal RBCs. This reaction (d) would induce destruction of the fetal RBCs (a^2; hemolysis). Progressive hemolysis in the fetus induces the recruitment of immature RBCs (erythroblasts) from its bone marrow. Overall reduction of the oxygen-carrying RBCs results in fetal anemia and possible fetal death. Transfusion of Rh positive blood to the newborn (containing no anti-Rh antibodies) is often necessary to prevent a fatality.

Color the subheading 2nd Pregnancy: Treated, and color the structures at the bottom of the plate.

Hemolytic disease of the newborn can be prevented by administering anti-Rh antibodies to the mother soon after conclusion of her first pregnancy (*^{3}). The anti-Rh antibodies will react with the Rh positive fetal RBCs that crossed the placenta before it was discharged and eliminate those RBCs from the woman's circulation. This will reduce the probability of her producing her own anti-Rh antibodies and will eliminate the hypersensitivity reactions in the developing fetus during the next pregnancy.

HEMOLYTIC DISEASE OF THE NEWBORN

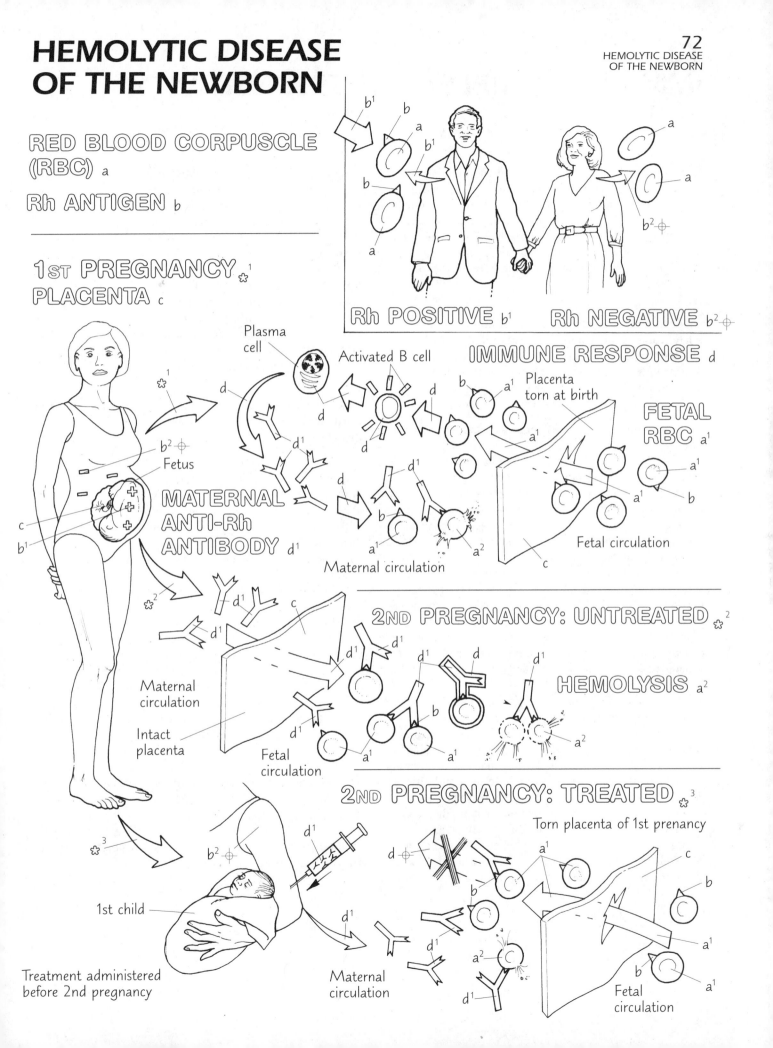

RED BLOOD CORPUSCLE (RBC) a

Rh ANTIGEN b

1ST PREGNANCY ☆¹
PLACENTA c

Rh POSITIVE b¹   Rh NEGATIVE b²⊕

Plasma cell

Activated B cell

IMMUNE RESPONSE d

Placenta torn at birth

FETAL RBC a¹

MATERNAL ANTI-Rh ANTIBODY d¹

Fetus

Maternal circulation

Fetal circulation

Maternal circulation

Intact placenta

Fetal circulation

2ND PREGNANCY: UNTREATED ☆²

HEMOLYSIS a²

2ND PREGNANCY: TREATED ☆³

Torn placenta of 1st prenancy

1st child

Treatment administered before 2nd pregnancy

Maternal circulation

Fetal circulation

73
IMMUNE COMPLEX HYPERSENSITIVITY

Immune complex-mediated hypersensitivity (type III) involves the formation of antigen-antibody masses in vascularized tissues. The immune complexes induce inflammatory responses in those tissues and permanently alter their structure and function. In this form of hypersensitivity, the mechanism of injury is an interconnected mass of antigens and antibodies ("complex") that influences ("mediates") the onset of destruction.

Color the titles and structures (a) through (g^1). Structures (c), (c^1), and (c^2) are composed of combinations of persistent antigens and antibodies (a) and (b). Use a light color for (c) through (c^2). Following the arrows at mid-plate, color the titles and structures (e) through (g^1). Stop before coloring the diseased tissues (g^2).

Most immune responses begin with the introduction of foreign antigens into the body. In some cases, the antigens may be "self-antigens" where certain parts of the body are considered antigenic by the immune system. Foreign antigens may be derived from bacteria, viruses, proteins in mold spores, inhaled dust, injection of serum, and so on. As the antigens encounter phagocytes, B lymphocytes are activated and plasma cells are formed (not shown; recall Plate 58). IgM and IgG are the predominant antibodies formed.

With subsequent or continued (chronic) exposures to antigens over time, the numbers of persistent antigens (a) in the body increase. IgG and IgM antibodies (b) are formed and they combine with these antigens. Both large and small masses of antibody-antigen complexes are formed; these masses are called immune complexes. Large immune complexes (c) are generally taken up in the liver (d) and spleen and broken down by phagocytes there (d^1).

Small immune complexes (c^1) remain in the circulation. In the presence of excess antigen, in areas of relatively high blood pressure and turbulent blood flow, small immune complexes may invade the vascular lining (e) of certain blood vessels, wedging between cells down to and sometimes through the basement membrane. Progressive accumulation and deposition of immune complexes (c^2) may occur.

The deposit of immune complexes irritates the tissues in the blood vessel wall. This irritation may attract and activate complement, resulting in a complement cascade (f) at the deposition site. Certain products of the cascade, notably C3a and C5a, enhance immune complex deposition and attract phagocytes, especially neutrophils. Acute inflammation (g) of the vascular wall is accompanied by the release of digestive enzymes (lysozymes) from the phagocytes. The resulting tissue damage from digestive enzymes may continue, destroying not only the lining cells of the vessels but also the nearby tissues of the organ in which the vessels reside.

Progressive (chronic) inflammation (g^1) induces the formation of new blood vessels, the continued digestion and erosion of tissue, and the formation of fibrous tissue to replace the injured lining tissues. Such areas of inflammation are called lesions (g^2). The progression of such lesions is enhanced by the recurrence of antigen, immune complexes, and complement at immune complex sites.

Color the lesions (g^2) and their title in the lowest set of illustrations. Only the lesions are to be colored in these drawings.

Destruction of cells and tissues of blood vessels and neighboring tissues significantly affects the functional capacity of the organ in which the destruction is taking place. Immune complex formation and related lesions may affect various parts of the body. In the heart, for example, immune complexes can localize on the heart valves, and subsequent inflammation can permanently erode and deform their structure. This condition is called endocarditis, and can significantly alter blood flow through the affected heart cavities. Species of streptococci contain antigens that can act as persistent antigens and can contribute to immune complexes in this heart condition.

Streptococcal antigens also contribute to immune complex formation in the kidney. The kidney consists of many microscopic tubules (nephrons) that conserve water and other molecules during the formation of urine. The immune complexes accumulate in these tubules and the accompanying tufts of blood vessels (glomeruli), inducing inflammation and dysfunction (glomerulonephritis).

Another manifestation of immune complex accumulation occurs in the joint cavities of the skeleton. Here, blood vessels in the joint lining tissues are affected, resulting in inflammation of the lining tissues and the erosion of articular cartilage. As a result, the surrounding bone may enlarge (hypertrophy) in response to the instability of the destructive process. This condition is called rheumatoid arthritis. Rheumatoid arthritis primarily affects the fingers and toes.

IMMUNE COMPLEX HYPERSENSITIVITY

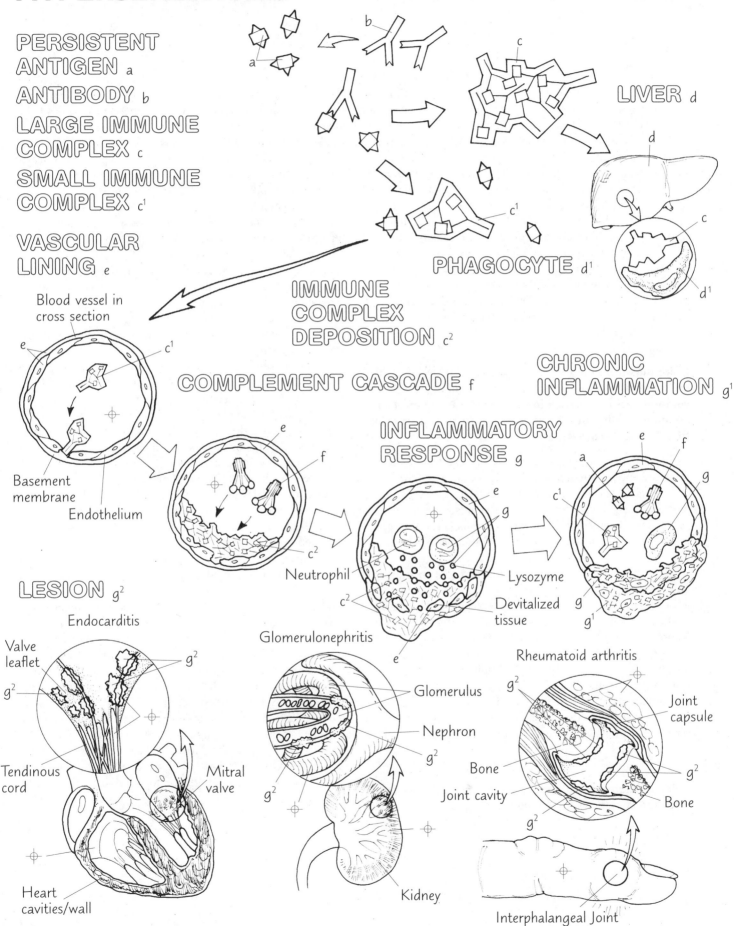

PERSISTENT ANTIGEN a
ANTIBODY b
LARGE IMMUNE COMPLEX c
SMALL IMMUNE COMPLEX c¹

VASCULAR LINING e

Blood vessel in cross section

LIVER d

PHAGOCYTE d¹

IMMUNE COMPLEX DEPOSITION c²

COMPLEMENT CASCADE f

CHRONIC INFLAMMATION g¹

INFLAMMATORY RESPONSE g

Basement membrane

Endothelium

Neutrophil

Lysozyme

Devitalized tissue

LESION g²

Endocarditis

Valve leaflet

Tendinous cord

Mitral valve

Heart cavities/wall

Glomerulonephritis

Glomerulus

Nephron

Kidney

Rheumatoid arthritis

Joint capsule

Bone

Joint cavity

Bone

Interphalangeal Joint

CELL-MEDIATED HYPERSENSITIVITY

Cell-mediated hypersensitivity (type IV) is an immune response associated primarily with T lymphocytes and phagocytes. Unlike other forms of hypersensitivity presented in Plates 71-73, this hypersensitivity does not involve activated B lymphocytes and antibodies; nor does it occur immediately. Indeed, another name for this type of immunity is delayed hypersensitivity. The cell-mediated hypersensitivity illustrated here is one seen with poison ivy, manifested as an allergic contact dermatitis (derm-, skin). The mechanism of injury is initiated by combining a small molecule (a hapten) with body proteins to create an allergen (antigen, recall Plate 54). The allergen sensitizes certain T cells to form memory cells on first exposure without signs or symptoms giving evidence that the sensitization took place. On subsequent exposure, the T cells are activated, and blisters form on the skin reflecting the tissue damage.

Color the subheading First Exposure, and the related titles and structures (a) through (f). Use a color for (c) that combines the colors used for (a) and (b); for example, the addition of yellow to red makes orange. Be sure to use colors that do not obscure detail.

Cell-mediated hypersensitivity can arise from exposure to a number of different potential antigens, such as those found on clothing, jewelry, and cosmetics. These chemicals or haptens include dyes, bacterial enzymes, protein fibers, formaldehyde, and other substances. In poison ivy, the hapten is urushiol, a small alcohol molecule found on the surface of the leaves.

The process of cell-mediated hypersensitivity begins with an exposure to leaves of the poison ivy plant. Hand contact with these leaves permits transfer of molecules of urushiol (a) to the skin surface. The urushiol, functioning as a hapten, combines with proteins (b) in the skin epidermal cells. These combined antigen-proteins, called allergens (c), are phagocytosed (d) and taken to T lymphocytes (e) in nearby lymphoid tissues. The T cells are sensitized by these encounters, and proliferate. Among the progeny are T memory cells (e^1). They leave the lymphoid tissues, enter the circulation (f), and are drawn to the site of first exposure. In this case, the site is the hand.

Color the subheading Second Exposure, and the related titles and structures (g) through (i).

Over a period of time, the sensitized T lymphocytes (the progeny of memory cells) gather in the skin tissues of the affected hand in increasing numbers. This time period may be weeks or months. At some point in time, the numbers of sensitized T cells reach a level where hypersensitivity is achieved. Now a second exposure to antigens is needed to set off clinical manifestations of this hypersensitivity.

If the person experiencing the first exposure to poison ivy encounters the poison ivy plant a second time, the hapten (urushiol) comes in contact with and combines with skin proteins. These allergens are taken up by phagocytes. The phagocytes present the allergen to sensitized T cells. The sensitized T cells secrete lymphokines (g) that induce phagocytes to aggregate, activate, and engulf the allergens. Ordinarily, these phagocytes would destroy the allergens and the crisis would be over. In cell-mediated hypersensitivity, however, the process is greatly exaggerated. Digestive enzymes (h) from the sensitized phagocytes pour into the local fluid spaces and cause cell destruction (lysis) and local fluid accumulation. These lesions (i; blisters) itch profusely and become red and granular ("raw"). These are the typical signs of poison ivy. As the urushiol molecules are removed from the area by the phagocytes, the blisters heal and disappear over a period of a few to several days.

The sensitized T cells will remain indefinitely in the skin tissues, and will induce another bout of cell-mediated hypersensitivity should the poison ivy plant be encountered again. These reactions are different from the types I, II, and III hypersensitivity responses (see glossary).

Contact dermatitis due to poison ivy is but one manifestation of cell-mediated hypersensitivity. Scaling and drying of skin (induration) can occur from exposure to a number of antigens, including shampoos, fabrics, metals, and so on. Cell-mediated hypersensitivity is also brought on by intracellular bacteria, such as *Mycobacterium tuberculosis*. Resulting tissue destruction followed by scarring (deposition of thick bundles of fibrous tissue), can significantly interfere with lung function. Other cell-mediated hypersensitivity responses may include destruction of insulin-secreting islets in the pancreas, or formation of granulomas characterized by persistent masses of highly active phagocytes encircled by sensitized T lymphocytes, and associated masses of blood vessels. These granulomas tend to occur at sites of antigen/T cell interactions, especially the lymphoid tissues (spleen and lymph nodes).

CELL-MEDIATED HYPERSENSITIVITY

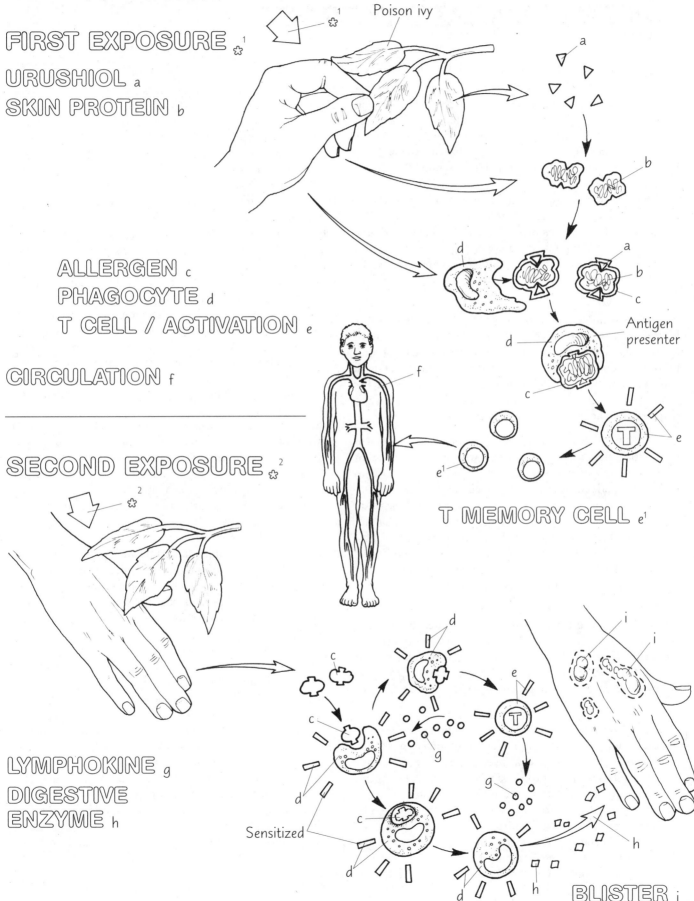

FIRST EXPOSURE ✿¹

URUSHIOL a
SKIN PROTEIN b

Poison ivy

ALLERGEN c
PHAGOCYTE d
T CELL / ACTIVATION e

CIRCULATION f

Antigen presenter

T MEMORY CELL e¹

SECOND EXPOSURE ✿²

LYMPHOKINE g
DIGESTIVE ENZYME h

Sensitized

BLISTER i

AIRBORNE BACTERIAL DISEASES: UPPER RESPIRATORY TRACT

Airborne bacteria gain access to the upper respiratory tract by way of the nose and mouth. Once in the tract, they become trapped in droplets of mucus and saliva. The upper respiratory tract includes the nasal passages of the external nose and the nasal cavity, the paranasal sinuses, the pharynx, the auditory tubes, and the larynx. The microscopic structure of the lining of these cavities offers a warm, moist environment that encourages bacterial growth. That same structure protects the lining cells from bacterial penetration and provides mechanisms for their destruction as well. Although airborne bacteria are often limited to the upper respiratory tract, they may infect additional organs if significant numbers access the blood supply.

Plan to use shades of blue or purple for the Gram-positive bacteria, and shades of pink for the Gram-negative bacteria. Use the same shade for both the disease and related bacterium. These colors are those elicited on Gram staining. For now, use light shades of the same color for the titles and structures (a) through (a². Do not color the trachea and the lower respiratory tract.

The lining tissue of the upper respiratory tract (a) is called respiratory mucosa (a¹). It is lined with a single layer of epithelial cells covered with hairlike cilia and covered with mucus (a²) derived from goblet-shaped cells in the mucosa. The mucus traps microorganisms that migrate into the nasal cavity with the airflow. The cilia move them into the pharynx for swallowing or expectoration. Under certain conditions, on which heavy concentrations of bacteria are superimposed, these bacteria penetrate the mucosa and populate and reproduce in the local tissues. Bacterial disease is initiated.

Color the titles and related bacteria at each of the five stations, starting with station 1. Use shades of blue or purple for (b) and (f), and use shades of pink (but not red) for (c), (d), and (e). Use a neutral color for (g). Color the arrows in the center of the plate as well.

Corynebacterium diphtheriae (b¹; station 1) is a club-shaped, Gram-positive rod that infects the nasopharynx. These bacteria multiply in the epithelial layer, and produce a powerful toxin that interferes with protein synthesis in the cells. As the cells die and accumulate in the inflamed tissues, a thick, leathery pseudomembrane (pseudo-, false) is formed, characterizing the disease diphtheria (b; *diphtheria, membrane*). This material can block the respiratory passages, causing suffocation. When these bacteria access the blood they cause a systemic infection. Diphtheria is rare in the United

States because most people are immunized against it with the diphtheria toxoid in the DPT (diphtheria-pertussis-tetanus) preparation (in the third of the three injections series administered, extracts of *B. pertussis* are used instead of the whole cell preparation; the extracts are called DTaP, referring to acellular pertussis).

Whooping cough (c; pertussis), is seen primarily in infants less than six months of age. It is caused by *Bordetella pertussis* (c¹; station 2), a small, nonmotile, encapsulated Gram-negative rod. *B. pertussis* often populates the nasal and pharyngeal mucosa. They are later carried to the lower airways, causing inflammation (bronchitis, bronchiolitis). The production of mucus in the irritated mucosa blocks the airway, inducing persistent coughing. These coughs often occur in rapid-fire succession, followed by quick inhalations that produce a high-pitched "whoop." *B. pertussis* responds to several antibiotics, and killed *B. pertussis* cells and extracts are used in the DPT and DTaP immunizations to develop immunity at an early age.

Meningitis is an inflammation of the meninges (g), the membranes that envelop the brain and spinal cord. Meningitis can be caused by an infection in which the infecting organisms came to the meninges via the blood stream from a primary infection of the upper respiratory tract. One cause of meningitis is *Neisseria meningitidis*, a small Gram-negative diplococcus (d¹; often called meningococcus). The disease is called meningococcal meningitis (d; station 3).

Meningitis can also be caused by a small Gram-negative rod *Hemophilus influenzae* (e¹; station 4). This organism inhabits the upper respiratory tract; unchecked, it enters the blood and causes inflammation of the meninges. Hemophilus meningitis (e) is seen most often in children six months to two years old ; it is characterized by stiff neck, severe headache, drowsiness, irritability, and other neurological signs. Less serious than meningococcal meningitis, it is treated similarly with antibiotics such as rifampin.

Streptococcus pyogenes (f¹; station 5) is a Gram-positive coccus occurring in chains. On reaching the mucosa, the streptococci multiply rapidly and induce the condition called "strep throat" (f). The local tissues become inflamed as the phagocytes attack the bacteria, and the regional lymph nodes swell as the immune system reacts to the bacterial presence. The surface of the mucosa becomes streaked with red, reflecting vasodilatation and the inflammatory process. These changes are often accompanied by a high fever. If significant numbers of the bacteria enter the blood, septicemia may result, and other organs may become infected. Some strains of streptococci, themselves infected by bacteriophages, release toxins in the blood that damage vessels, leading to blood leakage and the rash characteristic of scarlet fever.

AIRBORNE BACTERIAL DISEASES: UPPER RESPIRATORY TRACT

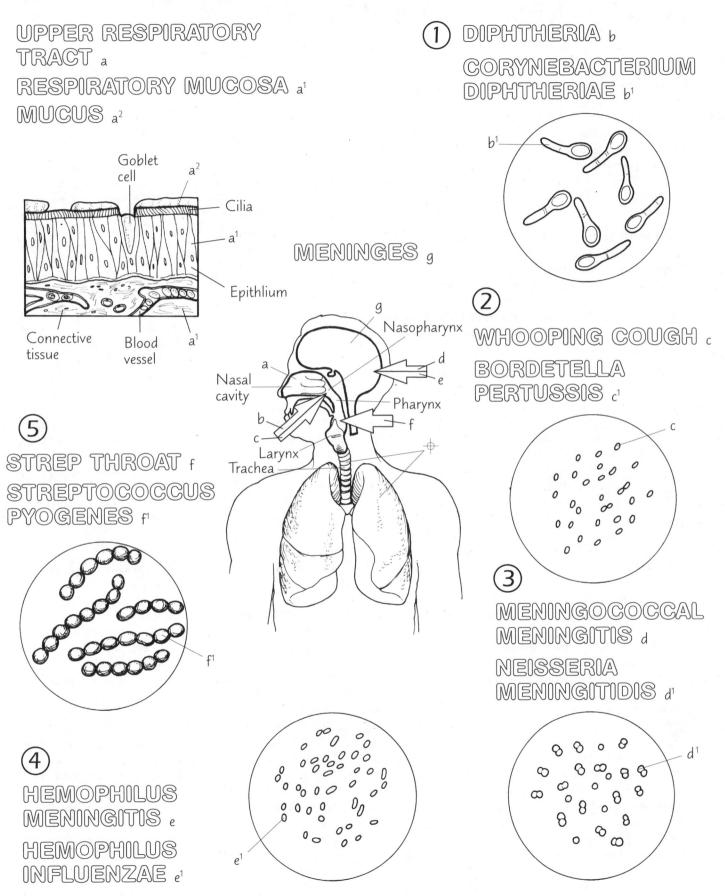

UPPER RESPIRATORY TRACT a
RESPIRATORY MUCOSA a¹
MUCUS a²

Goblet cell
a²
Cilia
a¹
Epithlium
Connective tissue
Blood vessel
a¹

MENINGES g

g
Nasopharynx
a
d
Nasal cavity
e
Pharynx
b
f
c
Larynx
Trachea

① DIPHTHERIA b
CORYNEBACTERIUM DIPHTHERIAE b¹

b¹

② WHOOPING COUGH c
BORDETELLA PERTUSSIS c¹

c

③ MENINGOCOCCAL MENINGITIS d
NEISSERIA MENINGITIDIS d¹

d¹

⑤ STREP THROAT f
STREPTOCOCCUS PYOGENES f¹

f¹

④ HEMOPHILUS MENINGITIS e
HEMOPHILUS INFLUENZAE e¹

e¹

AIRBORNE BACTERIAL DISEASES: LOWER RESPIRATORY TRACT

The lower respiratory tract includes the bronchi, bronchioles, and alveoli (air sacs). Certain airborne bacteria can transit the upper respiratory tract, bypass the defense mechanisms, and reach the lower tract. Multiplying there, they damage the alveoli in which gaseous exchange occurs between the outside environment (air) and the internal environment (blood). Thus, bacterial diseases of the lower respiratory tract can be serious and life-threatening.

Set aside red for the acid-fast bacteria (b¹), a shade of blue or purple for the Gram-positive bacteria (c¹), and a shade of pink for the Gram-negative bacteria (e¹). Use the same shade for the bacterium and the related disease. Color the lower respiratory tract (a) and the microscopic structure (a¹) through (a³). Then color the titles and bacteria at station 1, and continue through station 4. Color the arrows in the center of the plate as well.

The upper part of the lower respiratory tract (a) is largely lined with the same respiratory mucosa as that of the upper tract (Plate 75). The lower tract begins with the trachea in the lower neck and upper thorax. At the level of the upper sternum, the trachea divides into main bronchi (a¹). These bronchi dive into the lung tissue and divide into smaller branches. The smaller of these branches are called bronchioles (a²). The walls of bronchioles have a considerable amount of smooth muscle. As the bronchioles divide into small ducts and air sacs (a³; alveoli), the lining tissue thins to a layer of flat cells. Bronchioles are easily obstructed by infectious disease processes, making breathing difficult (dyspnea). The air sacs (a³) are particularly vulnerable because they are lined with exceedingly thin cellular walls that are immediately adjacent to thin walled capillaries (not shown). Disease of the air sacs compromises oxygenation of the blood and can be life-threatening.

Tuberculosis (b) is a lung disease caused by the rod-shaped *Mycobacterium tuberculosis* (b¹; station 1). This bacillus resists staining by the usual methods; however, it stains red in the acid-fast technique. Tuberculosis is often transmitted by airborne droplets of sputum from a tuberculosis carrier (one who harbors the microorganisms but does not exhibit symptoms or signs of the disease). Hence, tuberculosis can spread rapidly in dense populations. Once in the lungs, the bacteria localize in hard nodules called "tubercles" (hence, the disease's name). The pathogens tend to reside high in the lung at the apex. *M. tuberculosis* can enter the circulation and spread without symptoms, and over time it may reach a number of well oxygenated, blood-filled organs (e.g., bone marrow) and form chronic lesions (miliary tuberculosis). After some weeks or months following the initial infection, signs of tuberculosis emerge, including weight loss, fatigue, anorexia, and cough. The respiratory sputum becomes bloody as progressive deterioration of the lung occurs. Diagnosis is established by microscopic examination of sputum and the tuberculin skin test. Isoniazid is often an effective therapeutic agent, blocking DNA synthesis in the bacteria. Rifampin is also often effective, acting to inhibit bacterial RNA synthesis.

The term "pneumonia" refers to inflammation of the lungs with consolidation (air sacs filled with cell debris and particulate material called exudate). Although many different microorganisms can cause pneumonia, 90% of bacterial pneumonias are caused by *Streptococcus pneumoniae* (c¹; station 2). This Gram-positive chain of cocci are normally found in the upper respiratory tract. Young children and older people are particularly vulnerable to infection by this microorganism. Following trauma, surgery, or severe influenza, resistance is lowered and the bacteria invade the lung tissues and cause pneumococcal pneumonia (c). The polysaccharide capsule of the bacterium plays a significant role in its pathologic character by resisting phagocytosis. Pneumococcal pneumonia is characterized by chills and fever for about three days; in addition, bloody sputum, chest pain, and generalized muscle pain and weakness further characterize the disease. Aggressive and immediate penicillin therapy often resolves the uncomplicated disease; untreated, the disease may be fatal.

Another type of pneumonia, affecting people of all ages, is caused by small bacteria (about 0.2 micrometers in diameter) belonging to the subgroup Mycoplasmas (*Mycoplasma pneumoniae*, d¹; station 3). Mycoplasmas have no cell wall and therefore, no Gram reaction. The disease is called primary atypical pneumonia (d); "primary" because it occurs without any previous respiratory infection being present; "atypical" because the symptoms are milder than in other pneumonias. Often the disease is termed "walking pneumonia" because infected persons do not require extended hospitalization. Microbiological testing often includes complement fixation. Erythromycin is commonly used for therapy since penicillin has no effect on a bacterium without a cell wall.

Legionnaires' disease or legionellosis (e; station 4) was first observed in 1976 when an epidemic broke out among attendees at an American Legion convention in Philadelphia; 221 were affected, and 34 died of the disease. The disease is caused by *Legionella pneumophila* (e¹), a Gram-negative rod. Presenting as pneumonia, the disease is accompanied by fever, fatigue, chills, diarrhea, weakness, muscle pain, and a nonproductive cough. The bacteria exist where water collects (air conditioning units, water cooling towers, stagnant pools), and are apparently airborne by wind currents. Erythromycin is commonly used to treat Legionnaires' disease.

AIRBORNE BACTERIAL DISEASES: LOWER RESPIRATORY TRACT

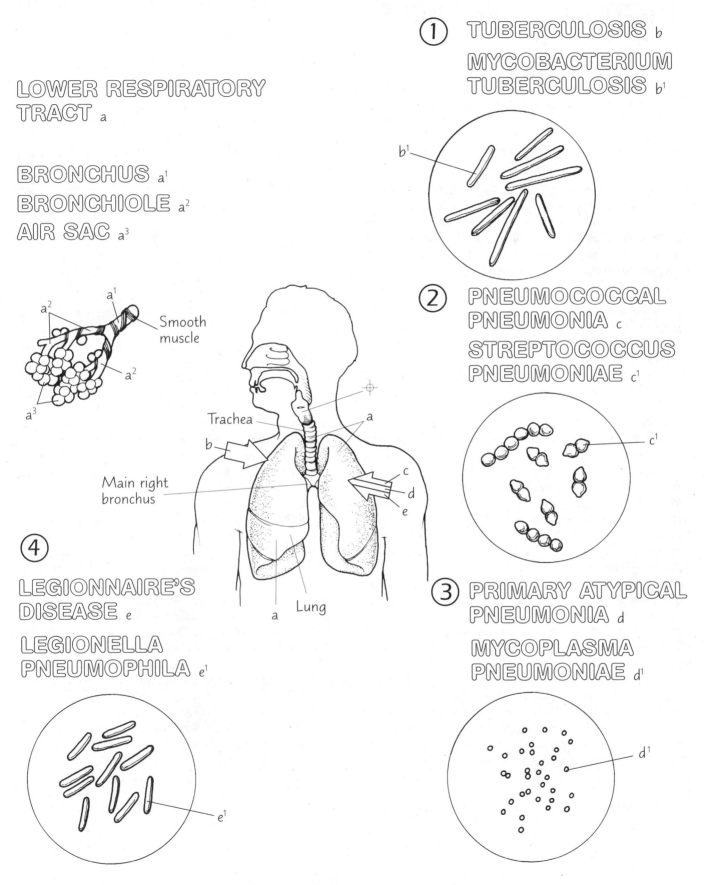

LOWER RESPIRATORY TRACT a

BRONCHUS a¹
BRONCHIOLE a²
AIR SAC a³

① TUBERCULOSIS b
MYCOBACTERIUM TUBERCULOSIS b¹

② PNEUMOCOCCAL PNEUMONIA c
STREPTOCOCCUS PNEUMONIAE c¹

③ PRIMARY ATYPICAL PNEUMONIA d
MYCOPLASMA PNEUMONIAE d¹

④ LEGIONNAIRE'S DISEASE e
LEGIONELLA PNEUMOPHILA e¹

Smooth muscle

Trachea

Main right bronchus

Lung

FOODBORNE / WATERBORNE BACTERIAL DISEASES

Foodborne and waterborne bacteria enter the body by way of the digestive tract. Many remain in this tract, but others invade the circulatory system and cause serious blood problems. In this plate, some significant pathogens of the digestive tract and the diseases they cause are presented.

Color the title Digestive Tract and its structure (a) in the center illustration; note the parts. Then color the titles and bacteria at station 1; continue through station 4. As in the previous two plates, use a shade of blue or purple for the Gram-positive bacteria (b^1), and shades of pink for the Gram-negative bacteria (c^1), (d^1), and (e^1). Color the arrows in the center of the plate as well.

The mouth, pharynx, and esophagus, all organs of the upper digestive tract (a) share a multi-layered lining that resists bacterial invasion. Because food and water usually move through these parts rapidly, there is little opportunity for infection. The stomach and upper small intestine, lined by a single layer of cells, offer an acidic and hostile environment for bacteria. Much of the small and all of the large intestine, also lined by a single layer of cells, present a more suitable environment, which is less biochemically hostile. Also, the contents move through at a slow rate conducive for infection.

Staphlylococcal food poisoning (b; station 1) affects millions of people annually and is caused by Gram-positive clusters of cocci called *Staphylococcus aureus* (b^1). These bacteria normally populate the skin surface. They constitute part of the normal skin flora, and they gain their nutrition by feeding on sloughed, dead, or decaying cells. *S. aureus* contain a number of enzymes and toxins that make the bacteria agents of infection when they cross the skin or mucosal lining of the digestive pathway. The toxins include coagulase, hyaluronidase, enterotoxins, and hemolysins (recall Plate 49). These factors tend to overcome the nonspecific and immune resistance mounted by the body. When the toxins are consumed in sufficient quantity, they irritate the intestinal mucosa lining; within hours, the toxicity induces vomiting, intestinal muscle contractions (cramps), and diarrhea. Staphylococcal food poisoning is usually self-limiting; more serious sequelae (shock, septicemia) are rare. The enterotoxins of *S. aureus* are produced at room temperature. Refrigeration at temperatures below 45° F prevents bacterial growth and inhibits toxin production.

Typhoid fever (c; station 2) is a disease carried and transmitted only by humans. The etiological agent, *Salmonella typhi* (c^1), is ingested with contaminated water or food. Contamination occurs from contact with feces or urine containing the organism. Sanitation workers at sewage treatment plants are particularly vulnerable to infection. Transmission of the organism world-wide is also facilitated by food handlers. *S. typhi* is a Gram-negative rod that enters the circulation by way of the small intestine. Bacteremia (increased concentrations of the organism in the blood) occurs about one to three weeks following contact. Extremely high fever, weakness, sweating, anorexia, vomiting, and diarrhea are common clinical features of the disease. In some cases, other organ systems are involved, causing coughing, enlarged spleen, urinary tract symptoms, and meningeal inflammation. Blood and tissue culture will identify the organism. Treatment with chloramphenicol, co-trimoxazole, or amoxicillin-ampicillin is usually successful in eradicating the disease.

Salmonella enteritidis (d^1; station 3) is one of many hundreds of species of the Gram-negative *Salmonella* genus, and one of several of these organisms pathogenic to humans. The disease state caused by this organism is called salmonellosis (d). It is transmitted to humans by contaminated food as well as by human to human transmission via fecal/urine contamination of food and water. The skin of uncooked chicken and turkey have been shown to carry the organism; washing the hands and work surfaces after contact with these plucked birds will prevent spread of the pathogen to other foods. The organisms induce formation of ulcers among the intestinal lining cells, often producing bloody stools. Within 48 hours following *S. enteritidis* infection, the signs and symptoms of gastroenteritis appear: abdominal pains, nausea, vomiting, fever, diarrhea, and chills. The disease is self-limiting and generally resolves within two or three days.

Cholera (e; station 4) is an extremely serious disease characterized by severe fluid and electrolyte loss through diarrhea. It is caused by the Gram-negative spiral (comma-shaped) bacillus *Vibrio cholerae* (e^1). The disease is endemic in India; epidemics have occurred throughout Africa and Asia. The active portion of the pathogen is an enterotoxin that induces substantial leakage of water and electrolytes from cells in the intestinal mucosa. The bacteria do not invade the mucosa; they remain on its surface. Stool specimens generally reveal the organism. Treatment is fluid and electrolyte replacement; failure to replace the lost water and electrolytes may result in shock associated with dehydration and acidosis.

FOODBORNE / WATERBORNE BACTERIAL DISEASES

DIGESTIVE TRACT a

① STAPHYLOCOCCAL FOOD POISONING b
STAPHYLOCOCCUS AUREUS b¹

② TYPHOID FEVER c
SALMONELLA TYPHI c¹

④ CHOLERA e
VIBRIO CHOLERAE e¹

③ SALMONELLOSIS d
SALMONELLA ENTERITIDIS d¹

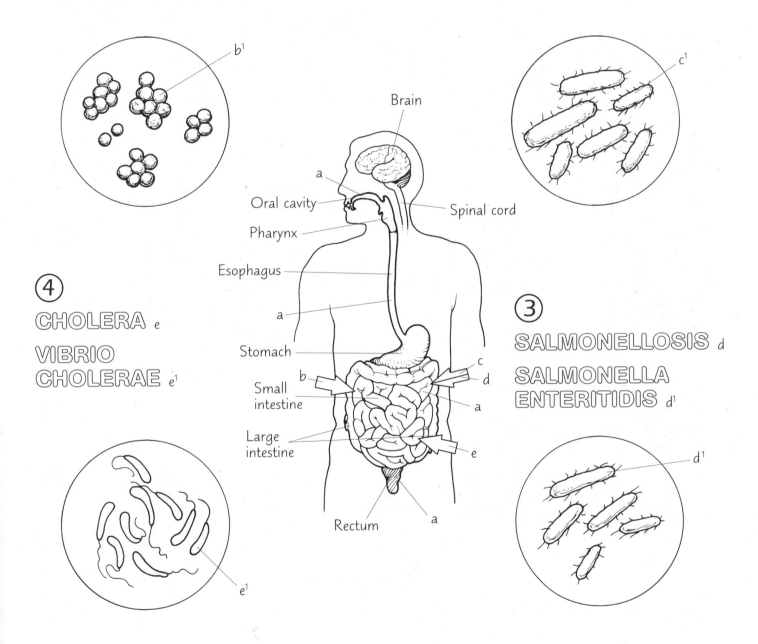

Brain
a
Oral cavity
Pharynx
Spinal cord
Esophagus
a
Stomach
b
Small intestine
c
d
a
Large intestine
e
Rectum
a

b¹

c¹

e¹

d¹

SOILBORNE BACTERIAL DISEASES

Microorganisms occupy numerous niches in various environments on this planet. Soil consists of dirt and rocks of various sizes, all containing a number of elements essential to life. Mixed within the soil are huge numbers of organisms, including great varieties of microorganisms. Several of these are pathogenic to humans; many can enter the body from the soil by way of wounds on the feet.

Color the title and soil bacteria (a) at upper right. Color the title of the disease Anthrax (b) and the related bacteria (b¹). Color the parts of the body (b²) and (b³) manifesting the disease state in the central illustration, and the related disease signs listed at right. Continue with the following two diseases and related bacteria (c) through (d¹). As in the previous three plates, use shades of blue or purple for these Gram-positive bacteria (b¹), (c¹), and (d¹). Use colors light enough not to obscure detail.

Anthrax (b) is a soilborne disease that affects mainly cattle, sheep, horses, and wild animals, but is transmissible to humans. It is caused by *Bacillus anthracis* (b¹) an aerobic, Gram-positive sporeforming rod often occurring in chains. Spores are ingested by animals while grazing; infection in humans occurs by introduction of spores through skin breaks. People employed where animal products are processed or animals are handled are at some risk for this disease.

Endospores of *B. anthracis* enter the body, and induce reddish skin eruptions or vesicles (b²) which tend to develop into craters or ulcers by the sloughing of necrotic skin cells. The spores revert to vegetative bacteria and may invade the underlying subcutaneous tissue (cellulitis), a potentially life-threatening condition. Should the organism invade the bloodstream (about 20% of those infected become septic), hemorrhagic lesions (b³) may develop in organs such as the liver, spleen, and kidneys. *B. anthracis* can be identified by skin biopsy, and/or culturing the blood or infected tissue. Aggressive antibiotic (penicillin) therapy is effective.

The agent of tetanus (c) is *Clostridium tetani* (c¹), an anaerobic, Gram-positive, sporeforming rod. Spores of this organism, like those of the *Bacillus anthracis*, survive long periods under difficult environmental conditions. Both the spores and the organism may enter skin wounds. In pus-filled, necrotic wounds (c²) where there is no oxygen, the organisms multiply and release an exotoxin that is remarkably lethal in very small quantities (0.000001 mg/kg of body weight). It should be noted that these organisms cannot multiply in oxygenated environments, such as clean wounds debrided of devitalized (dead or necrotic) tissue.

The exotoxin formed in necrotic wounds is a neurotoxin; it diffuses into the circulation where it reaches the spinal cord (c³) and blocks inhibitory neurotransmitters. Sustained excitatory impulses are sent to muscle cells, resulting in involuntary, unremitting muscle spasms (c⁴). Spasmodic contractions cause the body to become rigid. The chewing muscles clamp the jaws shut (lockjaw). Indeed, lockjaw (trismus) is a prominent sign of the disease. Spasms of the respiratory muscles may require tracheostomy. Such spasms, in the absence of assisted ventilation, may lead to death.

The disease can be prevented by immunization with tetanus toxoid. Tetanus toxoid is used in the DPT (diphtheria-pertussis-tetanus) and DTaP (diphtheria-tetanus-acellular pertussis) preparations to yield immunity that lasts for ten years. Once the disease is established, passive immunity with tetanus antitoxin is required (recall Plate 62).

Soil contamination of a superficial (skin) or deep (muscle, bone) wound may lead to gas gangrene (d). Gas gangrene is a wound condition (d²) characterized by devitalized tissue (d³), a lack of oxygen or blood flow to the wound, and the presence of multiplying anaerobic bacteria in the wound.

Clostridium perfringens (d¹), Gram-positive, anaerobic, sporeforming rods, proliferate in necrotic tissue, producing gas bubbles that swell at the wound site, block the blood flow, and contribute to the anaerobic state and the sustenance of the bacteria. The gas bubbles can be seen in plain film x-rays of the site, and the wound is dark and smells putrid. Treatment includes hyperbaric oxygen therapy, aggressive penicillin therapy, and excision of the devitalized tissue. Delay in treatment can result in amputation or septicemia (d³).

SOILBORNE BACTERIAL DISEASES

SOIL
BACTERIA a

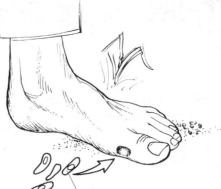

a

① ANTHRAX b
BACILLUS ANTHRACIS b¹

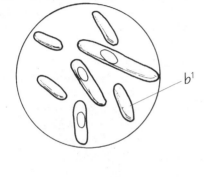

b¹

② TETANUS c
CLOSTRIDIUM
TETANI c¹

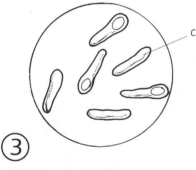

c¹

③ GAS GANGRENE d
CLOSTRIDIUM
PERFRINGENS d¹

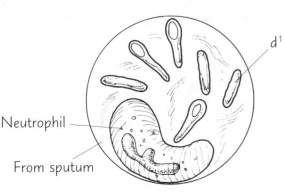

d¹

Neutrophil

From sputum

Spinal
cord

c⁴

Muscle

Liver

b³

Spleen

Kidney

c²

b²

Muscle

c⁴

d² d³

DISEASE SIGNS: ✿

b — SKIN VESICLES /
ULCERS b²
HEMORRHAGIC
LESIONS b³

c — WOUND /
NECROSIS c²
SPINAL CORD
DAMAGE c³
MUSCLE
SPASM c⁴

d — WOUND d²
GAS FORMATION
IN WOUND /
NECROSIS /
SEPTICEMIA d³

ARTHROPODBORNE BACTERIAL DISEASES

Arthropods are animals with jointed appendages and segmented exoskeletons. Representative arthropods include insects such as flies, fleas, lice, cockroaches, and bees, as well as noninsect spiders, mites, and ticks. Some of the smaller arthropods such as ticks and fleas carry bacteria (a) that are pathogenic to humans.

Color the title (a) at upper right, and the bacteria (a) harbored in the exoskeleton of the tick and the flea. Color the titles and related bacteria at station 1. Then color the disease signs associated with plague (b) and the designated part of the body in the center of the plate. Two signs are not shown (n.s.). Continue with stations 2, 3, and 4. As before, use shades of pink for the Gram-negative bacteria (b^1) and (d^1).

Plague (b) is one of the most well-known and historically important arthropodborne bacterial diseases. Historians write that the European population in the 1300s was reduced by one third because of this disease. The responsible microorganism is *Yersinia pestis* (b^1), an aerobic Gram-negative rod. The bacterium stains at the ends of the cell, giving an appearance under the microscope of a safety pin. Plague disease is now rare, but is seen in the southwestern United States, India, the Middle East, and Southeast Asia. The bacteria are carried by rat fleas that inhabit the wild rat and other rodents. The fleas that attach to human skin bite the flesh and introduce the bacteria into the human body. There are several forms of plague, including bubonic (bubo, inflamed or enlarged lymph node) plague, septicemic plague, and pneumonic plague.

The disease begins with fever and general malaise. The lymph nodes in the groin and the axilla (armpit) swell (b^2) and become tender. Left untreated bubonic plague develops to septicemic plague (b), with fatal results in 60% or more cases. Lung infections (pneumonic plague) are nearly 100% fatal. Cultures of infected lymph nodes or the blood, usually reveal the bacteria. Aggressive therapy with various antibiotics is the preferred treatment.

Lyme disease (c) is named after the Connecticut city where the disease was first described in the mid 1970s. The disease is due to a spiral bacterium (*Borrelia burgdorferi*, c^1) that has no Gram reaction. The organism is carried by ticks which are parasitic to white-tailed deer and field mice. Tick bites on human skin introduce the bacterium into the body. Characteristically, a rash (c^2) is produced at the infected site. Subsequent systemic involvement via the circulation causes fever, malaise, and weakness (c^3). Early treatment with tetracycline or penicillin is usually effective. Untreated for several months, the disease spreads to the joints (c^4), heart, and nervous system (c^5; meningitis, paralysis, and encephalitis). At that late stage, successful antibiotic treatment is more difficult.

Tularemia (d) is a disease similar to plague but not as lethal; it is caused by *Francisella tularensis* (d^1), a Gram-negative, endotoxin-containing pleomorphic rod that can reproduce in either anaerobic or aerobic environments (facultative bacteria). The organism is carried by ticks, fleas, and mosquitoes, all of which are parasitic on rabbits. Humans bitten by these arthropods can be infected with *F. tularensis*. People handling rabbits during food preparation are the most likely to be infected, either from inhalation of the bacteria or by way of small wounds on the hands. The transmission of this organism from human to human is not likely. In the case of wounds on the skin, skin ulcers (d^2) or cratering occurs at the portal of entry by the bacteria. The bacillus is taken up by phagocytes, and one of the initial signs of the disease is enlargement of the lymph nodes (d^3) in the region of the entry site. Once the organisms reach the circulation, there are more generalized symptoms such as fever, chills, malaise, and weakness. *F. tularensis* can be identified in agglutination tests; however, the organism is difficult to culture. Tetracycline, streptomycin, or gentamicin may be an effective treatment.

Rocky Mountain spotted fever (e) is caused by *Rickettsia rickettsii* (e^1) which is a parasite of ticks. Rickettsia are very small rod-shaped bacteria. They are invisible with the light microscope and so display no Gram reaction. Tick bites provide the means of transmission for rickettsiae in humans. About a week following the infection, local inflammation and capillary bleeding deep to the entry site cause a skin rash (e^2) which may spread quickly. Clinical signs associated with the infection include headache, fever, swollen upper and lower limbs (e^3), muscle pain, and neck stiffness. Untreated, the disease may go on to disseminated intravascular coagulopathy (DIC; e^4) in which there is uncontrolled bleeding. Early treatment with tetracycline or chloramphenicol is successful.

ARTHROPODBORNE BACTERIAL DISEASES

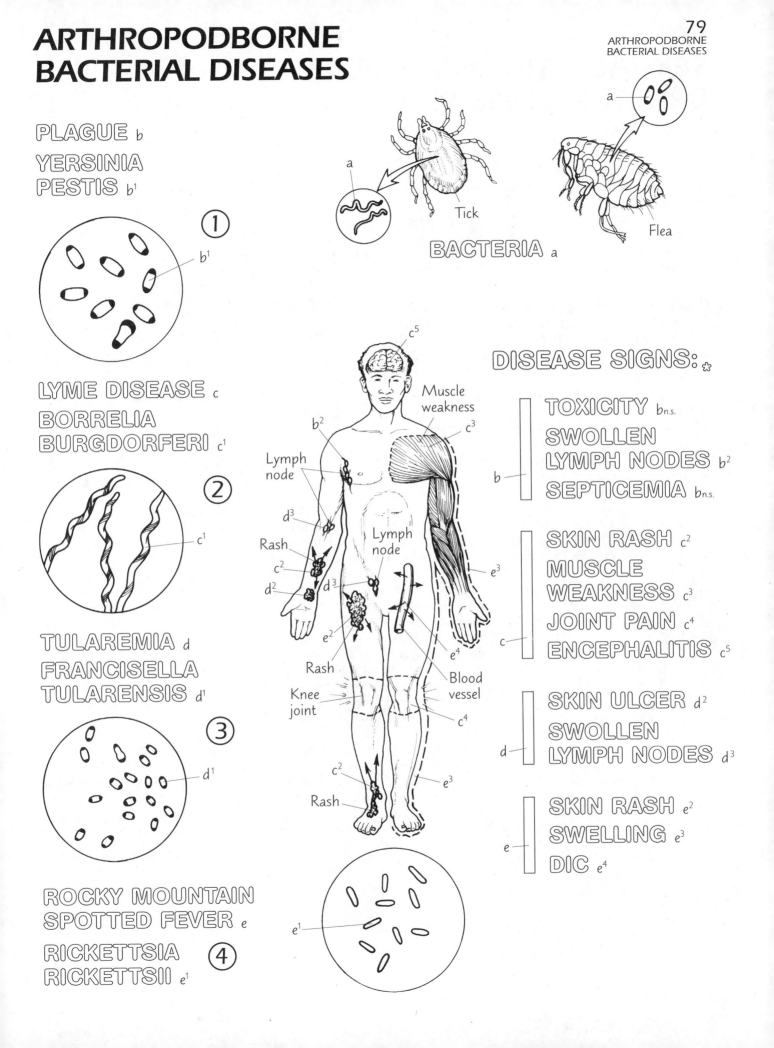

PLAGUE b
YERSINIA PESTIS b¹

① — b¹

LYME DISEASE c
BORRELIA BURGDORFERI c¹

② — c¹

TULAREMIA d
FRANCISELLA TULARENSIS d¹

③ — d¹

ROCKY MOUNTAIN SPOTTED FEVER e
RICKETTSIA RICKETTSII e¹

④ — e¹

BACTERIA a
Tick — a
Flea — a

Muscle weakness — c⁵
b²
c³
Lymph node
d³
Rash
c²
d²
d³
e²
Rash
Knee joint
c²
Rash
Lymph node
e³
e⁴
Blood vessel
c⁴
e³

DISEASE SIGNS: ✿

b {
TOXICITY b n.s.
SWOLLEN LYMPH NODES b²
SEPTICEMIA b n.s.
}

c {
SKIN RASH c²
MUSCLE WEAKNESS c³
JOINT PAIN c⁴
ENCEPHALITIS c⁵
}

d {
SKIN ULCER d²
SWOLLEN LYMPH NODES d³
}

e {
SKIN RASH e²
SWELLING e³
DIC e⁴
}

SEXUALLY TRANSMITTED BACTERIAL DISEASES

When microorganisms move directly from an infected person to an uninfected person during sexual activity, bacterial disease can develop in the uninfected partner. Sexual activity between persons involves contact with the external genitals and neighboring area, including the anus and surrounding perineal and perianal areas in both sexes. Here we present three of the most common of the sexually transmitted bacterial diseases.

Color the titles (a) through (c¹), the bacteria in each of the three circles, the diseased parts of the internal and external reproductive organs indicated, and the disease signs associated with each disease. As before, use shades of pink for the Gram-negative bacteria (b¹) and (c¹).

Syphilis (a) is caused by *Treponema pallidum* (a¹), a spiral bacterium, or spirochete, that has no Gram reaction. The bacterium functions in a low oxygen environment and is sensitive to dehydration. It invades the cellular lining of the urinary or reproductive passages, or the skin through small defects or breaks, and enters the circulation almost immediately. After about three weeks, the first cutaneous or mucosal lesion is a circular, surface wound called a chancre (a²). It appears at the site of bacterial entry and progressively changes to form an ulcer. Extragenital chancres (a⁴) may appear around the mouth or anus. The lymph nodes draining the site of entry (usually the inguinal, or groin, lymph nodes) become swollen (a³) as well. These signs manifest the first stage of syphilis (primary syphilis). The chancre heals without treatment; and the lymph nodes return to normal size without treatment. Over this approximate nine week period, the spirochetes replicate dramatically without symptoms. They are, however, quite sensitive to treatment by penicillin which destroys them and terminates the disease process.

Untreated, syphilis enters the second stage (secondary syphilis) at about two weeks to six months, a condition characterized by appearance of new mucosal or skin lesions, signs and symptoms of a generalized illness (fever, sweats, headache, sore throat, malaise, weakness, and joint pain), and lymphadenopathy. Liver disease and asymptomatic disease of the meninges and brain/spinal cord may also occur during this period. Within two to six weeks, the signs and symptoms of secondary syphilis disappear. Yet infiltration of many organ systems by the spirochetes continue.

Tertiary (third stage) syphilis reflects a general systemic infection that develops over the next several years. It is characterized by various expressions of brain dysfunction (dementia, paralysis, and generalized weakness), loss of movement-related sensation (tabes dorsalis), generalized neurological deficits, skin and bone lesions, and cardiovascular (aortic) disease. Syphilis-causing organisms are identified by complement fixation (Wasserman) test, VDRL testing, and the rapid plasma reagin test (RPR). Penicillin is the preferred treatment for all stages of the disease.

Gonorrhea (b) is due to a small Gram-negative diplococcus called *Neisseria gonorrhoeae* (b¹). It passes between the mucosal cells of the reproductive or urinary tract during sexual contact and multiplies in the urethra and cervix in females, and in the urethra in males, forming lesions (b²). The mouth and anus can also be sites of infection. Inflammation ensues, and a pus-laden discharge (b³) follows. Infection of the external genitals can migrate to the uterine tubes/ovaries in females and epididymis/ductus deferens in males. Scar tissue may form in these tubes during repair of tissues damaged by the organism. Tubal obstruction (b⁴) may result, inducing sterility. Common symptoms include pain on urination, discharge from the reproductive tract, and general discomfort. Gonorrhea is diagnosed in men by microscopic examination of urethral discharge; in women, cultures of cervical mucus/cells are examined. Penicillin and tetracycline are useful antibiotics. Untreated, gonorrhea may involve all of the organ systems, causing meningitis, cardiovascular disease, and arthritis.

Chlamydia (c), an infection caused by *Chlamydia trachomatis* (c¹), is often experienced as nongonococcal (or nonspecific) urethritis (c²). The infection is characterized by pain and discharge during urination, and cervicitis (c³; inflammation of the uterine cervix). It is transmitted during sexual contact. This species of bacteria, like other chlamydiae, has a very simple structure. It cannot be cultivated in the laboratory and therefore has no Gram reaction. It is cultivated only in living cells. Uncommonly, *C. trachomatis* infects the reproductive tracts of females and males in a manner similar to gonorrhea in which scarring of the tubal linings causes obstruction and sterility. Untreated, infection can progress to pelvic inflammatory disease (PID) in women. Chlamydiae are identified by taking vaginal or urethral swabs and combining them with antigens in an enzyme-linked immunoassay (ELISA) test. Tetracycline is the drug of choice for treatment.

SEXUALLY TRANSMITTED BACTERIAL DISEASES

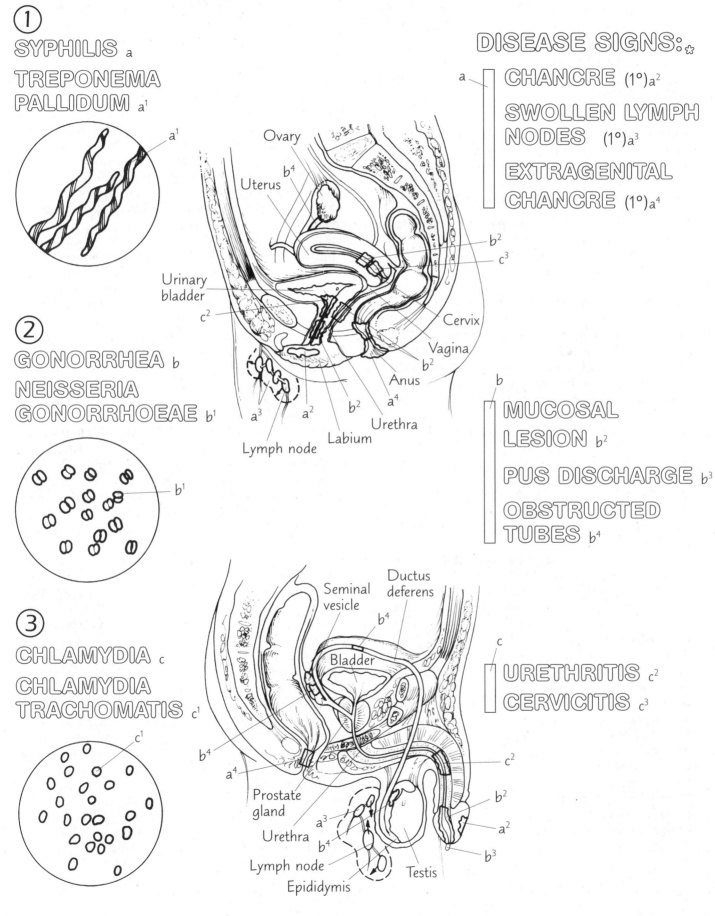

① SYPHILIS a

TREPONEMA PALLIDUM a¹

② GONORRHEA b

NEISSERIA GONORRHOEAE b¹

③ CHLAMYDIA c

CHLAMYDIA TRACHOMATIS c¹

DISEASE SIGNS:❋

a — CHANCRE (1°)a²
SWOLLEN LYMPH NODES (1°)a³
EXTRAGENITAL CHANCRE (1°)a⁴

b — MUCOSAL LESION b²
PUS DISCHARGE b³
OBSTRUCTED TUBES b⁴

c — URETHRITIS c²
CERVICITIS c³

Ovary
Uterus
b⁴
Urinary bladder
c²
b²
c³
Cervix
Vagina
Anus
b²
a³   a²   b²   a⁴
Urethra
Lymph node   Labium

Seminal vesicle
Ductus deferens
b⁴
Bladder
c
b⁴
a⁴
Prostate gland
Urethra
a³
Lymph node
b⁴
Epididymis   Testis
c²
b²
a²
b³

CONTACT BACTERIAL DISEASES

Contact bacterial diseases occur by interpersonal skin contact, such as shaking hands, random touching between or among people, and kissing.

Color the titles (a) through (d⁴), the bacteria in each of the four circles, the diseased parts, and the disease signs. As before, color Gram-positive bacteria (a¹) purple or blue, Gram-negative bacteria (c¹) pink, and acid-fast bacteria (d¹) red. Do not use colors that obscure detail.

Staphylococcus is a Gram-positive cluster of cocci normally resident on the skin surface (Plates 11 and 77). One of the pathogenic strains of this organism is *Staphylococcus aureus* (a¹). *S. aureus* enter the shafts of hair follicles and the ducts of oil (sebaceous) glands connected to the hair follicles. Here they reproduce, adhering to and irritating neighboring skin cells, and causing an inflammation to occur. Direct observation of the inflamed area often reveals "whiteheads" or "blackheads," reflecting trapped sebum or dead cells (from sebaceous glands). Unable to escape the obstructed hair shaft, masses of dead white blood cells (pus), bacteria, and inflamed tissue may accumulate, forming large abscesses called boils or furuncles (a³). In the eyelid, such obstruction and inflammation of gland ducts is called a stye (a⁴).

Laboratory diagnosis is usually performed by lancing the boil and cultivating the staphylococci on such bacteriological media as mannitol salt agar. Observation of Gram-positive staphylococci confirms the diagnosis. Surgical excision of the staph-laden, pus-filled abscess may be necessary. Treatment with penicillin, ampicillin, oxacillin, or other beta lactam antibiotics is usually indicated.

S. aureus can populate furrows of the skin created by abrasions or scratches, as occurs commonly with children. Commonly seen around the mouth and nose, the area may become crusty and enlarge or elevate, forming red-ringed vesicles. Such elevations may reflect impetigo contagiosum (a²); that is, patches of skin infected by *S. aureus* (especially the newborn). Penicillin and erythromycin are used to treat such skin infections.

Toxic shock syndrome (not shown) occurs when *S. aureus*, occupying the skin or mucosal lining of a cavity, such as the vagina, produces toxins that enter the blood stream and cause fever, vomiting, diarrhea, and a sunburnlike skin peeling. In the female, staphylococci may inhabit contraceptive devices or highly absorbent tampons that remain in contact with the body long enough to produce toxins that enter the circulation. These toxins, in some cases, can cause shock (rapid drop in blood pressure that prevents the body tissues from receiving normal amounts of oxygen, nutrition, and other important elements) with possible fatal consequences. Therapy is directed to relief of symptoms while the body repairs or replaces the damaged tissues.

A strain of *Chlamydia trachomatis* (b¹) can cause an eye disease known as trachoma (b). This disease is common in Mediterranean countries where it is transmitted to the eyes by hand contact. These tiny, round intracellular bacteria inhabit the inner lining of the eyelids, called the conjunctiva. Irritation and inflammation of the conjunctiva (b²; conjunctivitis) ensues. Chronic scratching trauma from the infected conjunctiva to the contiguous cornea may induce inflammation (b³; keratitis) and scarring and, often in association with secondary infections, progressive blindness. Personal transmission of *C. trachomatis* can be controlled and prevented by prophylactic treatment with tetracycline as is done in Mediterranean countries where trachoma nurses travel among towns administering doses of antibiotic to the eyes of all residents of the town.

Bacterial conjunctivitis (c) is an infection of the conjunctiva, a thin, transparent membrane covering the exposed eye and inner eyelids. The Gram-negative rod *Hemophilus influenzae,* type III (c¹; also known as *Haemophilus aegypticus*) is the microbe traditionally thought responsible for the condition. The inflamed conjunctiva becomes streaked with red or pink, giving rise to the name "pinkeye" (c²); the eyelids may be inflamed as well (blepharitis). The disease can be acquired by hands or face towels, long-wearing contact lenses, or simply by aerosol (airborne) transmission. Diagnosis is normally performed by isolating the bacteria on selective media and identifying the bacteria by morphology and Gram staining. Neomycin is commonly prescribed for therapy.

Hansen's disease (d; leprosy) is a condition of distorted skin and underlying deformities caused by infection with *Mycobacterium leprae* (d¹), an acid-fast rod without a Gram reaction. The bacterium is probably transmitted by discharge of nasal secretions that are passed from one person to another. Two principal varieties of Hansen's disease exist. The tuberculoid form is characterized by strong immune resistance to the organism, discrete skin nodules, and locally desensitized skin. The progressive or lepromatous form is characterized by low immune resistance to the organism with extensive distribution of the pathogen throughout the body; it is characterized by numerous nodules (d²) causing cosmetic deformities (d³), cutaneous nerve lesions (d⁴) resulting in local losses of sensation, a reduced ability to sweat, and certain motor nerve lesions resulting in hand and foot deformities. The establishment of the disease can be confirmed by observing the acid-fast bacilli in body fluids. Hansen's disease requires long term therapy; the drug dapsone, often in association with rifampin, has been used to curb its progression.

CONTACT BACTERIAL DISEASES

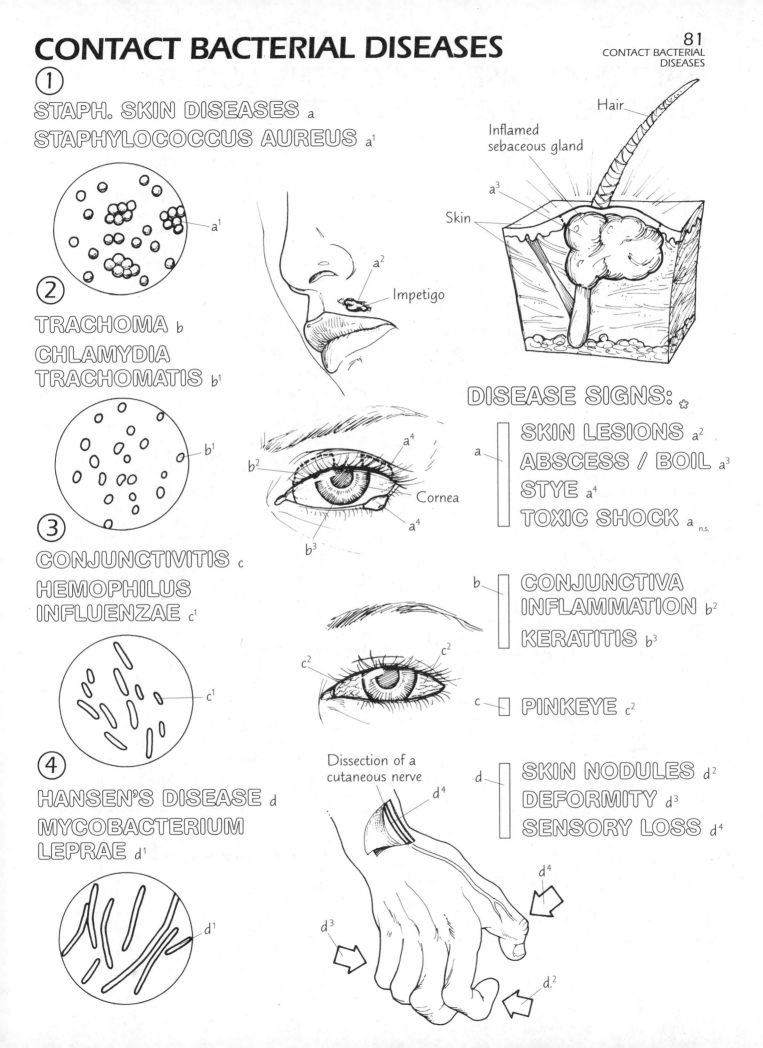

① STAPH. SKIN DISEASES a
STAPHYLOCOCCUS AUREUS a¹

② TRACHOMA b
CHLAMYDIA
TRACHOMATIS b¹

③ CONJUNCTIVITIS c
HEMOPHILUS
INFLUENZAE c¹

④ HANSEN'S DISEASE d
MYCOBACTERIUM
LEPRAE d¹

Hair
Inflamed
sebaceous gland
a³
Skin

Impetigo
a²

a⁴
b²
Cornea
a⁴
b³

Dissection of a
cutaneous nerve
d⁴

c²
c²
c²

DISEASE SIGNS: ✱

a | SKIN LESIONS a²
ABSCESS / BOIL a³
STYE a⁴
TOXIC SHOCK a n.s.

b | CONJUNCTIVA
INFLAMMATION b²
KERATITIS b³

c | PINKEYE c²

d | SKIN NODULES d²
DEFORMITY d³
SENSORY LOSS d⁴

d³
d⁴
d.²

BACTERIAL DISEASES OF THE LOWER RESPIRATORY AND DIGESTIVE TRACTS

Set aside a pink color for the Gram-negative bacteria (e¹) and (f¹). Color the title and structures of the lower respiratory tract (a) a light color. Color the titles, the organisms, arrows, and disease signs (b) through (c³).

Chlamydia psittaci (b¹; psittic-, parrot) belongs to a group of tiny bacteria called chlamydiae. These bacteria are larger than mycoplasmas but smaller than rickettsiae. Like rickettsiae, they are very simple and are cultivated only in living tissue. Strains of *C. psittaci* infect parrots as well as a great variety of wild and domestic species of birds, including chickens and seagulls. The bacteria are transmitted to humans by airborne infected particles, such as bird droppings and feathers, bird handling with open wounds on the hands, and skin-breaking bites of infected birds. The infection is called psitticosis (b; also called ornithosis).

The disease may be accompanied by headache, dry cough (b²), and other influenzalike symptoms and signs. Infection with *C. psittaci* yields infective particles that can migrate down the bronchi to the bronchioles of the lungs. Here the toxic particles can cause patches of lung destruction (b³; toxic fulminating pneumonia) severe enough to cause death in certain cases. Complement fixation testing and injection of infected sputum into mice corroborate the clinical diagnosis. Therapy with tetracycline is usually successful.

Q fever (c) is a bacterial disease of the lower respiratory tract caused by *Coxiella burnetii* (c¹), a rickettsia. Measuring only about 0.50 micrometers in length, *C. burnetii* has no spores, flagella, or capsule. The microorganism is cultivated only in living tissues, such as fertilized eggs. For many years the cause of the disease characterized by fever was unknown, giving rise to the term "Q" (for query) fever. The microorganism is carried by cattle ticks among other possible hosts. It is transmitted to humans in unpasteurized milk and by airborne particles in areas populated by large domestic animals. Dairy workers and large animal handlers are at risk for infection with *C. burnetii.*

The infection is associated with headache, high fever, dry cough (c²), fatigue, muscular aches and pains, and pneumonia (c³). A complication of the untreated disease is infection of the lining tissues of the heart cavities and valves (Q fever endocarditis). Tetracycline is usually used to treat Q fever, which is rarely fatal.

It may be helpful to examine Plate 76 for the other significant pathogens of the lower respiratory tract, and Plate 75 for major pathogens of the upper respiratory tract. All of these pathogens, including the ones colored here, share a common portal of entry (mouth, nasal cavity) by means of inhalation. The more extensive the infection, the more likely the lungs will become involved (pneumonia) and the greater risk of infection of the blood, vascular organs, the meninges, and the central nervous system.

Color the intestines representing the digestive tract (d) and its title. Color the titles, the organisms, arrows and disease signs (e) through (f³). Use shades of pink for (e¹) and (f¹).

Shigella sonnei (e¹; named after the Japanese physician Shiga and the Danish bacteriologist Sonne) is a waterborne and foodborne Gram-negative rod and one of four pathogenic species of the genus *Shigella.* It is ingested, reproduces in the small intestine, and its toxins damage cells lining the large intestine. Diarrhea (e³), bloody mucoid stools, and cramps (e²) characterize the disease shigellosis (e). Loss of body water may be substantial, and rehydration of the patient may be necessary. Cultures taken from rectal swabs generally assist in confirmation of the disease. Antibiotics are effective in limiting the disease. Recoverers generally become carriers and continue to shed the bacilli for may weeks.

Campylobacteriosis (f) is a foodborne and waterborne bacterial disease in which bacteria colonize the intestine and cause cramps (f²), diarrhea (f³), inflammation, and fever. The responsible organism is *Campylobacter jejuni* (f¹), a Gram-negative spiral bacterium. Contaminated drinking water, raw milk, and contaminated food have been implicated in recent outbreaks. The laboratory diagnosis is made by growing *C. jejuni* colonies cultured from stool samples. Most patients recover in several days. Erythromycin therapy hastens recovery.

Ingested waterborne and foodborne pathogens cause a great deal of human misery around the world, in the form of fever, abdominal cramps, nausea, vomiting, diarrhea, and dehydration. There are some similarities of signs and symptoms among salmonellosis, shigellosis, campylobacteriosis, typhoid fever, staphyloenteritis, cholera, and *E. coli* travelers diarrhea (recall Plate 77). The pathogens can be identified by stool culture. The diseases are generally self-limiting, but rehydration during recovery is important.

BACTERIAL DISEASES OF THE LOWER RESPIRATORY AND DIGESTIVE TRACTS

LOWER RESPIRATORY TRACT a

DIGESTIVE TRACT d

① PSITTICOSIS b

CHLAMYDIA PSITTACI b¹

(Greatly magnified)

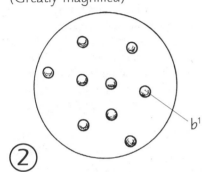

b¹

② Q FEVER c

COXIELLA BURNETII c¹

(Greatly magnified)

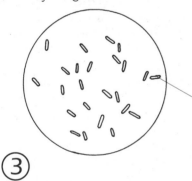

c¹

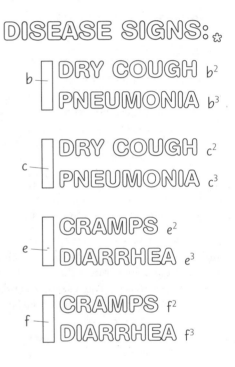

b²
c²
a
a
b³
c³
Lungs
Stomach
Intestines
d
d
e²
f²
Rectum
e³
f³

DISEASE SIGNS: ✿

b { DRY COUGH b²
    PNEUMONIA b³

c { DRY COUGH c²
    PNEUMONIA c³

e { CRAMPS e²
    DIARRHEA e³

f { CRAMPS f²
    DIARRHEA f³

③ SHIGELLOSIS e

SHIGELLA SONNEI e¹

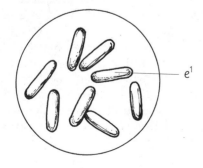

e¹

④ CAMPYLOBACTERIOSIS f

CAMPYLOBACTER JEJUNI f¹

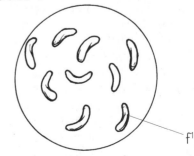

f¹

VIRAL DISEASES OF THE SKIN

When viruses enter and multiply within human cells, they kill the cells. The killing of cells by viruses and the related signs and symptoms constitute a viral disease. Viruses multiply within human cells by employing the cells' own metabolic machinery to reproduce. Viruses have no metabolic or reproductive structures. The massive numbers of developing viruses within the cell burst the cell membrane. Bacterial disease by contrast involves pathogens that have their own reproductive and digestive structures but use human tissue as a source of nutrition and substrate for multiplying. Bacterial infection, then, destroys human cells by consumption; toxins often accompany the bacteria, and these can destroy tissues and alter physiological processes in the body. No such toxins are produced by viruses.

Another problem with viral diseases is that there are no established methods for treatment. Antibiotics are ineffective because the viruses do not have mechanisms or structures with which antibiotics can interfere. Antiviral medications attempt to interrupt viral multiplication but this often has a deleterious effect on DNA metabolism in host cells.

Begin with station 1. Color the titles of the disease and the virus, and the related lesions or swellings on the body. Continue clockwise through station 6.

Chicken pox or varicella (a) is a highly contagious infection found mostly in children. It is characterized by groups of small, cutaneous, fluid-filled elevations (vesicles) over the face, neck, and back. The disease is caused by the icosahedral varicella-zoster virus (a¹), a part of the herpes family of viruses. The virus is spread by respiratory droplets. After passing through the respiratory tract, it enters the circulation, causes a brief fever to develop, and localizes in the skin tissues. Swelling of the lymph nodes at the back of the head (a) accompanies the formation of the skin vesicles (a). Interpersonal contact with fluid from these vesicles may transmit the virus. The vesicular rashes remain for several days, become filled with pus, and then form crusts that fall off without leaving scars. Laboratory diagnostic tests are based on nucleic acid probes (radioactively labeled DNA) and electron microscopy. Chicken pox is considered a benign disease in children; adults and a small population of immunosuppressed persons are at increased risk for a more severe form of the disease. A vaccine has long been under development and has recently been approved in the United States.

Herpes simplex (b) is an infection due to another strain of herpes virus (b¹; herpes simplex). Herpes is transmitted among individuals primarily by skin contact. Herpes simplex is experienced as cold sores (b) on the lips, gums, and in the mouth; and as lesions of the genital organs (genital herpes; not shown). Sexual contact may transfer the virus, and genital herpes is considered a sexually transmitted disease. Infection may also occur in the brain tissue (herpes encephalitis; not shown), particularly if a newborn acquires the virus from a woman who has genital herpes. Laboratory diagnostic tests

are based on nucleic acid probes and electron microscopy. The drug acyclovir may limit the severity of attacks.

The virus that causes measles (c; also called rubeola) is the RNA-containing helical, rubeola virus (c¹). It enters and passes through the respiratory tract, into the blood, and localizes in the skin tissues. Here it induces blood leakage from the capillaries, resulting in a red rash that breaks out at the hairline. The rash consists of small, elevated, round or oval spots (c) that form on the face and spread to the trunk and extremities. This very contagious disease, seen mostly in children two to six years old, is associated with fever, malaise, nasal discharge, and a dry cough. Complete recovery usually occurs within a week. Laboratory diagnostic tests are based on nucleic acid probes and electron microscopy. Immunization may be rendered by vaccination with attenuated measles virus injected with the MMR (measles-mumps-rubella) preparation. Rarely, in infants, adults, and immunocompromised persons, the disease may be associated with pneumonia.

German measles or rubella (d) is caused by the RNA, icosahedral, rubella virus (d¹). The virus is spread by respiratory droplets, as well as other body secretions and excretions. It affects the skin with dense groups of small, pale, quickly-fading, non-elevated skin rash (d). Clinically, fever and swelling of lymph nodes (d), along with nasal discharge, are seen in the young patients (generally three to nine years of age) with this disease. Laboratory diagnostic tests are based on nucleic acid probes and electron microscopy. In pregnant women, the virus can cross the placental barrier and spread to the fetus, causing hearing, vision, and cardiac defects. Immunization with attenuated rubella virus in the MMR preparation is recommended for all persons who have not already had the disease.

Mumps (e; also called parotitis) is an infection of the salivary glands, especially the parotid (e), and is seen mainly in children. Indeed, the swollen cheek, either unilaterally or bilaterally, is the characteristic sign of the disease. Also seen with the swelling are fever, headache, malaise, and anorexia. The causative agent is a helical RNA virus (e¹; mumps virus). Transmission of the agent from person to person is by respiratory droplets. Laboratory diagnostic tests are based on nucleic acid probes and electron microscopy. The disease is self-limiting and requires only supporting treatment. Immunization consists of an injection with attenuated virus in the MMR preparation.

Smallpox (f) has always been one of the great scourges of humanity. The responsible virus, containing DNA and a series of rods but no envelope, is called a pox virus (f¹). The virus is transmitted primarily by skin contact. Smallpox is characterized by elevated, fluid-filled, cutaneous vesicles. The vesicles later become pus-emitting ulcers and pitted scars (f; called pocks). Most victims either experience death or disfigurement. Prevention consisted of immunization with cowpox viruses (vaccination). However, no case of smallpox has been detected in the world since 1977, and the disease is now considered eradicated.

VIRAL DISEASES OF THE SKIN

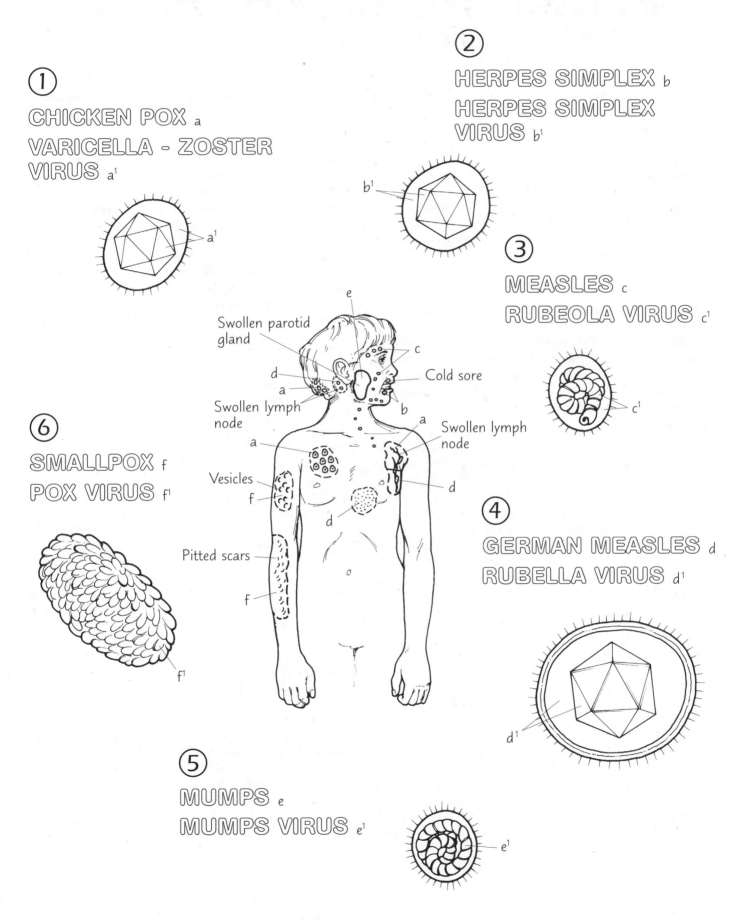

① CHICKEN POX a
VARICELLA - ZOSTER
VIRUS a¹

② HERPES SIMPLEX b
HERPES SIMPLEX
VIRUS b¹

③ MEASLES c
RUBEOLA VIRUS c¹

④ GERMAN MEASLES d
RUBELLA VIRUS d¹

⑤ MUMPS e
MUMPS VIRUS e¹

⑥ SMALLPOX f
POX VIRUS f¹

Swollen parotid gland
Swollen lymph node
Cold sore
Swollen lymph node
Vesicles
Pitted scars

84
VIRAL DISEASES OF THE DIGESTIVE AND LYMPHOID ORGANS

Viruses, like bacteria, can infect the organs of the digestive system, including the gastrointestinal tract, liver, gall bladder, and pancreas. The viruses are taken into the body by oral ingestion. Viral gastroenteritis (inflammation of the gastrointestinal tract) is a common disorder, especially in infants and young children. The disease often involves a course of diarrhea which, in these young people, can have devastating dehydrating effects that can result in death. These viruses often reach the liver and the lymphoid organs (lymph nodes and spleen) by the circulatory route after infecting the respiratory tract. Here we present several viral diseases, and their causative agents, that have a significant impact on humans.

Each of the viral disease states and their related virus are arranged as stations around the central anatomic drawing. Color the disease title and the virus title, the drawing of the virus, and then the arrow or site on the body that is particularly affected by that virus. Start with station 1 and continue through station 5. Use colors that do not obscure the illustrative detail of the viruses.

Yellow fever (a) is a mosquitoborne disease caused by an RNA icosahedral virus (a^1; yellow fever virus). The disease is endemic in South America, Central America, Africa, and the Caribbean. The virus is transmitted from blood to blood in humans by the mosquito *Aedes aegypti*. Once in the blood, the virus infects the vascular lining cells and passes into the liver as well. It incubates in the liver, disrupts the cells there, permitting pigmented bile to spill over in to the blood and discolor it. The skin, as a consequence, develops a yellow color. This condition, called jaundice (jaundice, yellow) is a reflection of liver damage (a^4), and is best seen in the sclera (white areas) on either side of the iris-ringed pupils of the eyes. Headache, fever (a^2), nausea/vomiting (a^3), and muscular aches (a^5) accompany the disease. Severe liver impairment, and hemorrhage into the gastrointestinal tract, can cause death. A preparation of live attenuated viruses can be used in a vaccine to promote immunity.

Dengue fever (b) is caused by an RNA icosahedral virus called dengue virus (b^1). As with yellow fever, the virus is transmitted to humans primarily by the mosquito *Aedes aegypti*. Once in the circulation, the virus infects vascular lining cells and monocytes of the blood. The normal flow of bloodborne infected monocytes brings the virus to multiple organs, such as the spleen (b^4). Fever (b^2), malaise, headache, nausea and vomiting (b^3), and joint pain (b^5) accompany the infection. Indeed, the disease is also known as "breakbone fever." Fever fluctuations give the disease the alternate name of "saddleback fever." Symptoms generally last a week.

Complications are uncommon; in endemic areas (Southeast Asia, India, and other tropical areas), a more severe hemorrhagic disease may ensue.

Infectious mononucleosis (c) is transmitted worldwide by interpersonal contact, mucus droplets and fomites (inert particles that carry the virus). The disease is caused by the DNA icosahedral Epstein Barr virus (c^1) of the herpes family of viruses. The virus infects the pharynx, often causing whitish patches. Sore throat, mild fever, swollen lymph nodes (c^2), a swollen spleen (c^3), and generalized weakness (c^4) accompany the disease. A high count of white blood cells with atypical lymphocytes (hence, mononucleosis) may be observed. The diagnostic test for "infectious mono" is called the heterophile antibody test. It consists of combining serum with horse erythrocytes to see if agglutination takes place. The disease is self-limiting and generally resolves over several weeks.

Hepatitis A (d) is an infectious disease of the liver caused by an RNA enterovirus (d^1). The virus is discharged in the feces; it is ingested orally into the body (fecal-oral route). The disease is common where there are unsanitary conditions and habits, contaminating water and food. Persons infected with hepatitis A have headaches, muscle pain (d^4), and nausea and vomiting (d^2). Some liver cell damage (d^3) occurs, and jaundice is an occasional feature of the disease. It is a self-limiting disease and has no chronic form. There is no specific treatment for the disease, but hepatitis A immunoglobulin can be given to provide short term protection.

Hepatitis B (e) is another liver disease caused by a DNA virus (e^1; hepatitis B virus or HBV). The virus has been observed in three forms, all of them called "infectious particles." These particles include the complete virus with DNA and envelope, and long, filamentous forms, and round, small forms. All contain antigens that bind with specific antibodies. The virus is transmitted from person to person by bodily secretions (blood, semen, and possibly other fluid discharges). These facts put at risk the administration of drugs by the intravenous (iv) mode, especially where antiseptic technique is not employed (as with many iv drug abusers). The amount of blood required to infect is small enough to be invisible to the unaided eye. The virus infects the liver (e^2), lymph nodes (e^3), and spleen (e^4). Liver damage may result in jaundice; symptoms of infection include headache (e^5), joint pain (e^6), and anorexia. Most infected persons discharge the virus, although some people, usually immunosuppressed, remain infected for life. Chronic, symptomatic hepatitis B, or cirrhosis of the liver is not a common sequela to the acute disease. There is no specific treatment for the condition, however, a genetically engineered hepatitus B vaccine is available for prevention.

VIRAL DISEASES OF THE DIGESTIVE AND LYMPHOID ORGANS

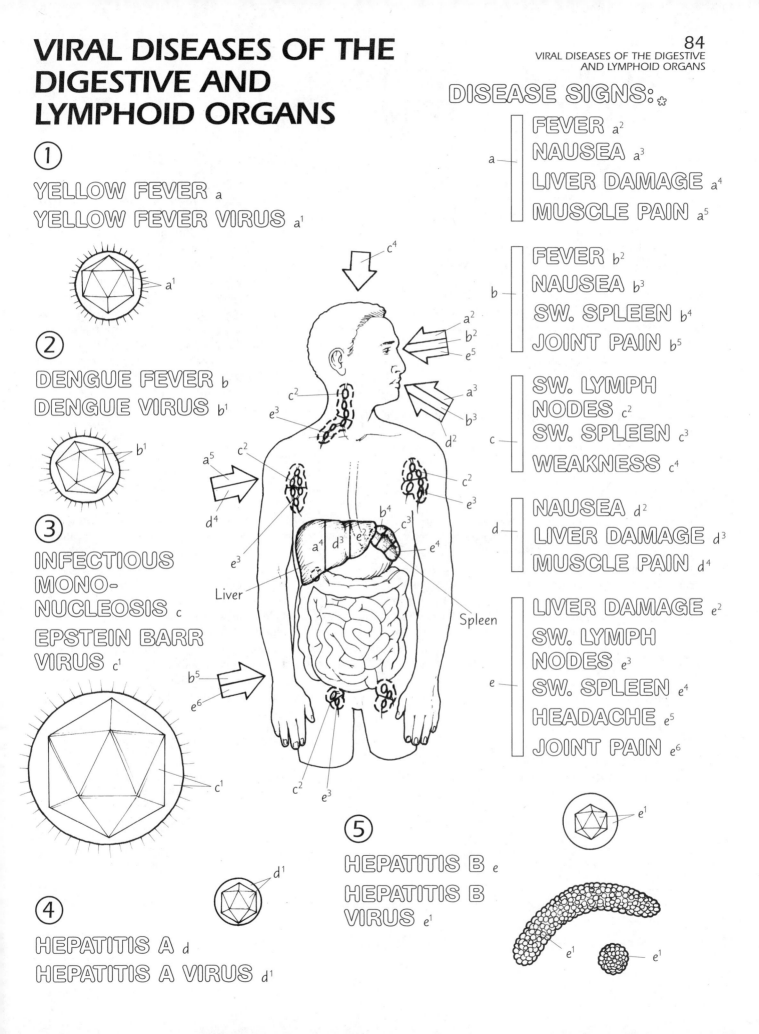

① YELLOW FEVER a
YELLOW FEVER VIRUS a¹

② DENGUE FEVER b
DENGUE VIRUS b¹

③ INFECTIOUS MONO-NUCLEOSIS c
EPSTEIN BARR VIRUS c¹

④ HEPATITIS A d
HEPATITIS A VIRUS d¹

⑤ HEPATITIS B e
HEPATITIS B VIRUS e¹

DISEASE SIGNS:

a —
FEVER a²
NAUSEA a³
LIVER DAMAGE a⁴
MUSCLE PAIN a⁵

b —
FEVER b²
NAUSEA b³
SW. SPLEEN b⁴
JOINT PAIN b⁵

c —
SW. LYMPH NODES c²
SW. SPLEEN c³
WEAKNESS c⁴

d —
NAUSEA d²
LIVER DAMAGE d³
MUSCLE PAIN d⁴

e —
LIVER DAMAGE e²
SW. LYMPH NODES e³
SW. SPLEEN e⁴
HEADACHE e⁵
JOINT PAIN e⁶

Liver

Spleen

85
VIRAL DISEASES OF THE RESPIRATORY SYSTEM

Viral diseases of the respiratory tract affect the system concerned with breathing. Structures of this system conduct air into the tiny air sacs (alveoli) of the lungs and conduct exhaled carbon dioxide out into the environment. Viruses affect this system by multiplying in the tissues and destroying the host cells. The accumulation of necrotic cells leaves the system open to bacterial infection that, in the extreme, can result in death. In the great influenza pandemic of 1919-1920, an estimated 20 million individuals died of the effects of the flu due mainly to secondary bacterial infections by species of *Staphylococci* and *Streptococci*. Here we present some of the common viral diseases and their etiologic agents.

Select light but contrasting colors for this plate. Color the upper and lower respiratory tracts in the center of the plate, and the related titles (a) and (b). Then color the titles of the diseases and related viruses, the viruses and the disease signs, beginning with station 1.

Influenza (c; station 1) is an acute contagious disease mainly of the upper respiratory tract (a). It is caused by a helical RNA virus (c^1). Chemical molecules in the envelope of this virus can be altered by mutation, making possible antigenic variations. Numerous strains of influenza virus are therefore possible, and recovery from disease due to one strain does not ensure protection from other strains. In other words, the new antigens react weakly with antibodies induced by previous strains. Three types of influenza, types A, B, and C (not shown) are known to exist. In each case, the patient suffers an abrupt onset of fever (c^2) and chills, fatigue (c^3), headache (c^4), muscle pain (c^5), and joint pain (c^6). Droplets of nasal mucus and saliva are the most frequent mode of transmission among individuals. Complement fixation testing can often make the diagnosis. The disease is usually short-lived and requires only palliative or supportive therapy. Bacterial infection superimposed on the viral infection (superinfection) can complicate and lengthen the disease process, especially in the very young and the very old. Influenza vaccines are available.

The helical RNA respiratory syncytial virus or RSV (d^1) causes a disease of the lower respiratory tract, primarily in children under two years of age. This disease, viral pneumonia (d), begins with inflammation of the nasal cavity and pharynx, causing related symptoms. The virus moves down into the bronchioles and alveoli, and infects the lining cells there (d^2; pneumonia). Inflammatory exudate and swelling blocks the airway, giving rise to respiratory difficulty (d^3; dyspnea). Other viruses may also cause the condition of dyspnea so characteristic of RSV infection. The term "respiratory syncytial" refers to the behavior of infected cells in culture in which the cells fuse to form a single mass of giant cells called a syncytium. It is thought that the bronchiolar inflammation may be the result of an immune reaction. The disease usually runs its course by seven days. Supplemental oxygen may be required; antibiotic treatment is only used in the event of bacterial superinfection.

Rhinoviruses are a group of over 100 different RNA viruses, all having icosahedral symmetry and no envelope (e^1). The viruses inhabit the upper respiratory tract (a) where they cause inflammation of the lining cells, giving rise to the symptoms of the common cold (e): headache (e^2), sneezing (e^3), a dry, scratchy throat, a "runny" nose (e^4), and fatigue (e^5). Droplets of sputum and nasal mucus are a mode of transmission among people, as is skin contact. The disease is self-limiting and only supportive therapy is required.

Adenoviruses (f^1) are a collection of over 30 types of icosahedral, double stranded DNA viruses lacking an envelope. They are very contagious, having a penchant for the lining cells of the upper respiratory tract (a). They are a cause of the common cold (f), the symptoms of which have been presented. Severe respiratory illnesses, including pneumonia, may also follow the cold in immunosuppressed persons, the very young, and the very old. Adenoviral infection sometimes occurs in the eye and in the meninges of the brain, accompanied by inflammation and related symptoms and signs there.

VIRAL DISEASES OF THE RESPIRATORY SYSTEM

UPPER RESPIRATORY TRACT $_a$
LOWER RESPIRATORY TRACT $_b$

DISEASE SIGNS: ✳

①

INFLUENZA $_c$
INFLUENZA VIRUS $_{c^1}$

c — FEVER $_{c^2}$
FATIGUE $_{c^3}$
HEADACHE $_{c^4}$
MUSCLE PAIN $_{c^5}$
JOINT PAIN $_{c^6}$

d — PNEUMONIA $_{d^2}$
DYSPNEA $_{d^3}$
FATIGUE $_{d^4}$

②

VIRAL
PNEUMONIA $_d$
RESPIRATORY
SYNCYTIAL
VIRUS $_{d^1}$

e — HEADACHE $_{e^2}$
SNEEZING $_{e^3}$
NASAL DRIP $_{e^4}$
FATIGUE $_{e^5}$

f — HEADACHE $_{f^2}$
SNEEZING $_{f^3}$
NASAL DRIP $_{f^4}$
FATIGUE $_{f^5}$

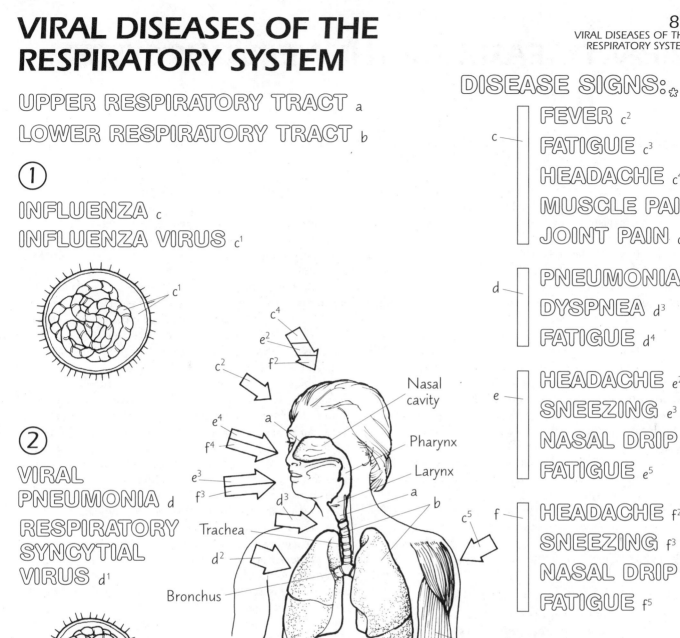

Nasal cavity
Pharynx
Larynx
Trachea
Bronchus
Lung
Muscle

③

COMMON COLD $_e$
RHINOVIRUS $_{e^1}$

④

COMMON
COLD $_f$
ADENOVIRUS $_{f^1}$

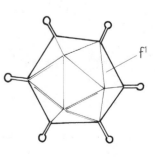

The portals of entry into the human body by viruses include the oral-fecal route, the nasal cavity, and defects or breaks in the skin. Viruses, like bacteria, may infect and cause symptoms in the upper respiratory tract, and then subsequently reach the nervous system by way of the blood and the lymph. The primary mechanism of injury to the cells by viruses is by direct cell damage due to intracellular viral proliferation. Such damage often kills the cells. This is a serious problem in the central nervous system where regeneration (mitosis) of neurons is very unlikely. Once the neurons are destroyed in the motor tracts, recovery of muscles supplied by those tracts is not likely even after the viruses have been eliminated.

The human nervous system consists of the central nervous system (CNS) and peripheral nervous system (PNS). The CNS includes the brain (b) and spinal cord (c). The meninges (a) envelop the brain and spinal cord. The brain consists of a cerebrum (cerebral hemispheres), the brain stem, and the cerebellum. The spinal cord is an extension of the brain stem and is located in the vertebral column (not shown). The PNS consists of nerves that flow out of the brain and spinal cord in specific patterns and routes, and reach every tissue, organ, muscle, and bone in the body. Here we show nerve roots (c^1) in relation to the spinal cord.

The main cell of the nervous system is the neuron. When microorganisms infect the nervous system, the neurons are their primary host. Significant loss of neurons causes dysfunction of some part of the nervous system.

Color the titles and related meninges (a), brain (b), and spinal cord (c) in the center of the plate. Then begin with station 1 and color the titles and names of each disease and related organism, and the related arrows. Use light colors.

Poliomyelitis (d; polio) is caused by a small icosahedral RNA polio virus (d^1) that has no envelope. The virus enters the body in contaminated food and water and through the nasal cavity. The infection begins with a sore throat as the virus multiplies in the pharynx and gastrointestinal tract. Over the next several days, the virus passes into the circulation, and localizes on the meninges (a), in the brain (b), and/or spinal cord (c). Once in the nervous system, the polio virus targets the motor cells in the spinal cord (anterior horn motor neurons). Destruction of these neurons causes degrees of paralysis in various parts of the body. The virus tends to localize in motor neurons high in the spinal cord; if motor cells in the upper cervical and thoracic cord are damaged, the muscular basis of respiration will be affected and death may occur from respiratory failure in the absence of mechanical ventilatory support.

Three types of polio virus are known to exist, and the preparations for immunity utilize all three. The Salk vaccine (introduced in 1955) contains the three types inactivated with chemicals. The Sabin oral vaccine (introduced in 1958) contains the three types attenuated by multiple passages through tissue culture (recall Plate 34).

Rabies (e) is due to a helical RNA virus (e^1) with no envelope. Once the body has developed symptoms and signs from rabies infection, the disease is nearly always fatal. The virus can be transmitted to humans by the infected saliva of warm-blooded animals including dogs, cats, bats, raccoons, skunks, and wolves. On entering the body following animal bites through the skin, the virus multiplies in the tissues, and in an about two to eight weeks the body experiences symptoms of headache, sore throat, fever, anorexia, and nausea. As the virus invades the brain and destroys neurons, the infected person becomes anxious and irritable. Muscle weakness or paralysis sets in at various sites. Psychological stress secondary to hydrophobia (fear of water) is followed by loss of consciousness (coma), respiratory paralysis, and death. Immunofluorescence of the rabid animal's brain makes the diagnosis. The immunization methods include vaccines and injection of immune globulin. New vaccines are under development.

Numerous viruses transmitted to humans by arthropods (ticks, mosquitoes) can cause inflammation of the brain (f; arboviral encephalitis) and possibly the spinal cord (f^1; arboviral myelitis). The viruses are icosahedral RNA arboviruses (f^2). Symptoms of infection begin with fever, chills, and headache. Seizures and altered consciousness or coma may follow. Types of arboviral encephalitis include St. Louis, California, LaCrosse, and Japanese encephalitis.

VIRAL DISEASES OF THE NERVOUS SYSTEM

MENINGES a
BRAIN b
SPINAL CORD c
SPINAL NERVE ROOTS c¹

① POLIOMYELITIS d
POLIO VIRUS d¹

③ ENCEPHALITIS f
MYELITIS f¹
ARBOVIRUS f²

② RABIES e
RABIES VIRUS e¹

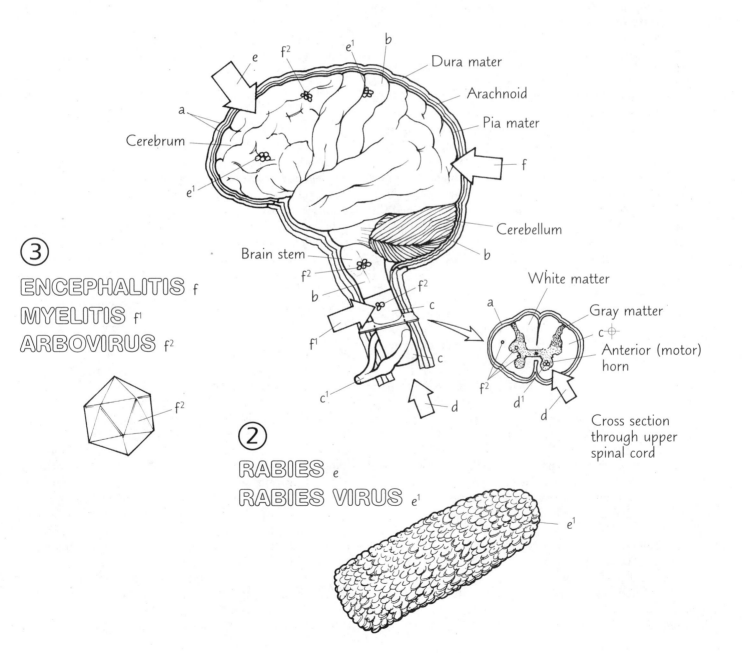

Dura mater
Arachnoid
Pia mater
Cerebrum
Cerebellum
Brain stem
White matter
Gray matter
Anterior (motor) horn
Cross section through upper spinal cord

ACQUIRED IMMUNE DEFICIENCY SYNDROME (AIDS)

Acquired immune deficiency syndrome (AIDS) has been the subject of intensive research in laboratories throughout the world since it was first recognized in 1981. The disease strikes the T lymphocytes of the immune system, and other cells as well. It commonly leaves the body defenseless against opportunistic infections due to diminished T lymphocytes and sharply depressed cell-mediated immunity. Since knowledge about AIDS is increasing daily, some of the information may have changed by the time you read these paragraphs and color the plate.

Look over the plate and plan your use of colors carefully. You can use the same color for titles/structures sharing the same subscript if necessary. Color the titles and parts of the HIV at the upper part of the plate. Use a light color for (a).

AIDS is caused by a helical virus known as the human immunodeficiency virus, or HIV (a). The virus is composed of two strands of RNA (b) surrounded by a core of protein (d), enclosed in another protein layer (not shown). The protein-RNA combination is surrounded by an envelope (a^1). On the surface of the envelope is a series of projections called spikes (a^2) composed of at least two glycoproteins called gp120 and gp41. These spikes fuse with the cell membrane during cell infection. The gp120 and gp41 molecules in the spikes change frequently to yield new strains of the virus. Associated with the viral RNA are two molecules of the enzyme called reverse transcriptase (c). This enzyme makes possible the use of RNA as a template to synthesize DNA, a reversal of the normal flow of biochemical information in a cell.

Color the subheading T Cell Infection and the titles and related illustrations in the center of the plate. Start at the left of the cell where the HIV has attached, and work to the right where the HIV is explosively discharged.

When an HIV infects a T lymphocyte, the envelope of the virus fuses with the cell membrane (e). The envelope spikes make the critical connection in the fusion as their gp120 and gp41 molecules unite with the CD4 receptor site on the membrane of the T lymphocyte. First the gp120 molecule makes contact, then the gp41 molecule induces the fusion of the envelope and membrane. The viral protein core enclosing the RNA now enters the cytoplasm of the cell; the envelope and spikes are left outside the cell. The protein core dissolves within the cell cytoplasm. The viral enzyme reverse transcriptase synthesizes DNA using the viral RNA as a template. The new viral DNA (h; also called provirus) enters the cell nucleus (f) and incorporates itself into the cellular DNA (i). The new viral DNA becomes a blueprint for the manufacture of viral parts and the formation of multiple copies of the HIV. The viral DNA transcribes strands of viral RNA

(j) that migrate to the ribosomes (k). The ribosomes turn out the protein for the viral parts (a^3). The viral parts are integrated into complete viruses. Viral RNA strands and reverse transcriptase are enclosed within the protein core as each virus is formed.

The multiplication of viruses within the T cells does not necessarily immediately follow the infection of the T cells. Long delays (called latencies) are common between infection and the onset of early AIDS, e.g., months to many years.

Infected T cells can transmit the HIV from one person to another. Person to person transmission occurs in direct sexual contact or through inoculation with or transfusion of contaminated blood. Interpersonal transmission of the virus by other means is unlikely.

Color the subheading (l) gray. Color the titles (l^1) and (l^2) which have no corresponding structures to color. Then color the subheading (m) gray, and the related titles and illustrations (m^1) through (m^5). The titles of the diseases (m^2 through m^4) receive the same color as the causative microorganisms.

The person whose cells are infected with HIV does not necessarily have AIDS. It is believed that most, if not all, HIV-infected persons will ultimately develop AIDS. The time for that development to occur is highly variable. The onset of AIDS begins with the biochemical signal that induces the viral DNA to program viral replication. Soon the infected T lymphocytes are producing hundreds of new viruses. As these HIV exit the cells, they usually destroy their host cells. As the disintegration of T cells progress, the symptoms of AIDS develop.

In the early stages there is fever, headaches, weight loss (l^1), diarrhea (l^2), nightsweats, lymphadenopathy, and intense fatigue. The onset of these signs can be mistaken for influenza or mononucleosis. This collection of signs is AIDS-related complex (l; ARC) or "early AIDS."

Full-blown AIDS occurs when the number of T lymphocytes is so low (m^1) that the body cannot mount a defense to a variety of bacterial, fungal, and protozoan infections. These infections include protozoan *Toxoplasma gondi*-induced toxoplasmosis (m^2), probably fungal (or protozoan) *Pneumocystis carinii* pneumonia (m^3), the fungus Cryptosporidium-induced cryptosporidiosis (m^4), and a skin cancer called Kaposi's sarcoma (m^5). Any of these diseases can be fatal with AIDS in the background. AIDS often involves the nervous system, and may be associated with personality changes, dementia, senility, or intellectual incapacity. In some individuals there is a wasting syndrome where the body gradually loses tissue and fluid.

Diagnostic tests depend on the observation of AIDS antibodies in the patient's serum. Azidothymidine (AZT) can be used for therapy even though it has severe side-effects in the bone marrow. AIDS is considered to be a fatal disease.

ACQUIRED IMMUNE DEFICIENCY SYNDROME (AIDS)

HUMAN
IMMUNODEFICIENCY
VIRUS (HIV) a
ENVELOPE a^1
SPIKE a^2
RNA b

REVERSE
TRANSCRIPTASE c
PROTEIN CORE d

T CELL INFECTION ✲
T CELL MEMBRANE e

NUCLEUS f
REVERSE
TRANSCRIPTION g
NEW VIRAL DNA h

CELL
DNA i
NEW
VIRAL RNA j
RIBOSOME k
VIRAL PART a^3

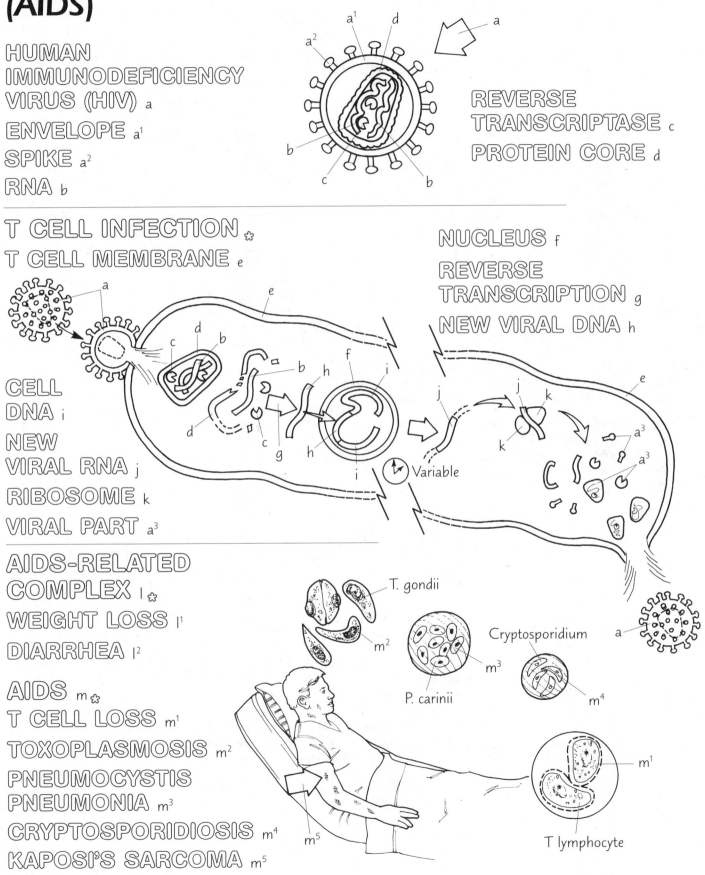

Variable

T. gondii
Cryptosporidium
m^2
P. carinii
m^3
m^4
a
m^1
T lymphocyte

AIDS-RELATED
COMPLEX l ✲
WEIGHT LOSS l^1
DIARRHEA l^2

AIDS m ✲
T CELL LOSS m^1
TOXOPLASMOSIS m^2
PNEUMOCYSTIS
PNEUMONIA m^3
CRYPTOSPORIDIOSIS m^4
KAPOSI'S SARCOMA m^5

PROTOZOAN DISEASE: SLEEPING SICKNESS

Protozoans are eukaryotic, largely unicellular, aquatic organisms that are generally microscopic in size (recall Plate 40). Of the more than 60,000 species of protozoans, about 15% of them are parasitic.

Among the protozoa that cause disease in humans is a flagellated organism belonging to the genus *Trypanosoma*. This protozoan is the agent of trypanosomiasis, more commonly known as sleeping sickness. Sleeping sickness can occur in two forms: African and South American.

The latter form, called Chagas disease (not shown), is caused by a different species of *Trypanosoma* (*T. cruzi*) than that causing the African form of the disease. Further, *T. cruzi* is transmitted by a different insect, the triatomid bug. Because this bug feeds at night and bites where the skin is thin, such as on the lips, it is often called the "kissing bug." In Chagas disease, the heart or gastrointestinal tract is most often affected by chronic infection.

In this plate we consider the African form of the disease.

Color the titles and related structures (a) through (c¹), the disease signs (a²) through (a⁸), and the related parts on the drawing of the infected victim. Use red for (c) and (c¹).

The protozoan *T. brucei* (a¹), an elongated microorganism with a single flagellum, is responsible for African trypanosomiasis (a). It has a characteristic undulating membrane on its dorsal aspect that helps propel the protozoan. *T. brucei* is found in Central Africa where the disease is both endemic and epidemic.

The protozoan is transmitted among humans by the tsetse fly (b; *Glossina palpalis*). The tsetse fly bites an infected individual (not shown), sucks the blood and acquires the protozoan. The protozoa multiply in the insect, accumulate in its salivary gland, and pass with the saliva when the insect bites (b¹) its next victim.

Once in the body, *T. brucei* multiply at the bite site, causing an inflammatory lesion (a²) in some non-indigenous people, but rarely in those permanently residing in the endemic area. The organism gains entry into the extracellular spaces under the bite site. It may then access the lymphatic and the blood vascular circulation. As the protozoan pathogens multiply by binary fission in the body fluids, the host develops a fever. Once in the lymphatics, the parasite will infect the lymph nodes, causing them to swell (a³) with phagocytes and lymphocytes. Lymph nodes in the back of the neck are particularly involved. The spleen often becomes infected, and swells (a⁴) with protozoans, phagocytes, and lymphocytes.

The immune system produces antibodies against the antigens of the protozoan, but unless the pathogens are all destroyed in the first sweep, they reproduce with new antigenic determinants on their cell membranes. Then a new array of antibodies must be formed specifically for these antigens. Ultimately, the reproduction of the protozoans overwhelms the capacity of the immune system.

After about four weeks, the protozoans leave the circulation around the brain and enter and proliferate in the cerebrospinal fluid. The neighboring meninges become infected and inflamed (a⁵; meningitis). The brain, surrounded by the meninges, becomes infected and inflamed (a⁶; encephalitis) as well. Meningo-encephalitis is accompanied by headache, stiff neck, progressive alteration in mentation, weight loss (a⁷), seizures, and coma (a⁸; hence, the name "sleeping sickness").

Without treatment, infection by *T. brucei* is almost always fatal. The mechanism of death is usually pneumonia or septicemia. Should another tsetse fly bite the individual before death occurs, the disease can be transmitted to others. The trypanosome is remarkably infectious, and medical personnel handling the blood of infected patients are at risk for infection if care is not taken.

Identification of the organism can be made with microscopic observation of properly stained infected blood smears. ELISA testing will usually identify the antigens. The antigens of *T. brucei* induce formation of both IgM and IgG antibodies in the human host.

Treatment to be effective must be administered before the protozoans infect the meninges and brain. Such treatment consists of suramin delivered intravenously The variation of antigens makes it difficult to develop a vaccine that will provide a reasonable degree of immunity against the infecting protozoan.

Historically, sleeping sickness had always been a deterrent to extensive colonization of Africa (and South America). In the late 1800s, British researcher David Bruce (for whom the protozoan is named) identified the tsetse fly as the transmitting agent for African sleeping sickness. It then became possible to interrupt epidemics of the disease and encourage British colonies in Africa. However, sleeping sickness is still widespread on the continent and continues to exact a heavy toll during epidemics.

PROTOZOAN DISEASE: SLEEPING SICKNESS

TRYPANOSOMIASIS (AFRICAN) a
TRYPANOSOMA BRUCEI a¹
TSETSE FLY b

DISEASE SIGNS: ✿

LESION a²
SWOLLEN LYMPH NODE a³
SWOLLEN SPLEEN a⁴
MENINGITIS a⁵
ENCEPHALITIS a⁶
WEIGHT LOSS a⁷
COMA a⁸

BITE / INFECTION b¹

BLOOD VESSEL c
RBC c¹

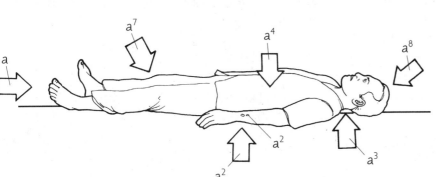

PROTOZOAN DISEASE: TOXOPLASMOSIS

Toxoplasmosis is caused by a species of protozoan known as *Toxoplasma gondii*, belonging to the class Sporozoa. It is a crescent-shaped organism without visible means of locomotion (no cilia, flagella, or pseudopodia). This agent of disease infects all mammals, and in humans can be ingested with raw or undercooked meat. It exists widely in nature, but its host in its infective stage is limited to the cat family, including domestic cats. The cat is the reservoir for this pathogen. The symptoms of the disease in the human are not often apparent and a significant disease state is unusual. The protozoa can cross the placental barrier in infected pregnant women, transmitting the disease to the fetus. Serious congenital damage can occur to the fetus in such cases.

Color the titles, related structures, and disease signs (a) through (l), beginning at station 1. Continue coloring through station 6. Use distinct shades for (a) through (a³).

Wild as well as domestic cats become infected with *T. gondii* by eating infected small animals, such as rats or mice. These rodents become infected by contact with materials contaminated with cat feces. The rat shown here carries the pathogenic protozoan in its asexual or pseudocyst (a) stage. The pseudocyst, filled with infectious forms of *T. gondii*, undergoes development in the cat's digestive tract (station 1). Such development includes the formation of oocysts (a¹) which are subsequently excreted with the feces. Each oocyst contains smaller sporocysts; and each sporocyst contains infective sexual forms of *T. gondii* called sporozoites (a²).

Field grasses contaminated with oocyst-containing cat feces can be consumed by cattle destined for human consumption. Eating raw or uncooked meat transfers the protozoan pseudocysts to humans. Another means of transmission from cat to human is by handling cat litter or other cat feces-contaminated material, and ingesting the oocysts with other food (c; station 2).

Once in the human intestinal tract, the walls of the oocyst are enzymatically digested, releasing the sporozoites into the lumen of the intestine (intestinal route, d; station 3). The sporozoites enter the intestinal lining cells (not shown), and are converted there into tachyzoites (a³). The tachyzoites migrate into the blood vessels near the lining cells of the colon (hematogenous route, e; station 4). Sweeping through the circulation, the tachyzoites rapidly multiply and infect, and often kill, cells throughout the body. The body responds to the infection with inflammation and immune responses.

Acquired infection with *T. gondii* (acquired toxoplasmosis) generally produces few signs and symptoms (station 5). However, there may be fever (f) and malaise, and the liver (g), spleen (h), and lymph nodes (i) may be inflamed, resulting in their mild enlargement (hepatomegaly, splenomegaly, adenopathy). Diagnosis generally consists of identifying the organism in tissue samples. Sulfonamide drugs hasten recovery.

The potential significance of this infection is in pregnant women and particularly the fetus (congenital toxoplasmosis; station 6). The mother may develop symptoms of toxoplasmosis that are mild (as above). The fetus may become infected by parasites that pass across the placental barrier (j) from the mother. In about 90% of cases of fetal infection, there are no significant signs or symptoms. In about 10% of cases, the fetus will experience mild to severe organ trauma from the parasites, especially during the second trimester of gestation (third through sixth month). In the worst of cases, the newborn will exhibit significant neurological damage (retardation, motor deficits). The characteristic lesions in these cases are calcifications of the cerebrum (k) where chronic inflammation has developed around the parasites, and retinochoroiditis (l) where chronic inflammation of the retina and choroid have resulted in necrotic lesions that progressively destroy sight.

In recent years, toxoplasmosis has been found in persons whose immune systems are suppressed. Persons receiving immune suppressing therapy to prevent organ rejection after transplant are at risk. Also at risk are individuals who have AIDS. In these people, toxoplasmosis represents an opportunistic infection that can cause cerebral lesions, seizures, and death.

PROTOZOAN DISEASE: TOXOPLASMOSIS

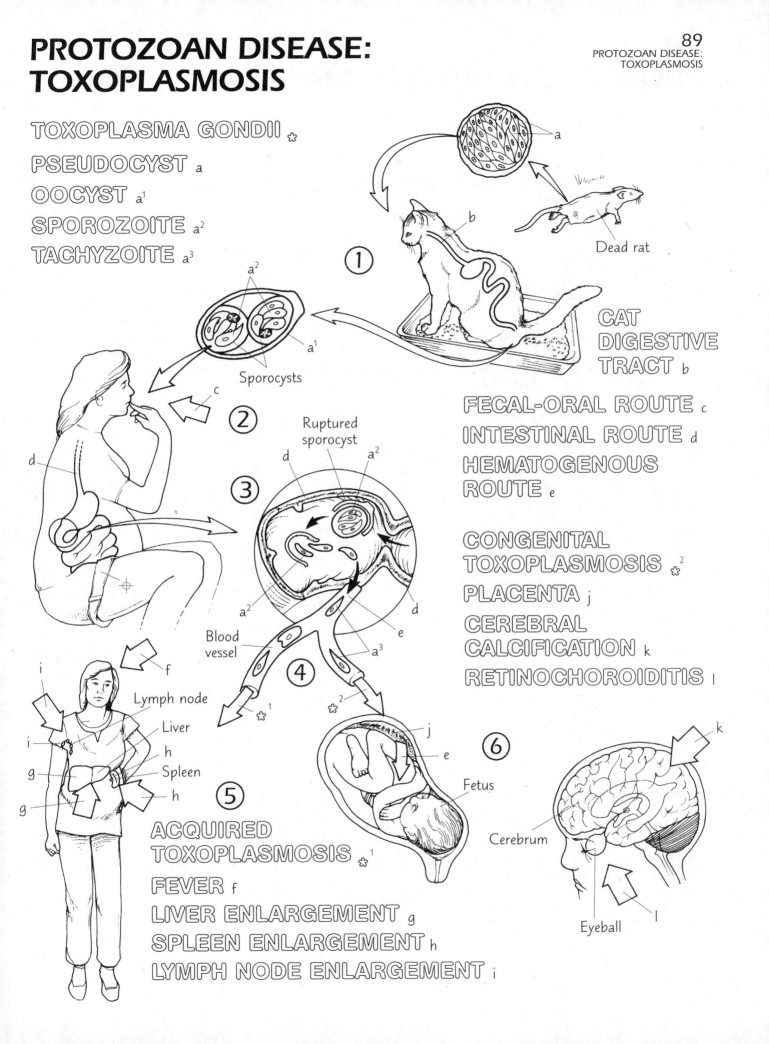

TOXOPLASMA GONDII ✿
PSEUDOCYST a
OOCYST a¹
SPOROZOITE a²
TACHYZOITE a³

CAT DIGESTIVE TRACT b

FECAL-ORAL ROUTE c
INTESTINAL ROUTE d
HEMATOGENOUS ROUTE e

CONGENITAL TOXOPLASMOSIS ✿²
PLACENTA j
CEREBRAL CALCIFICATION k
RETINOCHOROIDITIS l

ACQUIRED TOXOPLASMOSIS ✿¹
FEVER f
LIVER ENLARGEMENT g
SPLEEN ENLARGEMENT h
LYMPH NODE ENLARGEMENT i

Dead rat

Sporocysts

Ruptured sporocyst

Blood vessel

Lymph node
Liver
Spleen

Fetus

Cerebrum
Eyeball

90
PROTOZOAN DISEASE: MALARIA

Each year, more than 250 million humans suffer the debilitating effects of malaria. Public health officials consider malaria to be the most widespread infectious disease in the world and one of the most important global health problems. Malaria is caused by any of four species of *Plasmodium* (class Sporozoa). The life cycle is rather complex and involves both the mosquito and the human. The mosquito is a species of *Anopheles,* the only genus of mosquitoes that transmits malaria.

Color the subheadings Malaria: Plasmodium Species, and Human (Asexual) Cycle gray, and the related arrow ($*^1$). Then color the titles and related structures from (a) through (f^1) in the asexual or human cycle, following the arrows. Do not color the second infection yet. Be sure to use colors that do not obscure the details of the illustrations.

The human or asexual cycle of *Plasmodium* begins as the proboscis of the *Plasmodium*-infected mosquito penetrates the skin of an uninfected human (a; station 1). Saliva from the mosquito's salivary gland flows into the wound, and the sporozoites (b) of *Plasmodium* enter the tissue through the wound. The sporozoites, motile forms of the protozoan pathogen, enter the blood and migrate to the liver. In the liver cells (station 2), the sporozoites reproduce by asexual division (b^1), forming merozoites (c). Merozoites emerge from the liver into the circulation, and penetrate the red blood corpuscles (RBCs; station 3).

Within the corpuscles, the merozoites take on a ring-shaped form and are called trophozoites (d). The ring form takes only a portion of the space in the corpuscle (enlarged here for coloring). The early ring form trophozoite develops into a larger late ring form (d^1; station 4) which takes up almost all of the corpuscle, deforming it. While consuming the hemoglobin in the red blood corpuscles, the trophozoites divide asexually to become a mass of ameboid merozoites (c^1). These break out of the red blood corpuscles into the circulation to infect new corpuscles. In the process, many of the RBCs are destroyed. A significant number of these ameboid forms are destroyed by antibodies and other elements associated with the immune response (e; station 5). Some surviving merozoites go on to divide asexually, while others differentiate into male (f) and female (f^1) gametes. These gametes, enveloped in corpuscular membranes, continue development in the stomach and intestines of the mosquito.

Color the subheading Mosquito (Sexual) Cycle, and related arrow ($*^2$), the title and arrow (a^1), and the titles and related structures (g) through (i), as well as (b), (f), and (f^1), in the lower third of the plate.

While sucking the blood from the *Plasmodium*-infected human (station 6), the mosquito acquires the *Plasmodium* gametes enveloped in the membranes of the RBCs. In the mosquito's stomach, the gametes are freed from their envelopes. The male gametes develop flagella (station 7) and fuse with the female gametes to form fertilized eggs called zygotes (g). The nucleus of each zygote undergoes multiple reproductions; the zygote transforms and attaches to the gut wall as an oocyst (h). Masses of sporozoites are formed that burst from the oocyst to enter the mosquito's circulation (station 8). These sporozoites migrate to the tissues of the mosquito and some accumulate in its salivary gland. The mosquito is now capable of transmitting the parasite with the next skin penetration by its proboscis.

The symptoms and signs of malaria in humans begin with the rupture of infected red blood corpuscles. Anemia is a common finding. The release of toxic substances induce fever, sweating, and chills. Episodic signs and symptoms, including fever, sweating, chills, muscle pain, and headache, are characteristic. Each episode generally lasts for several hours and then remits. When the disease is untreated, these attacks continue periodically until the body's immune system can destroy the pathogens. This may take years, depending on the species of *Plasmodium.*

In some cases with children, the early attacks may be lethal. Diagnosis often depends on direct observation of blood smears and the examination of parasites. Where sophisticated laboratories are available, diagnosis may be performed with ELISA tests, fluorescent antibody tests, gene probe tests, or high technology procedures.

Quinine has been used for centuries to kill *Plasmodium.* Scientists have also synthesized derivatives of quinine such as chloroquine and primaquine. The effectiveness of these preparations may require their administration for a number of years. Worldwide efforts to control malaria have revolved around drug use and mosquito control. A vaccine is currently being developed.

PROTOZOAN DISEASE: MALARIA

MALARIA: PLASMODIUM SPECIES ✿
HUMAN (ASEXUAL) CYCLE ✿[1]

MOSQUITO / INFECTION a
SPOROZOITE b
ASEXUAL DIVISION b[1]
MEROZOITE c
RING TROPHOZOITE d
LATE TROPHOZOITE d[1]
AMEBOID
MEROZOITE c[1]
IMMUNE
RESPONSE e
MALE GAMETE f
FEMALE
GAMETE f[1]

SECOND
INFECTION a[1]

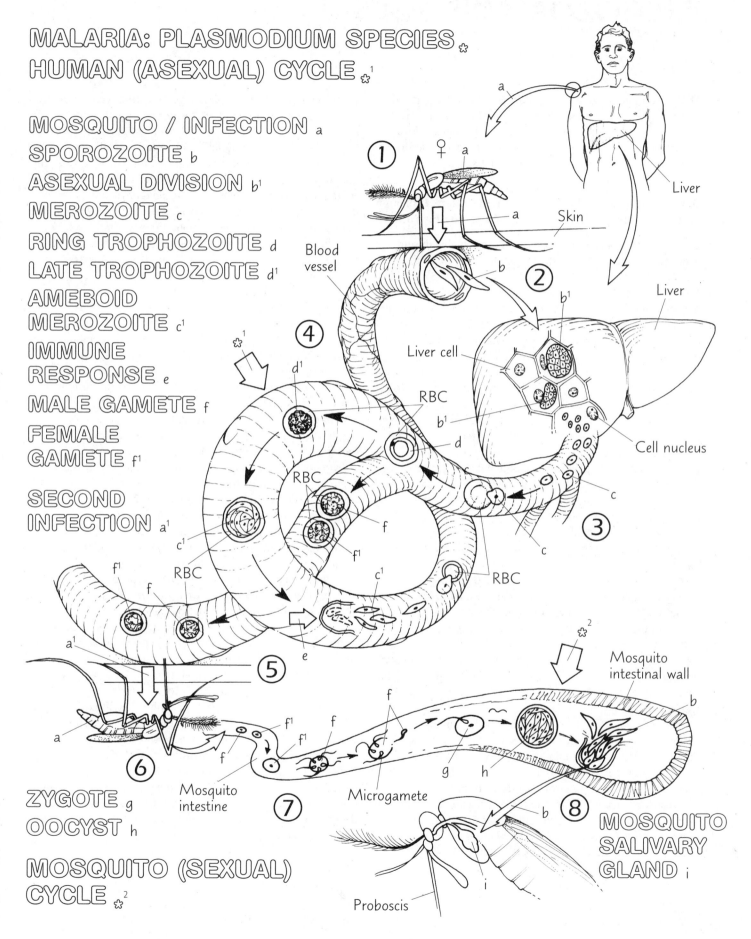

ZYGOTE g
OOCYST h

MOSQUITO (SEXUAL)
CYCLE ✿[2]

PROTOZOAN DISEASES

There are many diseases of protozoan origin that can affect humans. We have considered a number of significant pathogenic protozoans in the preceding plates. Here we illustrate six more disease-producing protozoans and briefly review the characteristics of the related disease states.

Color the titles and related organisms at stations 1 through 6, as well as the related arrows in the central illustrations.

Amebiasis (a; station 1; known also as amebic dysentery) is primarily an infection of the lower intestinal tract (a; colon, rectum). The pathogen that causes this disease is *Entamoeba histolytica* (a^1), an ameboid protozoan. The organism is ingested in contaminated water and food and passes through the stomach and duodenum in a cyst form. By the time it reaches the large intestine, the cyst wall is ruptured and the infectious trophozoites (a^1) emerge. They enter and disrupt the lining cells of the colon and rectum (a), inducing an inflammatory response. In some cases, these protozoans migrate into the blood supply and localize in the liver (a; hepatic amebiasis). Untreated, the intestinal lining cells may be destroyed, and the wall of the intestine can be eroded (ulceration) to the degree of perforation. Amebiasis is characterized by sharp (colicky) abdominal pain, bloody stools, and high fever. Protozoa eliminated in the feces can spread the disease. The diagnosis of amebiasis is made by identifying cysts or trophozoites in the stools.

Giardia lamblia (b^1) is the cause of an intestinal disease known as giardiasis (b). It is a multiflagellated organism with large paired nuclei. The reservoir for this pathogen includes dogs and some wild animals. The protozoan enters the body in feces-contaminated food and water. *G. lamblia* multiplies in the cells lining the colon, disrupting the absorption of water and inducing malabsorption. The characteristic sign of giardiasis is chronic, sometimes periodic, diarrhea, accompanied by anorexia, loss of weight, and debilitation. Gastric cramps are also symptoms of the disease. Diagnosis of the disease can often be made by observation of cysts or trophozoites in stool samples.

Leishmaniasis (c) is prevalent in tropical, subtropical, and certain temperate zones around the world (Africa, India, South America, Asia). Three major expressions of the disease have been determined based on the site of infection (skin, skin-mucous membrane junctions, and lymphoid organs). The organism responsible for liver and spleen disease (the lymphoid or visceral form of leishmaniasis, also called kala azar) is a complex protozoan named *Leishmania donovani*. The initial infective form of this protozoan is the promastigote (c^1; characterized by an elongated, flagellated form with distinctive nucleus), and is transmitted to humans by the bite of the *Leishmania*-infected sandfly. The promastigotes are taken up by phagocytes, and are quickly converted to oval forms called amastigotes (c^1). The amastigotes multiply within the phagocytes, ultimately destroying the host cells. The cellular fragments and the amastigotes are phagocytosed again, and the cycle repeats. The infected phagocytes eventually reach the lymphoid organs (spleen, bone marrow) and liver. Often associated with some other symptomatic illness, visceral leishmaniasis is accompanied by fever, malaise, and hepatomegaly, splenomegaly, and adenopathy.

Balantidium coli (d^1) is an oval, ciliated organism with a large kidney-shaped nucleus. The protozoan is worldwide, but the disease, balantidiasis (d) seems limited to certain tropical areas. Resident in numerous wild and domestic animals, the protozoan is acquired by humans from feces-contaminated food and water. It multiplies in the human intestinal tract (d) and causes extensive diarrhea and some ulceration. The pathogen is identified by examination of the stools. Balantidiasis is uncommon in the United States.

Trichomoniasis (e) is the only sexually transmitted disease caused by a protozoan. The responsible organism is *Trichomonas vaginalis* (e^1), a multiflagellated protozoan. The organism resides in the urogenital tract of *Trichomonas*-infected males and is transmitted to the female during sexual intercourse. The organism thrives in the reproductive tract, infecting the lining cells of the male urogenital tract and those of the female reproductive tract. Symptoms of the disease include pelvic discomfort, painful urination, and a thick vaginal discharge. The diagnosis is made by microscopic observation of the protozoa in samples of urine or vaginal discharge. Treatment is effective with metronidazole.

Pneumocystis carinii (f^1) is believed to be a complex protozoan possibly belonging to the class Sporozoa. There is some biochemical evidence that the organism may be a fungus, but, at present, the prevailing view is that it is a protozoan. *Pneumocystis* exists in three forms: cyst (not shown), sporozoite (not shown), and trophozoite. The life cycle of the microorganism is not yet known. It is commonly found in the human lung without consequence. In persons with AIDS, however, the protozoan multiplies rapidly in the air cells (alveoli) of the lung, causing a life-threatening pneumonia called pneumocystosis or *Pneumocystis carinii* pneumonia (PCP). The drug pentamidine isethionate can be used to limit the growth of the protozoa. Pneumocystosis is a common cause of death in AIDS patients.

PROTOZOAN DISEASES

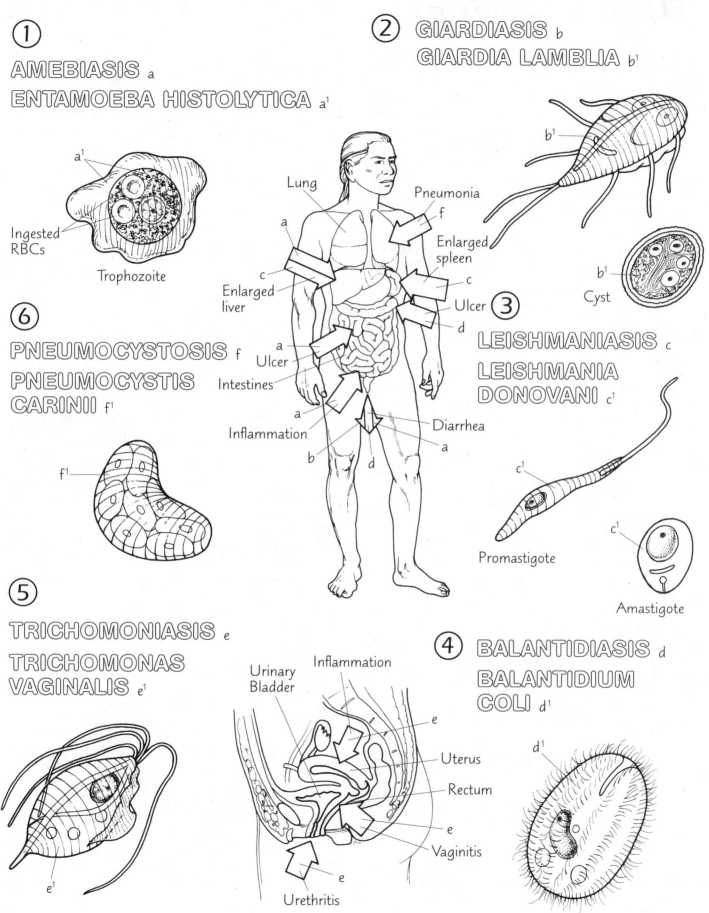

① AMEBIASIS a
ENTAMOEBA HISTOLYTICA a¹

a¹

Ingested
RBCs

Trophozoite

② GIARDIASIS b
GIARDIA LAMBLIA b¹

b¹

b¹ Cyst

⑥ PNEUMOCYSTOSIS f
PNEUMOCYSTIS
CARINII f¹

f¹

③ LEISHMANIASIS c
LEISHMANIA
DONOVANI c¹

c¹

Promastigote

c¹ Amastigote

⑤ TRICHOMONIASIS e
TRICHOMONAS
VAGINALIS e¹

e¹

④ BALANTIDIASIS d
BALANTIDIUM
COLI d¹

d¹

Lung
Pneumonia
f
a
Enlarged
spleen
c
c
Ulcer
d
Enlarged
liver
a
Ulcer
Intestines
a
Inflammation
Diarrhea
a
b
d

Urinary
Bladder
Inflammation
e
Uterus
Rectum
e
Vaginitis
e
Urethritis

Although most fungi pose no threat to humans, several pathogenic species exist. Fungal diseases are known as mycoses. Fungi may infect skin or mucosal surfaces (causing superficial mycoses) or they may enter the body through wounds or by inhalation and localize in deeper tissues (deep mycoses). Some potentially pathogenic fungi reside in or on the body as members of the normal flora without producing symptoms. Once the fungal growth is particularly dense, or if the person is immunocompromised, the infection may be accompanied by signs and symptoms.

Start at station 1, coloring the titles and related structures at that station. Then continue with stations 2 through 4.

Cryptococcus neoformans (a^1) is a yeastlike fungus that reproduces by budding (station 1). The fungus is usually found in the soil and is inhaled into the human respiratory tract. Infection by this organism is called cryptococcosis (a). Localized in the lungs (a^2), the fungus may cause a mild pneumonia. The fungus may enter the circulation and infect the brain and meninges (a^3), causing meningoencephalitis. This condition often has a rapid onset, and is accompanied by cerebral swelling (edema). Mental alteration, motor deficits, coma, and death are the possible sequelae. Immunosuppression encourages fungal proliferation in the body, especially in persons with AIDS where cryptococcosis is commonly an opportunistic infection. Blood and cerebrospinal fluid (CSF) cultures generally demonstrate the pathogen. Chemotherapy (with amphotericin B and/or itraconazole) may prevent the disease's progression.

Candida albicans (b^1; station 2) appears in both yeast and hyphal forms. Here are shown budding cells (blastospores, chlamydiospores) arising from chains of elongated cells that look like hyphae (pseudohyphae). Blastospores arise at the junctions of the pseudohyphae (nodes); chlamydiospores arise at the tips of the terminal pseudohyphae. Although *Candida albicans* exists as part of the normal flora on the skin and mucosal surfaces, it can cause both superficial and deep infections. Human-to-human contact is the usual vehicle of transmission. Candidiasis (b), the name given to *Candida* infections, is extremely common and can take many forms. One form is a superficial mycosis called thrush. Thrush appears as milkwhite, "cheesy" flecks or plaques on the gums and mucosal surfaces of the oral cavity. Another form occurs in the female reproductive tract where it is accompanied by burning, internal pain, and a white "cheesy" discharge (vulvovaginitis).

Candida can be transmitted between humans during sexual contact. Deep mycoses (systemic infections) can develop by way of the hematogenous route, involving the lungs, liver, spleen, and other vascular organs. The immune system can usually control growth of the normal *Candida* flora, but immunosuppressed persons such as those with AIDS commonly exhibit candidiasis as an opportunistic infection. Identification of the fungal infection requires isolation and culture of the fungus from infected tissues. Nystatin, ketoconazole, clotrimazole, and itraconazole are effective against *Candida albicans*. Azole suppositories are effective in vulvovaginitis candidiasis.

Species of *Trichophyton* (c^1; station 3) and several other fungi grow on the skin causing an infection called dermatophytosis. *Trichophyton* occurs in a filamentous, branching hyphal form (c^1). Asexual spores, both large (macroconidia) and small (microconidia), develop at the ends of the hyphae. Trichophytan species grow in the hairs and the keratin layer of skin. Fragments of the hyphae can be picked up from a towel, comb, hat, or the floor of a shower room. From such sources, the fungus attaches to the host skin. *Trichophyton* infection involves the scalp ("ringworm"), the foot (c; "athlete's foot"), the nails, and the skin in other moist body parts. The infections are sometimes known as tinea infections (tinea, worm); it was once thought that worms existed beneath the skin of infected persons. Laboratory identification for *Trichophyton* includes examination of infected skin scrapings with a potassium hydroxide preparation. Therapy for superficial mycoses include antifungal preparations. Significant skin lesions are often treated with griseofulvin.

Coccidioides immitis (d^1; station 4) is a species of imperfect fungi that grows in the soil in hyphal form (d^1). Its spores are released from sporangia and dispersed with the air currents. From these currents, the spores (d^2) can be inhaled. This is a particular risk to workers in fields or wherever soil dust is being inhaled. Once in the respiratory tract, the spores enter into the lungs' air cells. Here the spores enlarge to become oval, encapsulated structures called spherules (d^3). Endospore development occurs in each spherule. Eventually the spherule bursts, freeing more spores. Each of these form spherules, and the reproductive cycle continues. In about two weeks, chest pain and cough are experienced in less than 50% of those exposed. The disease, pulmonary coccidioidomycosis (d), is usually limited to the lungs, and may be relatively benign with influenzalike symptoms. Complications can occur in immunocompromised persons. The laboratory diagnosis can be made with the complement fixation test.

Two other fungal diseases (not shown) deserve mention. Species of *Aspergillus* consist of spore-bearing hyphae. These spores are carried by wind currents. Entering the ear, they cause an infection called otomycosis; entering the lungs by way of the respiratory tract, they cause an infection called aspergillosis. One species of *Aspergillus* infects food products and produces toxins called aflatoxins. Once ingested, the toxins can cause severe liver damage.

Sporothrix schenkii is also a spore-bearing fungus in the soil. Entering through breaks in the skin, as with rose thorn punctures, these fungi cause purple skin lesions at the site of the skin puncture (sporotrichosis).

FUNGAL DISEASES

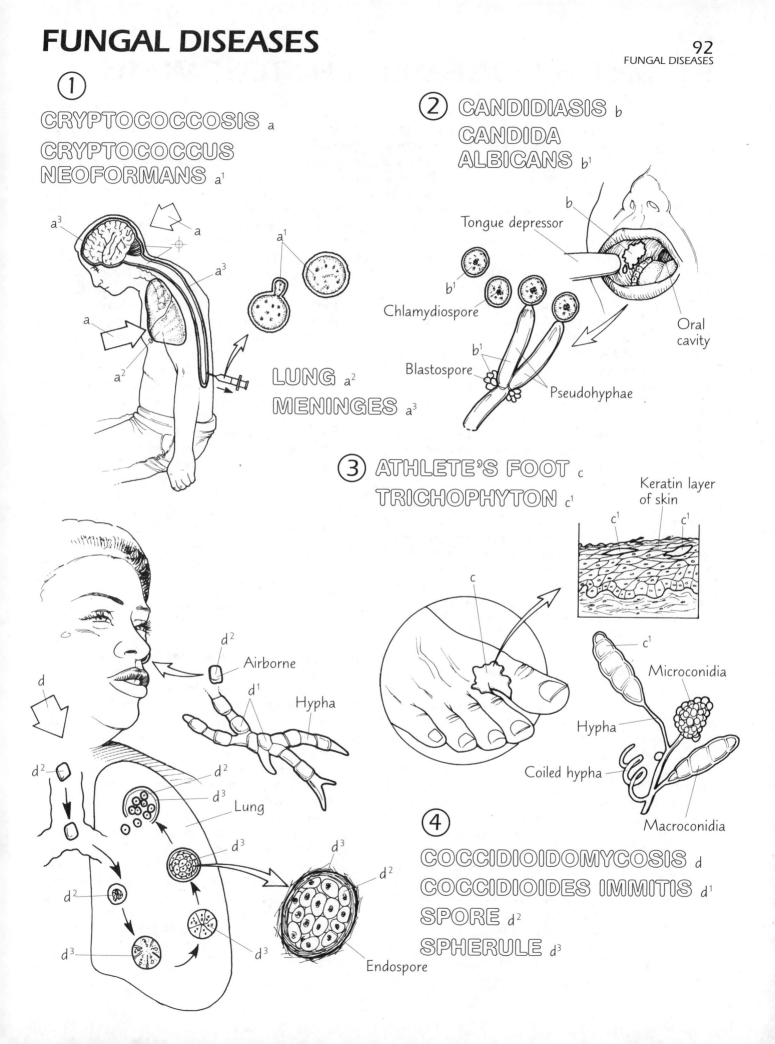

① CRYPTOCOCCOSIS a
CRYPTOCOCCUS NEOFORMANS a¹

LUNG a²
MENINGES a³

② CANDIDIASIS b
CANDIDA ALBICANS b¹

Tongue depressor
Chlamydiospore
Blastospore
Pseudohyphae
Oral cavity

③ ATHLETE'S FOOT c
TRICHOPHYTON c¹

Keratin layer of skin
Microconidia
Hypha
Coiled hypha
Macroconidia

Airborne
Hypha
Lung
Endospore

④ COCCIDIOIDOMYCOSIS d
COCCIDIOIDES IMMITIS d¹
SPORE d²
SPHERULE d³

HELMINTHIC DISEASE: SCHISTOSOMIASIS

Humans can be infected by parasitic worms (helminths). Helminthic parasites include the trematodes, or flukes (Plates 93, 94, 95), the cestodes, or tapeworms (Plates 95, 96), and the nematodes, or roundworms (Plates 97, 98). The flukes and tapeworms are flatworms classified in the phylum Platyhelminthes. Roundworms are members of the phylum Aschelminthes (also called Nematoda). Here we consider the parasitic blood fluke *Schistosoma mansoni*.

Schistosoma mansoni is one of three pathogenic species of the genus *Schistosoma*. These species of flukes are region-specific, limited by the species of snail that acts as a host for the transmission of the parasite. *S. mansoni* is distributed throughout Africa and South America. *S. japonicum* occurs mainly in the Far East. *S. haematobium* is found in Africa. It is estimated that hundreds of millions of people are infected by one of these species, suffering from a disease called schistosomiasis (also called bilharziasis in some regions of the world). Certain species of *Schistosoma* penetrate no farther than the skin. In these cases, young schistosomes cause dermatitis in the skin and a condition commonly called swimmer's itch. The body's immune system keeps the parasites in check, but the worms release allergenic chemicals that cause the itching.

Color the title *Schistosoma mansoni* (blood fluke), and the copulating flukes (a) at upper left. Begin with station 1. We enter the cycle there with the fluke eggs-contaminated human feces (m). Note that we leave the cycle at station 8 with the arrow (m) representing the discharge of egg-contaminated feces. Color the titles and related parts at each station. Note that the titles are arranged clockwise from upper right around to upper left. Color them in this order.

S. mansoni flukes (a) are elongated, structurally complex worms with suckers at one end. Unlike many other species of this genus, they occur as male and female rather than as hermaphrodites. The adult worms exist connected in a copulatory relationship throughout their life. The female produces eggs continuously, and the male fertilizes them. The worms have digestive and excretory systems that permit them to feed on cells of their host.

The hosts for the blood flukes' life cycle are the snail and the human. The fertilized eggs (b) of *S. mansoni* are contained in the feces of an infected human. In rural agrarian environments, human feces are often deposited in fresh water environments (station 1). The eggs soon hatch, releasing the cases or "shells," and the resultant forms are called miracidia (c; station 2). The miracidia are subsequently taken up by the snail (d; *Biomphalaria)*. The asexual phase of the fluke's life cycle takes place in the snail (station 3).

Within the digestive system of the snail, miracidia develop into sporocysts (e). The sporocysts divide and proliferate, each developing into a cercaria (f; station 4). Cercariae have tadpolelike forms with head pieces connected to bodies with forked tails. Greatly multiplied in number within the snail, the cercariae leave their host and enter the water. Swimming on the water's surface, the cercariae contact the skin of a human standing or sitting in the water. In a short time, the cercariae find hair follicles on the skin surface and enter them, losing their tails in the process. Each altered cercaria becomes a schistosomulum (a^1). It penetrates the connective tissue and finds a vein (station 5). Entering the vein (g), the schistosomulum flows with the currents into and through the heart (h; station 6), into the arterial circulation (i) and into the liver (j) by way of the hepatic artery. It is in the human host that the fluke completes the sexual phase of its life cycle.

The schistosomula never leave the circulation while in the liver; they mature and reproduce in the small veins there, and become adult schistosomes (a; one pair of copulating worms is shown in the liver, and is greatly enlarged for coloring). The schistosomes migrate from the small veins in the liver, against the flow, into the large portal vein and its tributaries draining the intestines (k; station 7). Here, in the small tributaries of the superior and inferior mesenteric veins, the blood flukes live. Eggs are produced in these veins and are swept away to be taken up into other organs, including the liver.

The adult flukes feed on hemoglobin from the red blood corpuscles (not shown). It is the immune and inflammatory reaction of the body's organs and tissues in response to the deposition of eggs that creates the signs and symptoms of the disease in humans.

Many eggs will reach the intestines (l) where they will be discharged with the feces (m; station 8), completing the sexual cycle.

While infecting the circulation of the human host, the parasites cause chills, fever, and sweating to occur. The liver, spleen, and lymph nodes generally enlarge. With dense infections, the population of eggs throughout the body induces chronic inflammatory changes and fibrous reactions surrounding the eggs. These reactions eventually cause some degree of obstruction in blood flow, killing of cells, and general scarring in the organs infected. The liver can become badly obstructed. The brain and spinal cord can be seriously affected, causing severe encephalitis and myelitis. The urinary tract can become obstructed due to fibrosis and thickening of tissues around egg deposits.

The diagnosis of schistosomiasis is confirmed by the identification of the eggs of *Schistosoma* in the microscopic examination of the victim's stools or urine. Biopsies of lower intestinal tract mucosae may reveal the pathogens or their eggs. The ELISA test is also effective in determining the presence of the flukes. Treatment against live worms can be effected by praziquantel. Once the fibrotic conditions and obstructions have set in, no treatment other than specific surgical responses are likely to be effective.

HELMINTHIC DISEASE: SCHISTOSOMIASIS

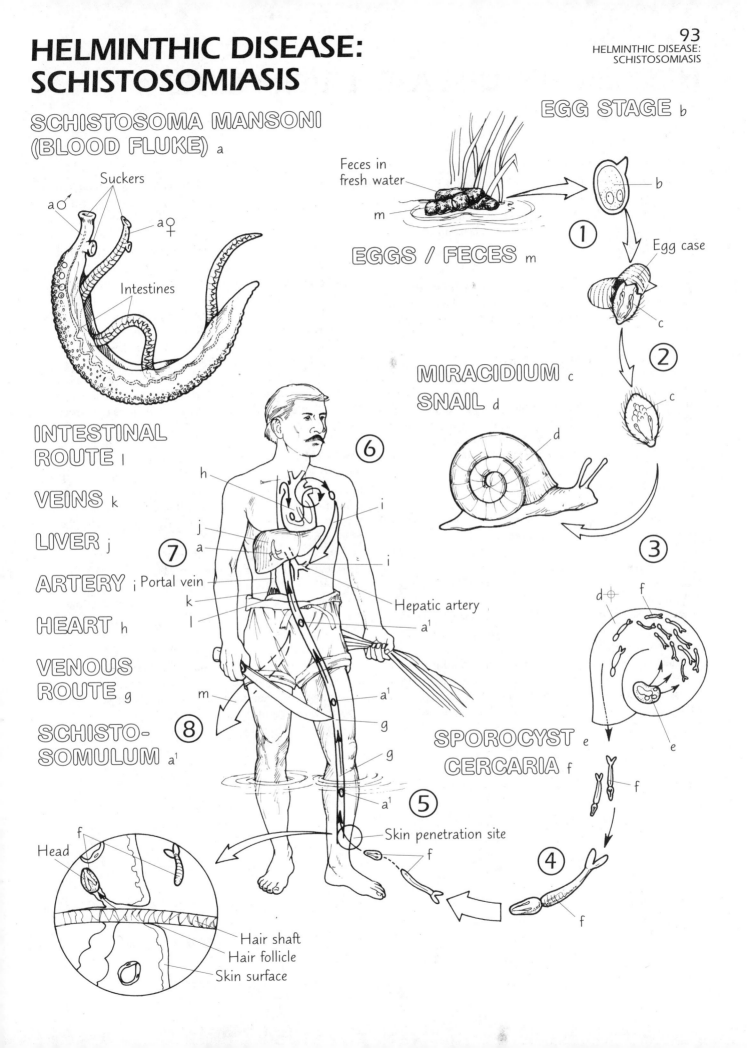

SCHISTOSOMA MANSONI
(BLOOD FLUKE) a

Suckers

a ♂

a ♀

Intestines

EGG STAGE b

Feces in
fresh water

m

EGGS / FECES m

b

Egg case

c

①

②

MIRACIDIUM c
SNAIL d

d

c

c

INTESTINAL
ROUTE l

VEINS k

LIVER j

ARTERY i

HEART h

VENOUS
ROUTE g

SCHISTO-
SOMULUM a¹

h

j

a

k

l

Portal vein

Hepatic artery

a¹

a¹

g

g

a¹

m

⑥

⑦

⑧

⑤

i

i

③

SPOROCYST e
CERCARIA f

d

f

e

f

④

Skin penetration site

f

f

Head

f

Hair shaft
Hair follicle
Skin surface

HELMINTHIC DISEASE: LIVER FLUKE

Among the flatworms that inhabit humans is a flat, leaflike parasite known as *Clonorchis sinensis* (Chinese liver fluke). Like the blood fluke (Plate 93), this parasite inhabits fresh water, but its entry and mechanism of disease in the human body is somewhat different.

Color the subheading Chinese Liver Fluke Infection gray, and the related organism and its title (a). Color the titles and parts associated with station 1, and continue through station 6.

The Chinese liver fluke (a) infects the liver and is common in many regions of Asia, especially in China, Korea, Japan, and southern Asia. The fluke is flat and broad, with two suckers and a body structure similar to the intestinal fluke. Human feces, contaminated with fluke eggs (b), are deposited in a freshwater environment (station 1). Within the egg case (b), the egg develops into a ciliated larval form called a miracidium (c; station 2). The miracidium can remain alive only a few weeks in its egg case. For the life cycle to continue, the miracidium and its case must be eaten by a snail (d).

C. sinensis undergoes several transformations in the snail (station 3). First, the miracidium emerges from the egg case as a mature miracidium. This takes place in the snail's digestive tract. The miracidium then burrows into the tissues of the snail. Here it transforms itself into a sporelike form called a sporocyst (e). Within the sporocyst, numerous new larvae called rediae (f) develop. Soon the rediae mature and form new larvae called cercariae (g). The cercariae mature and leave the snail through its digestive tract and enter the water.

Each cercaria swims about in the water until it locates a fish belonging to one of several species. Once attached to the skin of a fish, the cercaria burrows deeper into the body wall of the fish and lodges in the muscle. Now it forms a cyst around itself in the fish muscle (h; station 4). Within the cyst, the cercaria becomes a new form, the metacercaria (g^1). The fish is thus a second intermediate host for the parasite which can remain in this state indefinitely.

C. sinensis will complete its life cycle by returning to its human host. This takes place when the fish is eaten (i) in a raw or inadequately cooked state by the human (station 5). Ingested metacercariae pass into the stomach to reach the duodenum where they detach from their cyst (station 6). The immature flukes migrate (j) into and inhabit the common bile duct and the pancreatic duct (not shown). The flukes mature here, reproduce, and lay eggs in the lining cells of these ducts. Some of the eggs leave the ducts (not shown), and following the flow of bile, enter the small intestine (duodenum). They are discharged from the body with the feces (not shown).

The effect of *Clonorchis sinensis* infection in humans can be substantial. Liver, biliary duct, and pancreatic damage (pancreatitis) can occur secondary to egg accumulation in the ducts. The blocking of the bile ducts (cholangitis) causes a backup of bile into the liver. This can cause pressure damage to the liver cells where the bile is produced.

Diagnosis of liver fluke infection can be made by gross and/or microscopic inspection of the stools. The drug praziquantel is effective in killing the fluke and ending the infectious part of the disease. In very dense infections, there may be permanent biliary or liver damage from fibrosis or scarring, and even death.

To preclude the possibility of infection by *Clonorchis sinensis,* public health officials recommend that fish be heated to a minimum of 50° C for 15 minutes before consuming. This temperature and time will destroy any cysts that may be in the fish muscle.

HELMINTHIC DISEASE: LIVER FLUKE

CHINESE LIVER
FLUKE INFECTION ⚹
CLONORCHIS SINENSIS a

① EGG / EGG CASE b

Feces
Fresh water

INGESTION i

FLUKE
MIGRATION j

⑤

Liver

j
i

a
j
Stomach

Gall
bladder
Duodenum
Common bile duct

⑥

Torn cyst

② c
b

SNAIL d
d

MIRACIDIUM c

ENCYSTED
METACERCARIA g¹

③

SPOROCYST e
REDIA f

d
e
g
f
e
g
g
c
b

④

INFECTED
MUSCLE TISSUE h

CERCARIA g

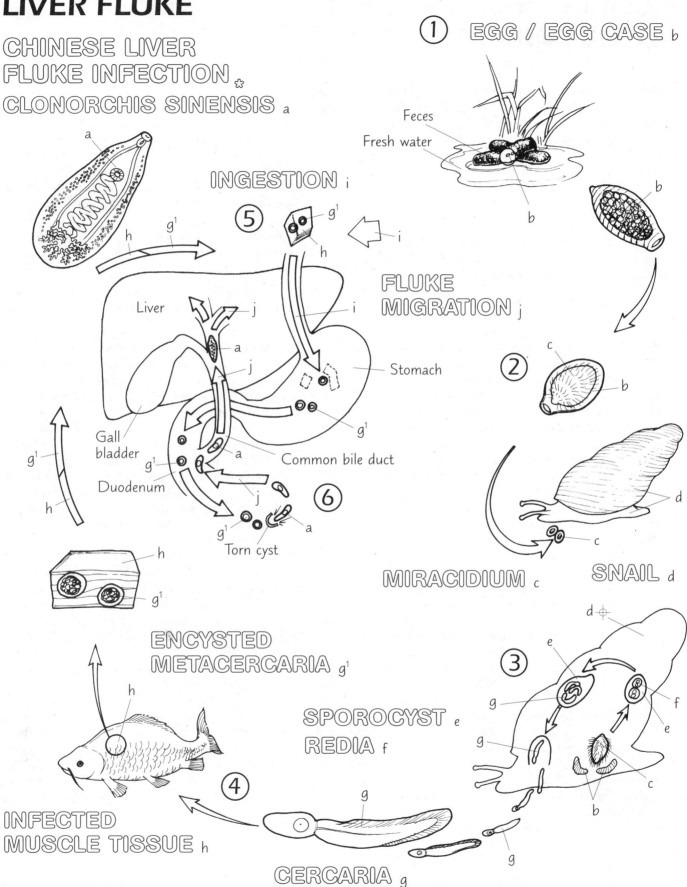

HELMINTHIC DISEASES: FLUKE AND TAPEWORM

Flukes and tapeworms both belong to the phylum Platyhelminthes (flatworms). There are significant differences between them. Flukes (trematodes) have a digestive tract; tapeworms (cestodes) do not. Flukes have a single body; tapeworms consist of detachable, recurring units called proglottids. Flukes require snails as intermediate hosts; tapeworms do not. In this plate, we finish with the flukes, and introduce the tapeworms.

Color the subheading Fluke Infections gray. Then color the titles and related organisms (a¹) through (c¹), using light colors. Color the arrows directed to the human body.

Fasciolopsis buski (a¹) has the typical shape of a fluke, with a broad, leaflike appearance. It is often called the intestinal fluke, since it is responsible for an intestinal disease (a). *F. buski* is a large fluke. Its eggs are found in the feces of its host. Deposited with feces in fresh water, the eggs pass through miracidium, sporocyst, redia, and cercaria stages. Certain species of snails are its intermediate host. Metacercariae form cysts in water plants. Ingestion of these plants passes the organism to the human host. The fluke lives in the small intestine for up to six months. During this time, it causes inflammation, ulcers, and abscesses on the intestinal mucosa, often resulting in bowel obstruction and/or malabsorption syndrome. The worms are identified in the stools of infected persons. Treatment for fluke infection consists of the oral medications praziquantel and niclosamide. A few seconds of boiling infected plants kills the metacercariae and prevents transmission of the disease.

Paragonimus westermani (b¹) is the fluke responsible for lung fluke disease (b; paragonimiasis) in humans and carnivores. Species of *Paragonimus* are found in Asia, Africa, and Central and South America. Eggs of these worms are excreted in human feces. Cercariae form in the fresh water snails that ingest the eggs. Crabs and other crustaceans eat the infected snails. Metacercariae encyst in the crabs which are a common human food. If undercooked, the crabs transfer the organism to the human. The larvae pass into the human duodenum where they lose their coverings. By way of the circulation, they migrate to the lung, where they mature and cause difficult breathing and chronic, often bloody cough. The eggs of the lung flukes are coughed up, swallowed, and excreted in the feces. The disease is diagnosed by complement fixation testing; it can be confirmed by the ELISA test.

The fluke responsible for liver infection (c) is *Fasciola hepatica* (c¹). *F. hepatica* resides in sheep and cattle. Eggs

reach the soil in feces of these animals, and the life cycle continues in snails that consume the eggs. Eventually the parasites are excreted from the snails, and become encysted metacercariae on water plants. Human ingestion of the infected plant (e.g., wild watercress) brings the parasites to the intestinal wall where they lose their cyst casings, enter the blood, and pass to the liver, gall bladder, and bile ducts. Tissue damage in the liver (inflammation, scarring, obstruction of ducts and vessels) can be substantial (liver rot) especially if the number of flukes is high. These flukes have a long life span (up to ten years). Repeat stool exams may reveal the organisms.

Color the subheading Tapeworm Infections, and the organisms and their titles (d) through (e²).

The beef tapeworm *Taenia saginata* (d¹; station 4) is similar to the tapeworm studied in another plate (Plate 96). Gravid proglottids are expelled to the soil in human feces. Feces-contaminated grasses are consumed by cattle after which tapeworm embryos travel to the cattle muscle and encyst. Consumption of poorly cooked or raw beef returns the parasite to the human. Intestinal infection (d; beef tapeworm disease) occurs in the human in which the tapeworm may reach a length of 8 meters (25 feet) without causing significant symptoms.

Hymenolepis nana (f¹) is a short tapeworm reaching a maximum of 25 mm in length. This "dwarf" tapeworm (station 5) is common in humans worldwide, infecting the human intestine where it holds fast with a complex set of sucker devices and hooks. Eggs released from the intestine contaminate foods and thereby spread to other individuals. No intermediate host is required. Dwarf tapeworm infection (f) occurs in the southeastern United States.

Echinococcus granulosus (e¹) is a small (about 5 mm long) tapeworm with three proglottids behind the scolex: one immature, one mature, and one gravid (station 6). The worm develops in the intestine of the dog and other canines; egg-laden gravid proglottids are deposited with the feces. The eggs reach humans via soil-contaminated plants or by contact with dogs. Ingested by the human, the eggs hatch in the small intestine, and travel by the blood to the liver, lung, kidney, brain, and/or other vascularized tissues where they form hydatid cysts (e²). These cysts are thick-walled structures enclosing a number of encysted larvae that can bud off the parent cyst. The cysts can cause obstruction, form emboli as well as abscesses, and may require surgery for removal.

HELMINTHIC DISEASES: FLUKE AND TAPEWORM

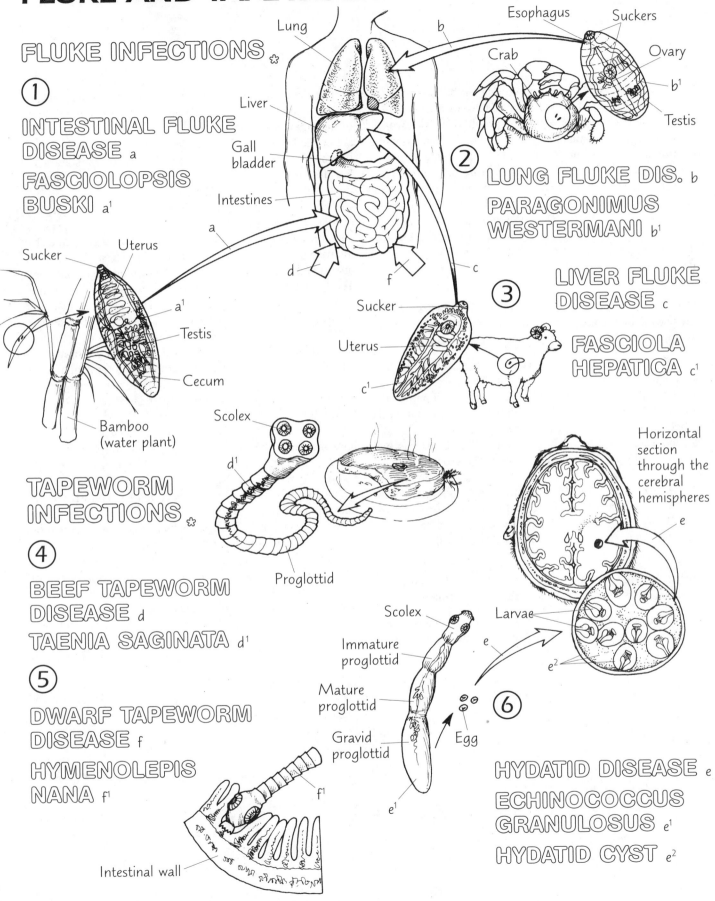

FLUKE INFECTIONS ✿

① INTESTINAL FLUKE DISEASE a
FASCIOLOPSIS BUSKI a¹

Sucker
Uterus
a¹
Testis
Cecum
Bamboo (water plant)

Lung
Liver
Gall bladder
Intestines
a
d
f

Esophagus
Crab
Suckers
Ovary
b¹
Testis
b

② LUNG FLUKE DIS. b
PARAGONIMUS WESTERMANI b¹

③ LIVER FLUKE DISEASE c
FASCIOLA HEPATICA c¹

Sucker
Uterus
c¹
c

TAPEWORM INFECTIONS ✿

④ BEEF TAPEWORM DISEASE d
TAENIA SAGINATA d¹

Scolex
d¹
Proglottid

⑤ DWARF TAPEWORM DISEASE f
HYMENOLEPIS NANA f¹

f¹
Intestinal wall

Scolex
Immature proglottid
Mature proglottid
Gravid proglottid
e¹
Egg
e

Horizontal section through the cerebral hemispheres
e
Larvae
e²

⑥ HYDATID DISEASE e
ECHINOCOCCUS GRANULOSUS e¹
HYDATID CYST e²

Tapeworms (class Cestoda, phylum Platyhelminthes) are parasites with long, flat bodies consisting of a head region and a ribbonlike series of segments (proglottids). Tapeworms are ingested and remain in the intestinal tract of humans while continuing to grow and lengthen. They are hermaphroditic, each proglottid fertilizing the neighboring proglottid. They have no digestive tract, and are dependent upon a host for survival, and a series of hosts for complete development. The fish tapeworm is named *Diphyllobothrium latum*, meaning "double-thin" and "wide." It is one of the longest tapeworms that can infect humans, attaining lengths up to 10 meters (32 feet) or more.

Color the subheading Fish Tapeworm Infection gray. Color the organism, the proglottid and eggs, and the related titles at upper left. Use colors that do not obscure the detail.

Human infection by the fish tapeworm is caused by eating uncooked or incompletely cooked fish. The disease occurs in areas worldwide where uncooked fish make up a part of the human diet. Examination of the adult fish tapeworm (a) reveals a body consisting of repeated structural units (proglottids) behind a head region called the scolex. The scolex has two longitudinal grooves, one on each side. The groove creates a suction on the host tissue to which it is attached. Behind the scolex is the neck region in which new proglottids are formed. The newly-formed proglottids push the more mature proglottids to the rear. The most distant ones, called gravid proglottids (a^1), are filled with fertilized eggs (b).

Begin the life cycle at station 1. Continue with the cycle, coloring titles and related parts through station 6.

The tapeworm grows in the small intestines of the human host after being ingested in larval form. The worm's presence during development causes little or no symptoms, except perhaps a mild diarrhea. About five or six weeks after ingestion of the tapeworm larvae, gravid proglottids are detached and are carried with the flow to the large intestine. There they fragment and eggs are discharged into the feces (station 1). For the life cycle to continue, the eggs must be deposited in fresh water where intermediate hosts are available.

Each egg will give rise to a larval form called a coracidium (c; station 2). This tiny, ciliated organism moves by means of cilia through the water. Some of these larvae are consumed by microscopic crustaceans (d; station 3). These function as intermediate hosts for the larvae (e) which undergo continued development within the body of the crustacean. When this small host is eaten by a small fish, such as a minnow (f; station 4), the minnow becomes a host for the developing larvae. The larvae proceed to migrate to the muscle tissue of the small fish. When a larger fish, such as a perch (g) consumes the minnow (station 5), the perch becomes an intermediate host for the tapeworm larvae. When the infected fish is caught by a human and eaten uncooked or inadequately cooked (station 6), the living larvae will be transferred. The larvae will pass through the human stomach and attach to the wall of the small intestine where they will develop into adult forms (a).

D. latum can maintain residence in the small intestine indefinitely without causing significant damage. In some cases, the tapeworm can retard the absorption of vitamin B^{12} in the human host, inducing a form of anemia. Presence of the adult tapeworm can only be determined by microscopic observation of the stools of the infected person. Treatment with niclosamide is effective.

In the United States, infections by *D. latum* are most common in the Great Lakes region. In some patients, anemia develops, possibly because the parasite consumes the body's supply of Vitamin B^{12} needed for red blood cell production.

HELMINTHIC DISEASE: FISH TAPEWORM

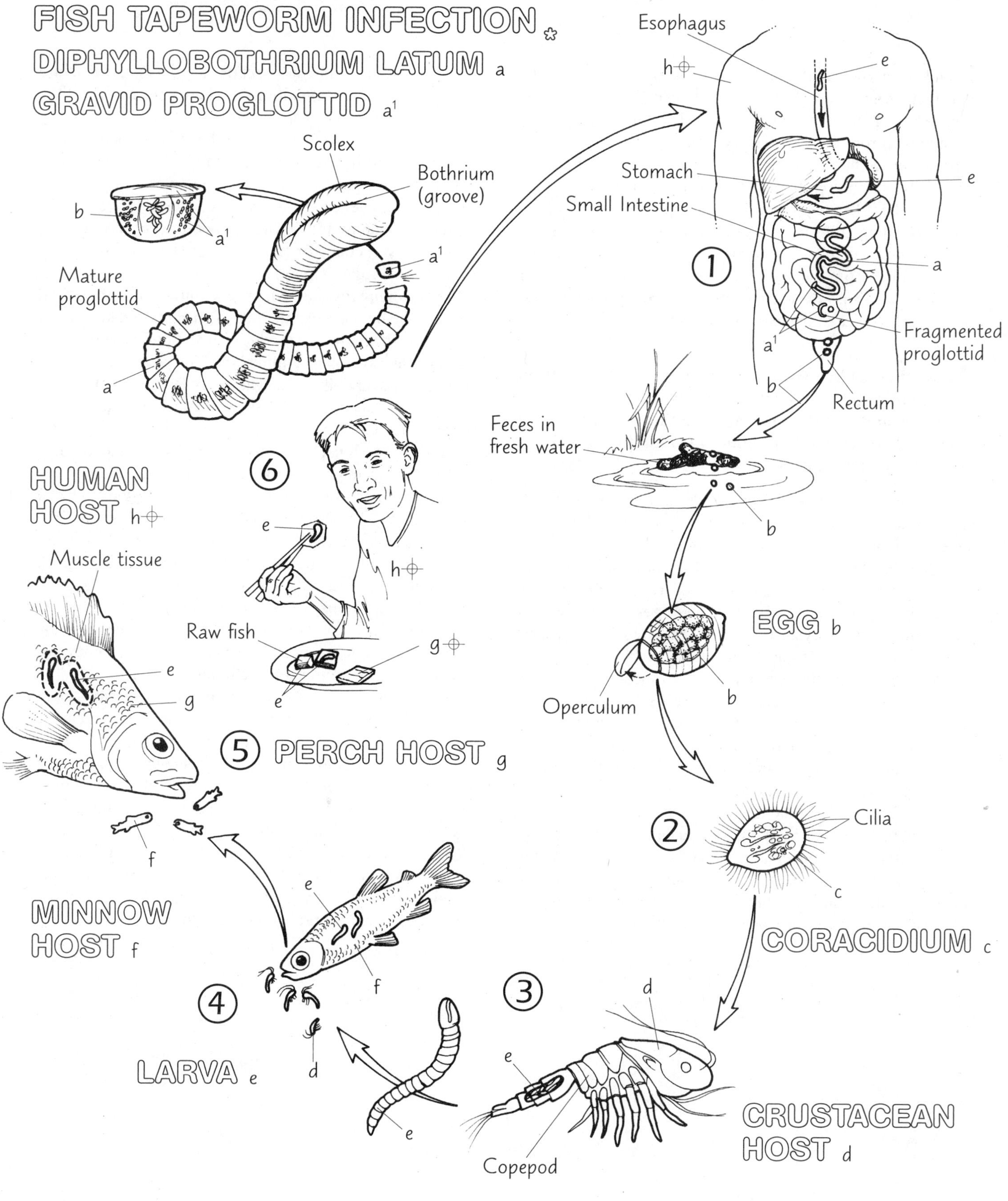

FISH TAPEWORM INFECTION ✿
DIPHYLLOBOTHRIUM LATUM a
GRAVID PROGLOTTID a¹

Scolex

Bothrium (groove)

b

a¹

a¹

Mature proglottid

a

Esophagus

h⊕

e

Stomach

Small Intestine

e

a

①

a¹

b

Fragmented proglottid

Rectum

HUMAN HOST h⊕

⑥

e

h⊕

Muscle tissue

Raw fish

e

g⊕

e

⑤ PERCH HOST g

e

g

Feces in fresh water

b

EGG b

Operculum

b

②

Cilia

c

CORACIDIUM c

MINNOW HOST f

f

e

f

④

LARVA e

e

d

③

d

e

Copepod

CRUSTACEAN HOST d

HELMINTHIC DISEASE: TRICHINOSIS

Roundworms (nematodes) are the third major helminthic group that include parasites pathogenic to humans. Recall that the "flatworms" include the flukes (trematodes) and the tapeworms (cestodes). From an evolutionary standpoint, nematodes are more advanced than the flatworms. Roundworms are threadlike organisms exhibiting bilateral symmetry, a lack of segmentation, and unlike the flatworms, a complete tubular digestive tract extending from mouth to anus. Roundworms range in length from a few millimeters to a meter. Roundworms occupy most habitats on Earth, including fresh and salt water environments. They exist in conditions ranging from the moist heat of the tropics to the extreme cold of the Arctic.

Color the subheading Pork Roundworm Infection (Trichinosis) gray. Color the organism (a) in the center of the plate, and its title. Then go to station 1 and color the titles and structures, continuing to station 6. Use colors that do not obscure detail.

Encysted larvae (a^1) of *Trichinella spiralis* (a) are found in the uncooked meat scraps commonly eaten by pigs, and other mammals (station 1). Pigs are important to humans as a common food source (e.g., ham, bacon, pork sausage, pork chops, and pork roasts). Once ingested by humans in undercooked pork muscle (station 2), the encysted larvae enter the stomach. The cysts (b) are fragmented by digestive enzymes (c) in the stomach, and the live larvae are freed (station 3). The motile larvae migrate to the small intestine where they penetrate the mucosal lining cells. Here they mature and reproduce, supported by the nutritive environment within the cells they inhabit (station 4).

Female roundworms produce live offspring; that is, they are viviparous (most nematodes are ovoviparous). Hundreds of these small larvae (about 0.8 mm long) are released into the intestinal mucosa by a single female over a number of days (station 4). The developing larvae, appearing as miniature adults, enter the lymphatics and small veins draining the intestinal circulation. By this vascular route, they migrate to skeletal muscle (d) or heart muscle (not shown).

The larvae enter the muscle cells and transform them into supporting "nurse" cells (station 5). These cells remain alive, providing a means by which the enclosed larvae can develop and mature. The surrounding tissue becomes inflamed, and blood vessels proliferate and inflammatory cells (neutrophils, macrophages, plasma cells) abound (station 6). These larvae-muscle cell complexes, called cysts (b), may survive months or years. Ultimately, the larvae die and the "cysts" become surrounded by cocoons of fibrous tissue that usually calcify in time.

When unencapsulated larvae populate the intestinal mucosa of the human host, the intestinal signs and symptoms of trichinosis begin to appear due to irritation of the mucosa. These signs and symptoms, mimicking food poisoning, include nausea, cramps, and diarrhea. They may last for weeks.

As the larvae invade the skeletal muscles, they induce inflammation (e), and muscle tenderness and pain (g). Depending on the muscles involved, serious dysfunction may occur, including muscle weakness (f), inflammation of the myocardium, and respiratory difficulties. Rarely, the brain and spinal cord may become infected with encysted larvae. The mortality rate may be high in humans if the disease is untreated.

A history of having eaten potentially infected pork suggests the possibility of trichinosis. Diagnostic tests for *T. spiralis* include biopsy of skeletal muscle believed to be infected, the ELISA, and the latex agglutination test. Therapy is limited largely to supportive measures, because encysted larvae are generally unresponsive to medication.

Trichinosis is generally familiar to those who eat pork or pork products because packages contain warnings to cook the pork thoroughly (not lower than 58° C/137° F). The U.S. Department of Agriculture has warned that microwave cooking may not kill encysted larvae. Pickling, heavy seasoning, smoking, and prolonged refrigeration at temperatures above -15° C, are not adequate substitutes for thorough cooking of pork or pork products. Approximately 150 cases of trichinosis are reported in the United States annually. Since routine inspection for *Trichinella* cysts is not practiced in slaughterhouses, the burden for prevention falls to the consumer.

HELMINTHIC DISEASE: TRICHINOSIS

PORK ROUNDWORM INFECTION (TRICHINOSIS) ✺

TRICHINELLA SPIRALIS a

T. SPIRALIS LARVA a¹

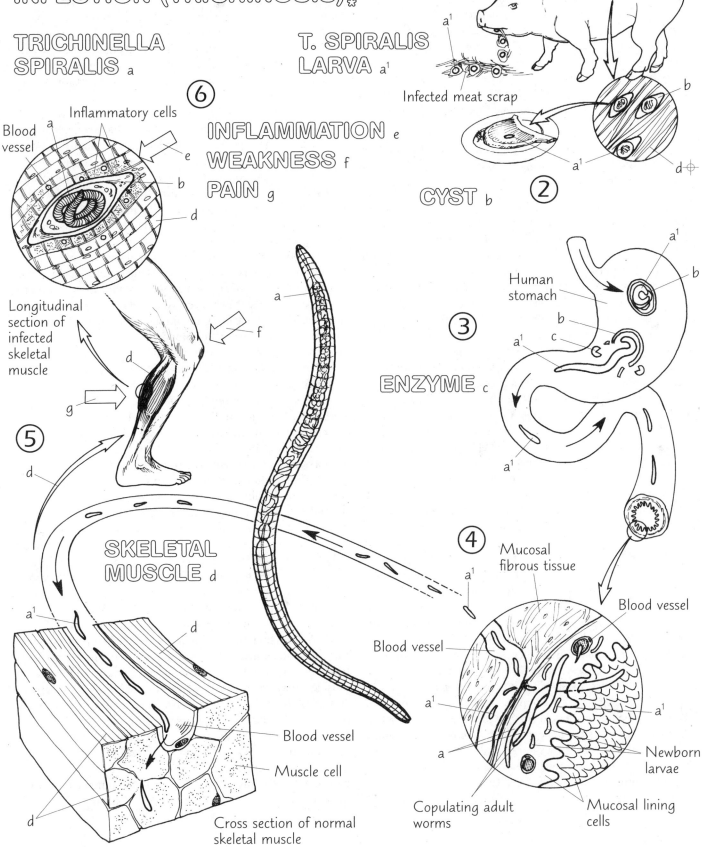

Infected meat scrap

CYST b

① ② ③ ④ ⑤ ⑥

Inflammatory cells

Blood vessel a

INFLAMMATION e
WEAKNESS f
PAIN g

Longitudinal section of infected skeletal muscle

Human stomach

ENZYME c

Mucosal fibrous tissue

SKELETAL MUSCLE d

Blood vessel

Blood vessel

Blood vessel

Muscle cell

Copulating adult worms

Newborn larvae

Mucosal lining cells

Cross section of normal skeletal muscle

HELMINTHIC DISEASES: ROUNDWORM INFECTIONS

Over one billion of the Earth's people are infected by one of several species of nematodes. *E. vermicularis, T. trichiura, A. lumbricoides,* and *N. americanus* are residents of the human intestine (intestinal nematodes). *W. bancrofti* and *Loa loa* are filarial nematodes; that is, they are threadlike in size and shape. They live in the blood and lymph of their human hosts, and in the connective tissues as well. *T. spiralis* (recall plate 97) is a tissue nematode, residing in skeletal muscle.

Begin at station 1, and color the titles, related organism (a¹), and the arrow (a). Use light colors. Continue coloring through station 6.

Enterobius vermicularis (a¹; station 1) is a roundworm known as the pinworm. It is probably the most common parasitic worm infection in the world, especially affecting young children. The worms live in the large intestine, and do not usually induce signs or symptoms that give evidence of their presence. At night, *E. vermicularis* females leave the rectum and migrate to the anus and surrounding perianal skin. Here eggs are deposited. When these eggs are picked up on the fingers, they can be transferred to other persons and to one's self by finger-to-mouth. Pinworm eggs are observed by microscopic study of a scotch tape preparation from the perianal area. If present, adult female worms (about one quarter of an inch or 10 mm in length) can be seen with the unaided eye. Several drugs are effective for controlling pinworms.

The whipworm *Trichuris trichiura* (b¹; station 2) is named for its long and slender anterior end that burrows into the villi of the small intestine. The larger, coiled end projects into the intestinal cavity. *T. trichiura* is distributed worldwide. The eggs of this worm are found in the soil. They are ingested by humans eating unwashed vegetation. The larvae are hatched in the small intestine, and migrate to the large intestine where they mature. The eggs of the female pass out of the tract with the feces, and can be directly observed on the fecal surfaces. Transmission of the worms to another individual can be effected by hand-to-hand contact; reinfection occurs by the fecal-oral route. Infection with these worms is associated with malnutrition, and is accompanied by chronic diarrhea. Diagnosis can generally be made by observing eggs or adult worms in the feces. Mebendazole is usually successful in treating the disease.

One of the longest (up to 30 cm or 12 inches) and most prevalent roundworm parasites is *Ascaris lumbricoides* (c¹; station 3). The eggs of this worm are discharged with the feces from an infected person and cling to plants in the soil. When the plants are ingested, the eggs are transferred to the human host. *A. lumbricoides* grows to maturity within the intestine. These worms contain protease enzymes that enable them to digest the proteins eaten by the host. Large numbers of *Ascaris* may block the intestinal canal or perforate the intestinal wall.

Migration into the blood by the larvae permits invasion of the lungs, liver, and other organs as well. Pneumonia, hepatomegaly, hepatitis, and toxemia may follow. The eggs can be identified in the feces during the intestinal phase of the disease. A number of medications are effective in killing adult worms, including pyrantel pamoate and mebendazole. Migratory larvae are more difficult to eradicate,

Necator americanus (d¹; station 4) and *Ancylostoma duodenale* (not shown) are hookworms that are parasitic in humans. The two species can be identified by the structure of their mouth parts. *N. americanus* has rounded cutting plates; *A. duodenale* has more conical projections called cutting teeth. Hookworm eggs are discharged in the feces of the infected human host. Larvae hatch and develop in the soil. They infect humans by entering the hair follicles, penetrating the skin, and entering the circulation. They can migrate via the blood to the lungs or heart. Hookworms localized in the lungs are coughed up into the pharynx and swallowed, ultimately reaching the intestine. The teeth of hookworm attach to the villi of the small intestine, macerate the surface tissue, and ingest the nutritional mucosa. This often causes abdominal pain, nausea, vomiting, diarrhea, anemia from the blood loss, and loss of protein (hypoproteinemia). Diagnosis is made by identification of eggs in the feces; treatment by administration of pyrantel pamoate and mebendazole is effective.

Wuchereria bancrofti (e¹; station 5) is a filarial roundworm transmitted by mosquitoes to human hosts. Larvae of *W. bancrofti* enter the circulation at the site of the insect bite. The larvae (microfilariae) of the worms mature in the lymphatic vessels, causing extensive deformation and blockage. After years of infestation and lymphatic obstruction, the upper and lower limbs became grossly deformed due to trapped lymph, a condition called elephantiasis (e). The microfilariae are returned to the mosquito with subsequent bites. Mosquito control is essential to interrupting the worm's life cycle. Microfilariae can be identified in blood smears and are eradicated by treatment with diethylcarbamazine. Adult worms are more difficult to kill.

Loa loa (f¹; station 6) is a roundworm parasite found in insects, such as deerflies and horseflies, in Africa. The worm is transmitted to humans with insect bites, and it resides in the fatty tissues under the skin of the new host. Larvae (microfilariae) enter the circulation. When temperatures outside are warm, microfilariae are attracted to the conjunctival surface of the eye (f; eyeworm; also called loaisis). The parasite returns to the deeper tissues as the skin temperature cools. Immune responses to the worm cause localized swelling and inflammation (calabar swellings) observable on the body surface. The microfilariae can be passed on to insects when they bite the host. Insect control helps reduce human infection. Microfilariae can be identified in blood smears.

HELMINTHIC DISEASES: ROUNDWORM INFECTIONS

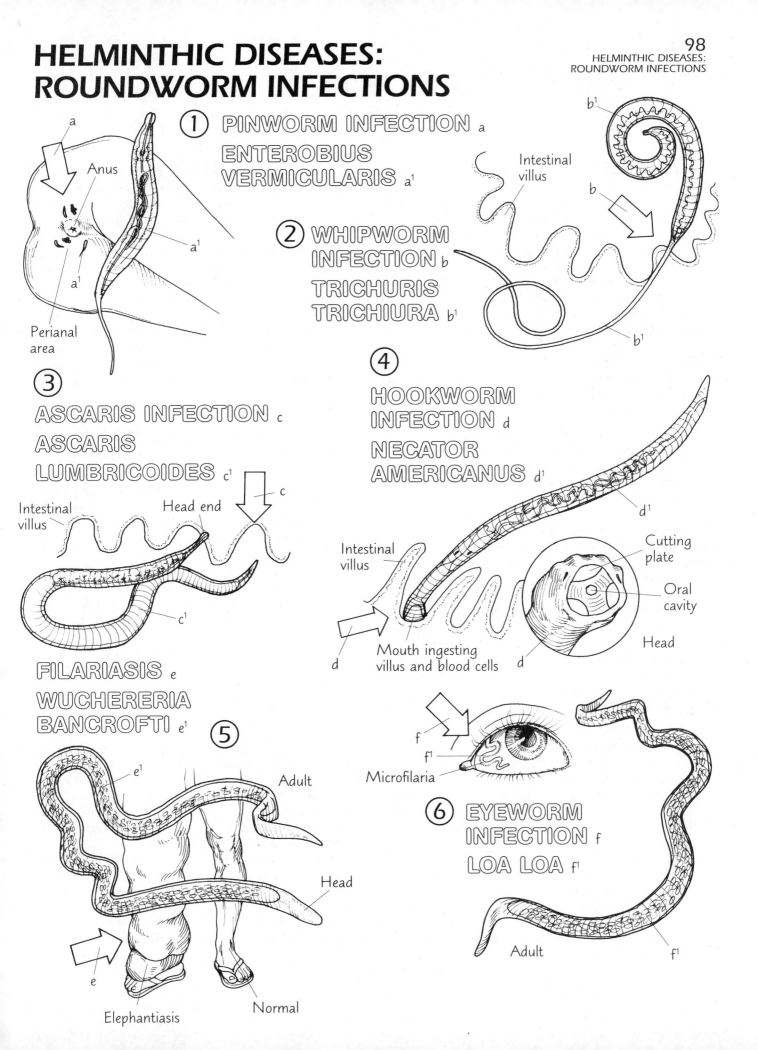

① PINWORM INFECTION a
ENTEROBIUS VERMICULARIS a¹

a
Anus
Perianal area
a¹
a¹

② WHIPWORM INFECTION b
TRICHURIS TRICHIURA b¹

b¹
Intestinal villus
b
b¹

③
ASCARIS INFECTION c
ASCARIS LUMBRICOIDES c¹

Intestinal villus
Head end
c
c¹

④
HOOKWORM INFECTION d
NECATOR AMERICANUS d¹

d¹
Intestinal villus
Cutting plate
Oral cavity
Head
Mouth ingesting villus and blood cells
d
d

FILARIASIS e
WUCHERERIA BANCROFTI e¹

⑤
e¹
Adult
Head
Normal
Elephantiasis
e

f
f¹
Microfilaria

⑥ EYEWORM INFECTION f
LOA LOA f¹

Adult
f¹

PASTEURIZATION OF MILK

Milk and milk products contain water, proteins, carbohydrates, fats, vitamins, and minerals, and provide an extremely nutritious medium for the growth of microorganisms. Milk can transmit such infectious diseases as Q fever, tuberculosis, campylobacteriosis, and salmonellosis. Pasteurization of milk destroys the microorganisms and enzymes, minimizing disease transmission and spoilage. Milk is sterilized in areas where refrigeration is unavailable or long time storage is necessary. Such milk must be stored in sealed, aseptic containers. Most developed countries around the world pasteurize their milk products. Due to the possibilities of widespread dissemination of pathogens in milk and milk products, the processing and pasteurization of milk is rigidly controlled by both state and federal agencies in the United States. In this plate we explore a common method for pasteurization: High temperature-Short time (HTST) pasteurization.

Careful attention to the notes here are critical to the successful coloring of the pasteurizing machine illustrated. We suggest you color the plate before you get into the text.

First, color the subheading (*) at upper left gray. Then color the following titles and structures in sequence: (b), a bright color; (c) and (d), light neutral colors; (e) and (e¹), warm and hot colors; and (h) and (h¹), cool and cold colors. Then start with the milk truck at upper right, and color (a) and its title a cool color. Plan on using three cool/cold colors for (a), (a³), and (a⁴); and use warm and hot colors for (a¹) and (a²). Follow the arrows in the "tubes." When you reach the holding tube (f), color it, and the valve (g). Color the diverted milk (a¹) a series of dash lines (- - -) with the color used for (a¹). Color the cold storage chamber (i) a cold color.

Raw milk is collected by tanker vehicles from dairy farms on the same day the milk is extracted. The cool raw milk (a; stored in the tanker at 45° F/7.2° C or less) is delivered to the processing plant and piped into storage tanks (silos).

Samples of the raw milk are taken on a regular schedule from the tankers as well as the silos for microbiological analysis. There should be no increase in numbers of microorganisms between samples taken from the tanker and the silos. Samples are incubated on a standard agar medium in Petri dishes for 48 hours at 89.6° F/32° C. Raw milk can contain a number of nonpathogenic as well as pathogenic micro-organisms (b), including *Escherichia coli, Streptococcus* species, *Staphylococcus aureus, Salmonella* species, *Pseudomonas* species, *Campylobacter jejuni,* and protozoa. The cleanliness of the dairy, the physiologic condition of the

milk animal, the condition of the udder and teats of the animal, and the cleanliness of the tanker truck are all factors that may contaminate the raw milk before processing. Serious contamination of raw milk is avoided by strict adherence to carefully regulated operational standards.

Raw milk is piped from the silos to a constant-level supply or balance tank (c). The milk is then piped into the regenerator section (d) of the HTST pasteurizer.

Within the pasteurizer, raw milk is conducted under suction between alternating pairs of thin, stainless steel, corrugated plates that are warmed by a counter-current of hot pasteurized milk (a²) passing between alternating pairs of those plates. Here we show the flow of milk in tubes for visual simplicity. The arrangement of plates can be visualized (by parallel lines) at the top of the pasteurizer.

The warmed milk (a¹) passes into the heater section (e) where it comes under pressure by a timing pump that provides the motive force for milk flow through the HTST pasteurizer. The heater section consists of warmed milk flowing between alternating pairs of stainless steel plates (not shown; here represented by tubes) heated by steam (e¹) or hot water passing between adjacent pairs of plates. The system is designed to heat the milk (a²) to a temperature of 161.6° F/72° C as it leaves the heater section to enter the holding tube (f).

The S-shaped holding tube holds the hot milk at a constant temperature for a period of at least 15 seconds. This ensures sufficient heat to kill the microorganisms. At the end of the holding tube, the hot milk passes by a sensing bulb of the thermometer and controller and then into a flow diversion valve. Insufficiently heated milk (a¹) is directed back to the balance tank for reprocessing. Microorganisms destroyed by pasteurization include species of *Escherichia, Brucella, Mycobacterium, Campylobacter, Salmonella, Listeria, Staphylococcus, Coxiella, Streptococcus, Yersinia,* and others.

Acceptably heated milk is directed forward into the regenerator section of the HTST pasteurizer. Here it heats the incoming raw milk (a) as described previously. The hot milk is then directed to the cooler section (h) where coolant (h¹; cold water) passing between adjacent pairs of steel plates brings the milk down in temperature. The cooled pasteurized milk (a³) is conducted out of the pasteurizer to cold storage tanks and then to packaging (a⁴). In many dairies, the milk is distributed the same day as it was collected.

Microbiological analysis of the pasteurized milk (taken at the bottling machine) generally reveals no significant bacterial content. A concern of the dairy microbiologist is the formation of a biofilm (layer of bacteria clinging to the smooth steel surface) on the interior surfaces of the stainless tubing; stringent cleaning practices eliminate this problem.

PASTEURIZATION OF MILK

HIGH TEMPERATURE-SHORT TIME (HTST) PASTEURIZATION ✿

COOL RAW MILK a
MICROORGANISM b
BALANCE TANK c
REGENERATOR SECTION d
WARMED MILK a¹
HEATER SECTION e
STEAM e¹
HOT MILK a²
COOLER SECTION h
COOLANT h¹

COOL PASTEURIZED MILK a³

COLD STORAGE i

PACKAGED MILK a⁴

HOLDING TUBE f

FLOW DIVERSION VALVE g

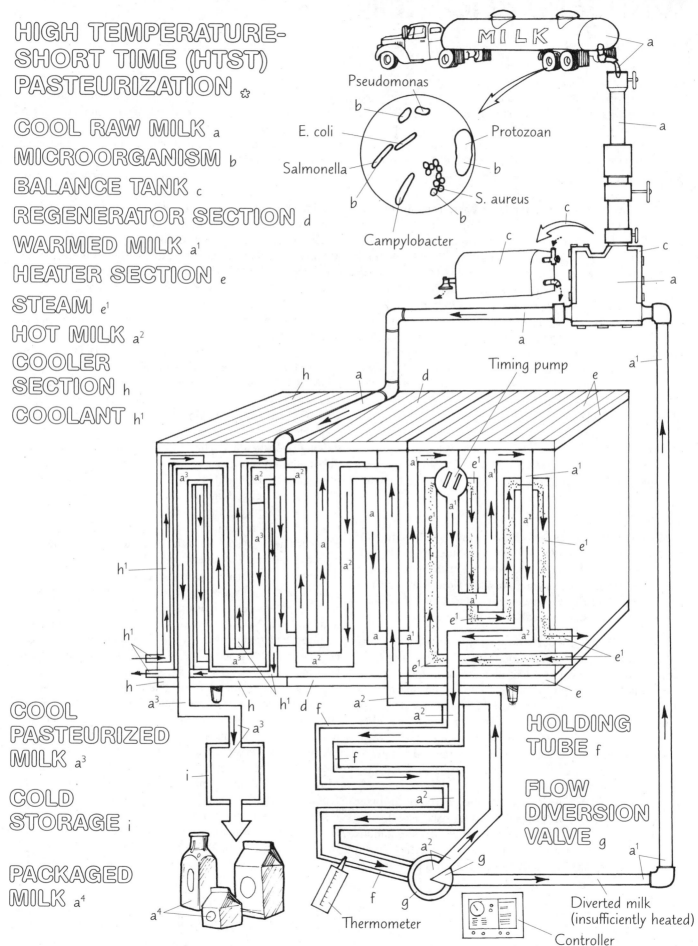

Pseudomonas
E. coli
Salmonella
Protozoan
S. aureus
Campylobacter

Timing pump
Thermometer
Diverted milk (insufficiently heated)
Controller

100
WATER PURIFICATION

Raw reservoir water (a) is subject to widespread contamination and requires purification for human consumption. Cows (b) grazing on hillside fields above a reservoir, shorebirds, ducks and other swimming birds (b), and humans (b) all represent potential sources of contaminants that can enter the water supply. Rain water coursing down hillsides into the reservoirs is the vehicle of other foreign material, including animal and human wastes.

These waterborne contaminants contain many micro-organisms (b[1]), including the bacteria *Shigella* (shigellosis), *Campylobacter jejuni* (gastroenteritis), *Salmonella* species (salmonellosis), *E. coli* (traveler's diarrhea), *S. typhi* (typhoid fever), *Yersinia enterocolitica* (enteritis), and *Vibrio cholerae* (cholera). Waterborne viruses include hepatitis A (liver infection), adenoviruses (respiratory infection), rotaviruses (infant diarrhea), Coxsackie viruses (common cold), and the Norwalk viruses (diarrhea). Waterborne protozoans include *Giardia lamblia* (giardiasis, a form of gastroenteritis), *Entamoeba histolytica* (dysentery), and oocysts of *Cryptosporidium* (cryptosporidiosis, characterized by diarrhea).

Cryptosporidiosis outbreaks, in particular, have occurred in a number of communities in the United States over the last ten years. The source of the pathogenic organisms in these cases was well, lake, river, and reservoir water. Purification of drinking water by a carefully monitored, multistep process that involves both chemical disinfection and physical treatment prevents such occurrences.

Set aside three progressively lighter shades of blue for (a), (a[1]), and (a[2]). Use shades of brownish-blue for sludge (j), (j[1]), and backwash water (j[2]). Color the titles and related parts (a) through (m) as you follow the flow of water from the reservoir to the water storage tank. Use light colors where there is risk of obscuring illustrative detail.

Water is stored for drinking purposes in water reservoirs (a), and aboveground or underground tanks. In reservoirs, large concrete towers rising 100 feet or more from the reservoir floor contain intake pipes covered with screens. The water flows by gravity to the bottom of the tower and is then conducted through a number of screens to the pumping station. Water for agriculture may be diverted into canals (not shown) before passing the screens. The water is pumped through one or more large diameter pipes (as much as 1 meter / 39 inches) to the treatment facility.

Initially, the raw water is discharged into a flash mixer (c) where it is sprayed (in approximately one second) with flocculating agents, such as cationic polymers (d) and solutions of alum or aluminum sulfate (e), and with chlorine (f) and potassium permanganate (g; $KMnO_4$). Cationic polymers attract negatively charged particles; alum is an agent that attracts microorganisms and particles that are positively charged. Chlorine is an oxidizing agent that reacts with microorganisms, killing them. $KMnO_4$ prevents formation of a carcinogen called trihalomethane. These chemicals are thoroughly dispersed in the partially treated water (a[1]) which is conducted to the flocculation basin through aerated passageways.

The chemicals continue to work the reservoir water in the flocculation basin (h). The flow of water is slowed, and is directed over and under plates, aided by rotating paddles. This water action enhances the formation of "floc": aggregation of small particles and the polymer / alum to form larger particles. These particles are carried with the water flow to the sedimentation basin.

The water moves slowly in the sedimentation basins (i), permitting the settling of the floc or large particle sediment to the basin floor. This sediment, called sludge (j), is discharged into a sludge basin (j[1]) or container from which water is returned to the reservoir and the sludge is conveyed to landfill areas or sludge lagoons. In some municipalities, the sludge is withdrawn from the sedimentation tank floor by a hose connected to a vacuum pump housed on a traveling bridge moving over the tank. The vacuumed sludge is conveyed to the sludge tank for disposal. The water flows through the sedimentation tank and is conveyed to the filter tanks (k).

As the water goes into the filter tanks, a non-ionic polymer (d[1]) may be added. Filter tanks contain one of a number of filters used for removing the finest particles from the water. These filters (k[1]) usually consist of anthracite (charcoal) over sand over stones. The filters are periodically back-washed to remove excess particulate matter and microorganisms; this waste (backwash) water (j[2]) is drained to the sludge tank.

Filtered water is chlorinated and treated with sodium hydroxide (l; NaOH) to raise the pH (lower the acidity) if required. The treated water (a[2]) is then stored in tanks (m) at, near, or some distance away from the treatment facility, for consumer use.

Water purification ensures that harmful microorganisms are not transmitted to consumers. The physical processing (flocculation, sedimentation, filtering) and the chemical treatment (chlorine and other chemicals) will kill most disease-producing organisms. Rapidly moving water also dilutes pathogens to the extent that their concentrations are harmless. Most municipalities monitor the quality of the water on a regular schedule at one or more of several sites, including the reservoir effluent, the sedimentation basin, and the inlet to the filter. Such monitoring includes pH, temperature, color, hardness, turbidity, coliform bacteria count, and so on.

WATER PURIFICATION

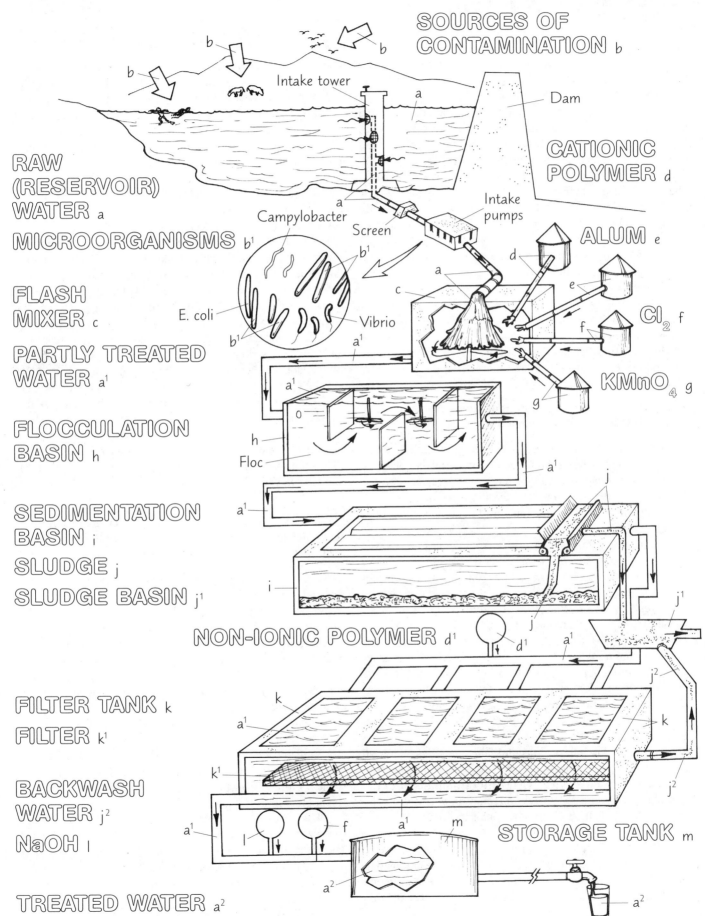

SOURCES OF CONTAMINATION b

RAW (RESERVOIR) WATER a

MICROORGANISMS b¹

FLASH MIXER c

PARTLY TREATED WATER a¹

FLOCCULATION BASIN h

SEDIMENTATION BASIN i

SLUDGE j

SLUDGE BASIN j¹

FILTER TANK k

FILTER k¹

BACKWASH WATER j²

NaOH l

TREATED WATER a²

CATIONIC POLYMER d

ALUM e

Cl_2 f

$KMnO_4$ g

NON-IONIC POLYMER d¹

STORAGE TANK m

Intake tower

Dam

Campylobacter

E. coli

Vibrio

Screen

Intake pumps

Floc

Metabolism is the process by which organic molecules are produced within an organism (anabolism), and organic molecules are broken down (catabolism). Catabolism is usually associated with energy release, stored in the form of adenosine triphosphate (ATP) and other electron acceptors, such as reduced NAD and FAD. This energy is then available for cell work. Metabolism takes place within the cells of an organism. In prokaryotes, it occurs within the cytoplasm and involves the cell or plasma membrane of the microorganism.

Fermentation is a catabolic process taking place within many microorganisms including certain bacteria and fungi. In fermentation, organic molecules, such as sugar, are converted to one of several possible end-products, including lactic acid and ethyl alcohol. These reactions generally occur in the absence of oxygen.

This plate introduces the biochemical process of fermentation that occurs in the bacterium *Lactobacillus* and the yeast *Sacchromyces cerevisiae*, and compares this anaerobic form of catabolism with the aerobic form (aerobic respiration) undertaken by most eukaryotic cells.

Guided by the arrows, color structures (a) through (d), and their titles. Use very light colors.

The organic molecule glucose (a) is a simple carbohydrate. Complex carbohydrates would be starch and glycogen, each of which consist of many linked glucose molecules. Microorganisms catabolize carbohydrates to generate energy. Catabolism of amino acids and fatty acids also generates cellular energy.

In microorganisms, chains of carbohydrates are broken down by enzymes to glucose and other simple sugars. Glucose can then be oxidized to form carbon dioxide (CO_2) and water (H_2O). Oxidation and reduction are the principal processes that break down molecules and capture energy. For every oxidation reaction, there has to be a reduction reaction.

During oxidation, electrons are removed from one molecule and transferred to another molecule. This is usually done by removing two hydrogen atoms (two protons and two electrons) from a donor molecule and transferring the electrons to an acceptor molecule. The donor molecule losing the hydrogens or electrons is said to be oxidized. The acceptor molecule is said to be reduced.

During reduction, a molecule gains electrons. Such an electron acceptor molecule is oxygen. Two other electron acceptor molecules are NAD (nicotinamide adenine dinucleotide) and FAD (flavin adenine dinucleotide). The reduced form of these electron-acceptor molecules (NADH, FADH) store energy.

When a molecule of glucose is oxidized, two molecules of pyruvic acid (c) are produced. The series of reactions that produces pyruvate is called glycolysis (b). The oxidation of a molecule of glucose also produces four molecules of high energy ATP, and two molecules of reduced NADH. This stored energy is used for synthesizing molecules during anabolic metabolism (cell work).

Pyruvic acid is the threshold molecule for two additional catabolic processes: respiration (d) and fermentation (e).

In aerobic respiration, pyruvic acid is converted to acetyl coenzyme A (d) which enters the Krebs cycle (d). The Krebs cycle consists of a group of enzymes that oxidize acetyl coenzyme A in a cyclic series of reactions. In one complete cycle, high-energy electrons are transferred to NADH and FADH, and high energy ATP is formed.

The NADH and FADH are then oxidized in a sequence of reactions called the electron transport chain. These reactions take place on the cell membrane of prokaryotes and the inner membranes of mitochondria of eukaryotes. As the electrons donated by NADH and FADH move along the chain or sequence of molecules, hydrogen ions (protons) build up outside the membranes. These protons then reverse their flow and flow back through the membranes (chemiosmosis, d). The movements of these protons drives the oxidation of NADH and FADH, resulting in the formation of high energy ATP. Oxygen is the final electron acceptor in aerobic respiration.

In anaerobic respiration (d), organic molecules such as glucose are oxidized, energy is released, and CO_2 and H_2O are produced, but oxygen is not involved. Anaerobic respiration uses nitrates and sulfates as final electron acceptors (Plate 105), and not oxygen.

Color the title and border (e). Color the fermentation route to lactic acid and its uses (f) to (f⁴), and the route to acetaldehyde and its uses (g) to (g²).

Fermentation (e) begins with the conversion of organic molecules to pyruvic acid. In the case of glucose, it is converted to pyruvic acid by glycolysis. The pyruvic acid is reduced to such end products as lactic acid, ethanol, acetic acid, butanol, CO_2, and H_2O. The final electron acceptor in fermentation must be an organic molecule. No oxygen is utilized; fermentation is essentially an anaerobic process. Fermentation end products are waste to the bacteria or yeast. They can, however, be of value to humans.

Lactobacillus (f), a Gram-positive anaerobic rod, works on glucose in bread dough, reduces pyruvic acid to lactic acid (f¹), and oxidizes NADH to NAD⁺. Lactic acid gives a characteristic flavor to sour dough breads (f²). *Lactobacillus* works on milk sugar to form lactic acid which gives unique flavors to yogurt (f³) and cheese (f⁴).

The yeasts *Sacchromyces cerevisiae* (g) break down grape sugars and grain sugars to form acetaldehyde (g¹) and CO_2. The acetaldehyde is reduced to ethanol (g²), accompanied by the oxidation of NADH. Ethanol is a form of alcohol, and is considered a desirable feature of wine (g³) and beer (g⁴).

FERMENTATION

GLUCOSE a

GLYCOLYSIS
PATHWAY b

PYRUVIC
ACID c

FERMENTATION e

RESPIRATION d

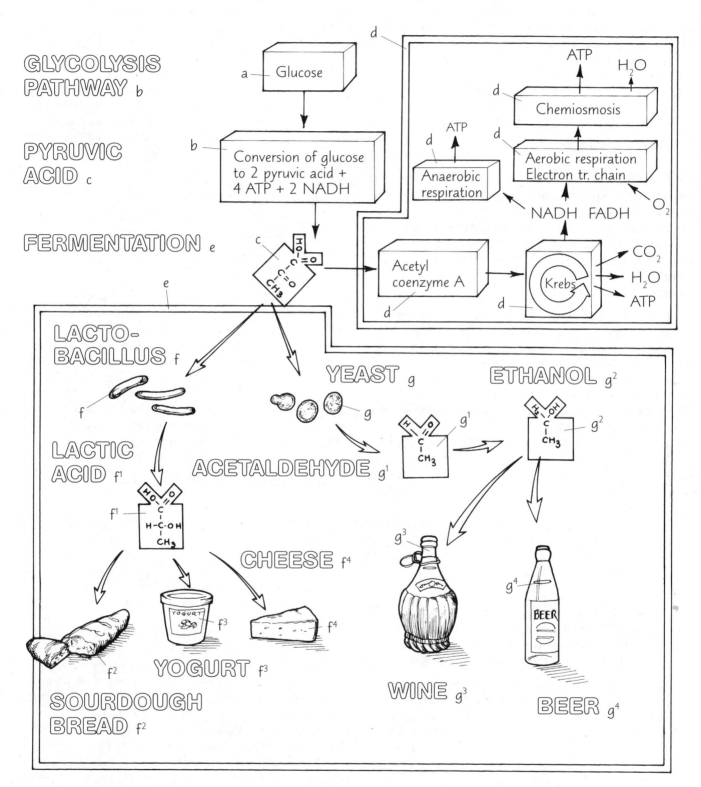

LACTO-
BACILLUS f

LACTIC
ACID f1

YEAST g

ACETALDEHYDE g1

ETHANOL g2

CHEESE f4

SOURDOUGH
BREAD f2

YOGURT f3

WINE g3

BEER g4

WASTEWATER TREATMENT

Wastewater (sewage) consists of liquid and solid waste products collected by drains and pipes called sewers. Wastewater contains microorganisms, some of which are pathogenic to humans. The proper treatment of wastewater prevents the dispersal of these pathogenic microbes, such as coliform bacteria, staphylococci, fungi, protozoa, and helminthic larvae. Most wastewater treatment plants incorporate three broad levels of processing: physical or primary treatment, biological or secondary treatment, and chemical or tertiary treatment.

Plan on using several shades of the same color for (d) through (d⁶). Color the titles and perimeters of the three processing modes (a), (b), and (c) with contrasting colors. At upper left, begin with wastewater (d) and its title, and color the flow through the physical processing units. Stop after coloring the primary clarifier. Use progressively lighter shades for (d) as the wastewater is processed, using the lightest shades for (d⁵) and (d⁶).

Subterranean conduits, called sewers, bring wastewater (d) to the grit chamber (e) in the treatment facility for preliminary physical processing. The wastewater passes through a bar screen which traps large debris. Insoluble material passing the screen, such as sand, eggshells, and gravel (d¹; grit) settles in the grit chamber and is moved out of the chamber by a rotating auger. This insoluble waste is often transferred to a drying field (f).

Ninety percent of the wastewater from the grit chamber flows into the primary clarifier (g), a type of settling tank. Here light, insoluble waste collects on the surface of the wastewater. This scum (d²) is drawn into v-notched collectors, called weirs, located around the outer, upper edge of the clarifier. The heavier solids (d³; sludge) settle and collect at the bottom of the clarifier. Both scum and sludge are transferred to the anaerobic digester. The wastewater is transferred to aerobic biological processing units.

Color the titles and structures associated with biologic processing (h) through (k). In the trickling filter, color the media (j), then color over that with the color used for (d), suggesting the wastewater flowing over the media.

Microorganisms are the principal actors in biologic processing. The anaerobic digester (h) is a closed tank in which anaerobic bacteria resident in the sludge digest or break down the organic matter present (dead organisms, insoluble vegetative matter, and so on). This fermentation process (d⁴) goes on for weeks. The end products of fermentation are converted by anaerobic bacteria to methane gas (CH_4) which collects at the top of the digester. This gas can be recovered and used by cogeneration units at the treatment plant to produce electricity. The unconverted indigestible sludge is pumped to drying fields, developed as a fertilizer, or put in landfills or on specified areas of ocean floors.

Aerobic digestion of organic matter occurs in the trickling filter (i). Here the wastewater is sprayed over fist-sized stones (j; called media) on which a microbial (zoogleal) film of aerobic bacteria and fungi develops. Aerated wastewater passes down through the rocks, creating a virtual food chain on the media. Bacteria and fungi feed on particulate organic matter; these small microorganisms are consumed by larger microorganisms (protozoa). As the amount of available nutrients decrease, the number of microorganisms decreases. In technical terms, there is a reduction of biological loading and biochemical oxygen demand (BOD). In time, the remaining organic matter in the wastewater is oxidized and mineralized, forming ammonia, nitrate, phosphate, sulfate, and carbon dioxide. The oxidation of wastewater may also occur in large ponds (not shown).

Another form of aerobic digestion of organic matter in the wastewater is the activated sludge system (k). Here wastewater, containing numerous microorganisms, is discharged into a tank in which air diffuses up from the bottom. The aerated, flowing water encourages the growth and reproduction of bacteria which necessarily consume the available organic matter. In effect, the activated sludge system speeds up the process that ultimately reduces the biological loading. Some of the wastewater cycles back to the tank (return activated sludge or RAS); eventually the wastewater is piped off to the secondary clarifier by way of a mixer.

Color the titles and structures associated with chemical processing.

Chemical processing of wastewater begins with the addition of chemicals, such as polymers (l), coagulants, and flocculents to the waste water in a rapid mixer tank (m). These chemicals bind the microorganisms in the wastewater, forming insoluble clumps (floc) in the secondary clarifier (n). The floc settles to the floor of the tank, accumulating as sludge. The sludge in the secondary clarifier may be transferred to an anaerobic digester or it may be transferred to drying fields

Wastewater is transferred from the secondary clarifier to the chlorine contact basin (o). Here sodium hypochlorite or chlorine gas (p) is injected into the flowing wastewater for disinfection. Filtration of the wastewater is sometimes used as a tertiary treatment method (not shown).

If the treated effluent (d⁵) is to be discharged into a lake, river or ocean, it must be dechlorinated (q) with sulfur dioxide or sodium bisulfite. If the effluent is to be used for reclamation (d⁶; e.g., irrigation), the chlorine is usually dissipated during conduction of the water in the miles of transmission pipes from the treatment plant to the reclamation site.

WASTEWATER TREATMENT

WASTEWATER d
GRIT CHAMBER e
GRIT d^1
DRYING FIELD f
PRIMARY
CLARIFIER g
SCUM d^2 /
SLUDGE d^3

ANAEROBIC
DIGESTER h
FERMENTED
SLUDGE d^4
TRICKLING
FILTER i
MEDIA j
ACTIVATED
SLUDGE
SYSTEM k

POLYMER l
MIXER m
SECONDARY
CLARIFIER n
CONTACT
BASIN o
CHLORINE p
TREATED
WATER d^5

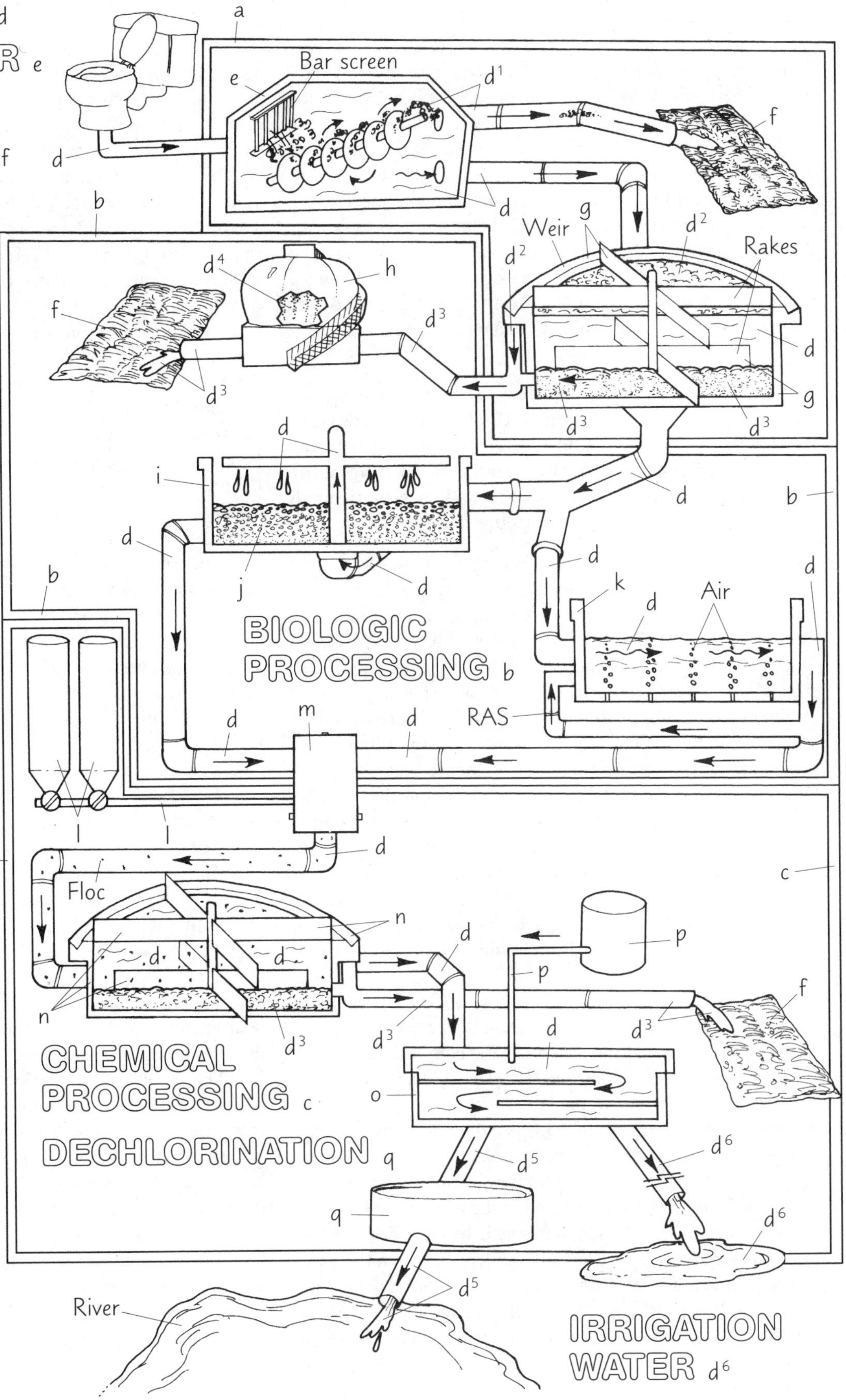

PHYSICAL PROCESSING a

Bar screen

Weir

Rakes

BIOLOGIC
PROCESSING b

Air

RAS

Floc

CHEMICAL
PROCESSING c

DECHLORINATION q

River

IRRIGATION
WATER d^6

Beer consists of water, barley malt, rice or corn adjuncts, hops, yeast, and alcohol. The word beer is derived from the Anglo-Saxon baere, meaning barley. The process of beer production begins with barley seeds which are collected and soaked in water, starting their germination. Germination produces carbohydrate-splitting enzymes called amylases that are essential to beer production. The germinated grains are dried, forming barley malt. The starch of the germinating grains is partially converted to sugar by the amylases during the malting process. Since yeasts cannot break down starch, the malt must be treated before the fermentation process to permit the amylase to continue converting the starch to fermentable sugar.

Set aside as many shades of one color as you have for (a) through (a^{11}). Consult the Using Color section to develop more shades and tints. Consider colors analogous to the color of beer, especially for (a^{10}) and (a^{11}). Starting at the top of the plate, color the titles and related parts (a) through (o). Use light colors so not to obscure detail.

Barley malt (a) is delivered from the storage silo (b) to a mill (c) and ground. It is then mixed with water (a^1) in a mash tank (d). Other grains, such as rice or corn, may be used as an adjunct (a^2); that is, as an additional source of fermentable sugar. The addition of rice also gives the beer lighter character. The adjunct grains are prepared in a cooker (not shown) with water and added to the mash tank. The rice and barley mixture or mash (a^3) is heated in the mash tank to enhance the enzymatic breakdown of the starch to simple sugars.

The mash is brought to the lauter tub (e) for filtering. The mash passes through a perforated plate (not shown), forming a grain bed (a^4) on top of the plate. This grain bed enhances the filtering of the liquid. The spent grains (a^5) are either discarded or used for animal feed. The filtered liquid, called wort (a^6), is rich in maltose (sugar) and will be the substrate for fermentation. The wort is recirculated through the lauter tub until the liquid is clear.

The wort is transferred to the brew kettle (f) or hopping tank. Here the sweet liquid is boiled and sterilized at 212° F/ 100° C, effectively halting the enzymatic conversion of starch to sugar. Dried petals of the vine *Humulus lupulus*, called hops (g), are added in stages to the brew kettle in a process called hopping. The essence of the hops is extracted during boiling in the kettle, giving the wort flavor, color (from the malt), and stability (from boiling). The spent hops are removed from the hop jack as the wort is transferred to the cooling tank.

The temperature of the hot wort is brought down in the cooler (h) to 52° F/11° C, an environment in which the yeast will soon metabolize the sugar. After passing through a settling tank (not shown) where it is further clarified by settling, the cooled wort is transferred to the fermentation tank (j) where yeast (i) is added. The process of putting the yeast in the fermentation tank is called pitching.

The fermentation of wort is effected by certain selected strains of the yeast *Saccharomyces cerevisiae* (Brewer's yeast). Other yeasts are considered contaminants ("wild yeast"). Yeasts that ferment rapidly accumulate at the top of the fermenter ("top yeasts") with the rise of carbon dioxide. The young beer taken off the top generally ferments at a higher temperature (about 65° F/18° C) after about six days and is used for making ale. Slower fermenters, producing less carbon dioxide, remain in suspension and flocculate, ultimately settling to the bottom of the tank ("bottom yeasts"), and produce a less alcoholic beer at a lower temperature (about 50° F/10° C) over a period of eight days or so.

Some brewers employ a second fermentation known as krausen, by using fresh pitched wort at the beginning of the lagering (aging) process. This enhances the maturation of the beer and carbonates it naturally. Carbon dioxide produced during fermentation may be collected and reintroduced later during carbonation (not shown).

The green beer (a^7) is pumped from the fermenter to a lager or aging tank (k) for storage of about three weeks. During this time, the flavor and aroma of the beer mature as the yeasts continue the fermentation process. The lager beer (a^8) is piped to a chiller tank (l) for further clarification. This process is called schoenering. Tannin or a silica gel may be added during this clarification process to pull out the undesired proteins, hops, and yeast. This mixture (floc) precipitates to the bottom of the tank and is removed.

The clarified schoene beer (a^9) is then sent through a tank (m) containing diatomaceous earth filters for final filtration. Some of the filtered beer (a^{10}) is then pumped to kegs (n) for shipping and consumption. This beer is not pasteurized and is called "draft beer." Because the kegs are chilled and kept cold, any further fermentation will be minimal.

Beer packaged in cans or bottles (a^{11}) is pasteurized (o) at 140° F/60° C for four to five minutes to kill remaining yeasts and bacteria (recall Plate 99). As an alternative to pasteurization, beer may be sterile filtered before canning or bottling, thus removing contaminants.

The alcoholic content of beer is about four percent by weight (or five percent by volume). The alcohol can be extracted from the beer to form a non-alcoholic brew.

If not rigidly controlled, microorganisms can destroy the quality of beer flavor, usually during fermentation or following filtration. Lactic acid-forming bacteria can cause clouding of the beer, and also give it an "off flavor." Wild yeast can cause bitterness in beer, and acetic acid-forming bacteria can cause souring of the beer. Rigid adherence to sanitary practices and process control will prevent these undesirable events.

BEER PRODUCTION

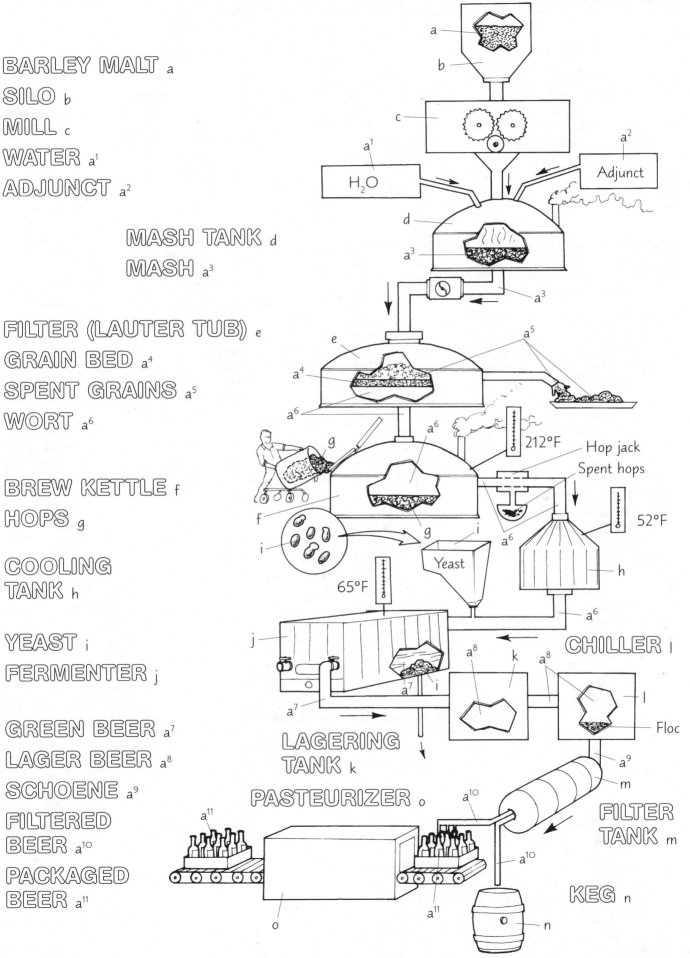

BARLEY MALT a
SILO b
MILL c
WATER a¹
ADJUNCT a²

MASH TANK d
MASH a³

FILTER (LAUTER TUB) e
GRAIN BED a⁴
SPENT GRAINS a⁵
WORT a⁶

BREW KETTLE f
HOPS g

COOLING
TANK h

YEAST i
FERMENTER j

GREEN BEER a⁷
LAGER BEER a⁸
SCHOENE a⁹
FILTERED
BEER a¹⁰
PACKAGED
BEER a¹¹

a
b
c
a¹ H₂O
a² Adjunct
d
a³
a³
e
a⁴
a⁵
a⁶
g
f
i
i
a⁶
212°F
Hop jack
Spent hops
g
i
Yeast
52°F
h
a⁶
65°F
j
a⁸ k
a⁸
CHILLER l
a⁶
a⁷ i
a⁷
l
Floc
a⁹
m
LAGERING
TANK k
PASTEURIZER o
a¹⁰
a¹⁰
a¹¹
FILTER
TANK m
a¹⁰
o
a¹¹
KEG n
n

Wine is essentially fermented grape juice, although other fruits can be used as sources of carbohydrate for the yeasts. The fermenting action of the yeast *Saccharomyces cerevisiae* converts the grape juice sugar to ethyl alcohol (ethanol). Fermenting of the juice of certain intact red grapes (with skins) results in red wine. Removing the skin of red grapes before fermenting can produce a "blush" or pink wine. Fermenting of the juice of white grapes results in white wine.

Starting at the top of the plate, color the titles and related structures (a) through (l^8). Choose or create shades and tints of one color for (l) through (l^8). Use colors that do not obscure the illustrative detail.

Grapes (a) used in the production of wine include species of *Vitis vinifera*, as well as hybrids of *V. vinifera*. The grapes are picked at a time when the sugar content is between 20 to 24 degrees brix (percent sugar by weight). The picked grapes, free of leaves but with stems and seeds, are placed in a hopper where an auger moves them into the destemmer and crusher (b). The stems (d) are separated from the grapes. The crushed grapes, and resulting juice, called must (c), are transferred to the fermentation tank (f; also called the fermenter).

Before being transferred to the fermentation tanks, the must is usually treated with sulfur dioxide (e; SO_2). This compound functions as an antioxidant, kills the natural (wild) yeasts, and inhibits spoilage by microorganisms. Most winemakers use their own preferred yeast strains that offer predictable results and are resistant to the lethal effects of SO_2.

The cultured yeast (g) is added to the fermentation tank (f) containing the grape juice (h) and skins and seeds (i). Here the yeast multiplies rapidly, using up all available oxygen. Large amounts of carbon dioxide (j; CO_2) evolve from the process and the must bubbles profusely. As the oxygen is used up, the yeast's metabolism shifts to fermentation. The grape glucose is now converted to ethyl alcohol (k; ethanol, ETOH), the product of fermentation (recall Plate 101). The temperature of the fermenting must needs to be controlled, as yeasts may die if the temperature is excessively high. The cloudy, fermented product, called free run wine (l), is transferred to a settling tank (n).

After the cloudy wine is transferred to the settling tank, the skins and seeds fill the bottom of the fermenter. These skins and seeds are then transferred to a press (m) where they are squeezed to elicit the press wine (l^1). This wine is transferred to the settling unit (n). The solid matter remaining in the press is called pomace and may be discarded.

In the settling tank, yeasts and the by-products of the fermentation process settle out of the wine and fall to the bottom of the tank. These solids (o; called lees) include proteins, yeasts, and other organic substances, and will be discarded, leaving clarified wine (l^2) in the tank.

The clarified wine is generally transferred to wooden barrels (p) for aging. As the wine ages in the barrel, sediments continue to fall out. The wine is transferred from barrel to barrel, in a process called racking, to further clarify the wine. The wine will mature (l^3) in the barrel and develop its distinctive flavor and aroma bouquet. Aging may go on for weeks, months, or years.

A secondary fermentation using a culture of malolactic acid bacteria (q) is generally promoted in the wine-filled barrels to anaerobically convert the natural malic acid in the grape to lactic acid and carbon dioxide. This process is called malolactic fermentation, and results in a smoother wine.

Once the aging process is complete, and providing the wine has no remaining fermentable sugars or malic acid, the wine may be bottled without additional processing if it has acceptable clarity. If the wine is to be bottled sweet or contains malic acid, it must be sterilized when bottled to prevent the possibility of undesired fermentation in the bottle. Filtration or pasteurization are the methods of ensuring sterile bottling.

The mature, fully fermented wine is bottled as "dry" (l^4). It may be altered before packaging by the addition of ETOH. These wines, such as port and sherry, are said to be "fortified" (l^5). An extra period of fermentation of wine in a burst-resistant container produces carbonation and a champagne or sparkling wine (l^6). When less than 3% sugar is added to wine, the product is called a sweet wine (l^7). If the sugar content is greater than 10%, the product is called a dessert wine (l^8).

Although the wine is considered mature at the time of packaging, in fact, the wine will continue to undergo subtle chemical changes in the container for years. Such changes are said to provide an added "bouquet" to the wine.

RED WINE PRODUCTION

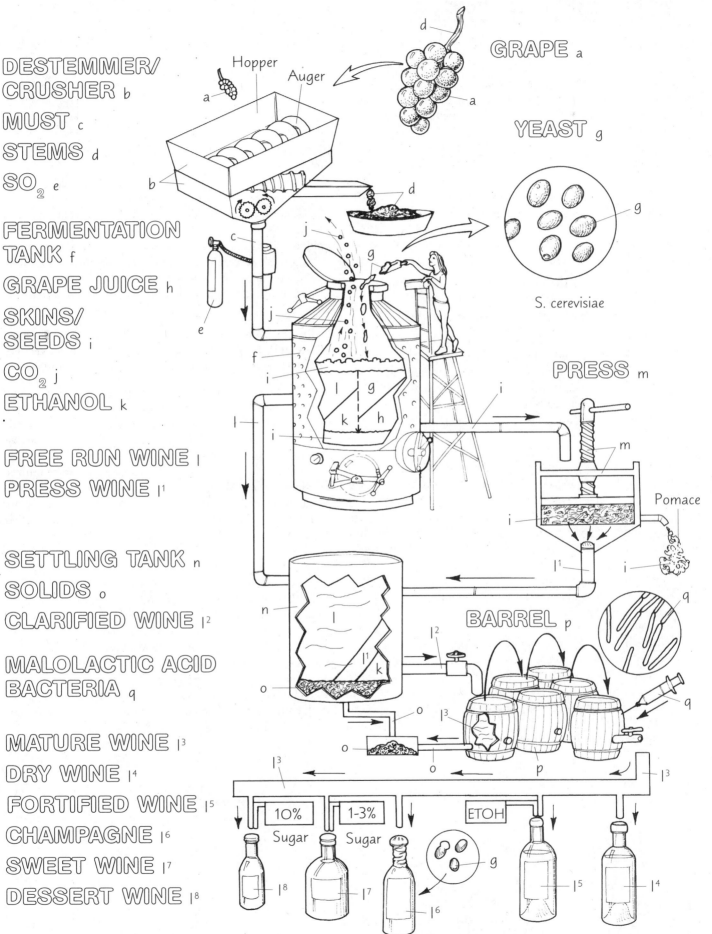

DESTEMMER/CRUSHER b

MUST c

STEMS d

SO$_2$ e

FERMENTATION TANK f

GRAPE JUICE h

SKINS/SEEDS i

CO$_2$ j

ETHANOL k

FREE RUN WINE l

PRESS WINE l^1

SETTLING TANK n

SOLIDS o

CLARIFIED WINE l^2

MALOLACTIC ACID BACTERIA q

MATURE WINE l^3

DRY WINE l^4

FORTIFIED WINE l^5

CHAMPAGNE l^6

SWEET WINE l^7

DESSERT WINE l^8

GRAPE a

YEAST g

Hopper

Auger

S. cerevisiae

PRESS m

Pomace

BARREL p

10% Sugar

1-3% Sugar

ETOH

Microorganisms play a positive role in nature by participating in several different cycles of elements in the soil, such as the nitrogen, carbon, and sulfur cycles. Nitrogen is essential to the formation of amino acids and nucleic acids, two key groups of organic compounds in animals. The most available form of nitrogen on the planet is nitrogen gas (N_2). But eukaryotes cannot synthesize nitrogenous compounds from nitrogen gas; they need the intervention of plants and microorganisms to do so. This plate reveals how nitrogen is fixed by bacteria and how free nitrogen is released by bacteria in what is called the nitrogen cycle. We enter the cycle with the animal's role.

Begin by coloring the animal and its title (a) at station 1. Continue counterclockwise around the cycle, following stations 2 through 10, finishing with arrows (k).

The animal (a) consumes organic matter and produces waste in two forms: urea (b) and indigestible protein (b^1). Urea is the principal nitrogenous component of urine. Indigestible protein, consisting of amino acids, is found in feces. The animal's body after death is also a waste material containing large amounts of nitrogenous matter. The principal molecule in these waste materials, with respect to the nitrogen cycle, is the amino group ($-NH_2$).

Indigestible protein and urea from animals, and protein from plants are broken down in the process of waste decomposition (c; station 2). Decomposition is the breakdown of complex molecules to simple ones. Decomposition is accomplished by a variety of aerobic and anaerobic soil bacteria near the surface or in deeper soil around plant roots. These bacterial decomposers (c^1) free amino groups (d^1; $-NH_2$) from amino acids (d) and urea, and form ammonia (e; NH_3; station 3). Ammonia is similar to an amino group, with an extra hydrogen atom.

Once ammonia has been formed in the soil, it may be utilized directly (e^1) by plants in synthesizing amino acids and proteins (station 4). In a process called nitrification (f), some ammonia is oxidized by the bacterial species *Nitrosomonas* (f^1) to form nitrite (g; NO_2^-; station 5) The nitrite is then converted to nitrate (h; NO_3^-) by soilborne bacterial species of *Nitrobacter* (f^2; station 6). Both genera of bacteria are soilborne Gram-negative rods.

Nitrites (g^1) and nitrates (h^1) can be extracted from the soil and utilized by the roots of the plants (station 7) by incorporating nitrogen into its amino acids, nucleic acids, and other nitrogenous compounds. The resulting plant structure becomes food (i; station 8) for animals. A portion of planetary nitrogen is thereby recycled in this way.

In a more common process, nitrate is denitrified (j; station 9) by soil and aquatic denitrifying bacteria (j^1), such as species of *Pseudomonas*. Denitrification of nitrate forms nitrogen gas (k; N_2). As the gas accumulates in the soil and water, it gradually diffuses into the atmosphere (station 10).

Fully 80 percent of the atmosphere is nitrogen. Nitrogenous compounds are essential for life; thus, for life to continue as we know it, a return trip into living things is required.

Color the titles for Nitrogen Fixation and the nitrogen fixers, and the bacteria and related arrows, at lower left. These, (l) and (l^1), should share shades of the same color.

The biologic process by which nitrogen gas is converted to ammonia is called nitrogen fixation (l; station 11). Nitrogen-fixing bacteria or fixers, such as *Clostridium* and cyanobacteria (l^1), play the essential role in this process because they possess the enzyme systems necessary to trap atmospheric nitrogen and convert it to ammonia. The ammonia can then be utilized by other soil bacteria, converting it to compounds essential to plants and animals. Nitrogen fixation can also occur, on a minor scale, during lightening discharges, and in certain other instances.

Some bacteria live in nodules on the roots of pod-bearing plants called legumes (not shown). Excess amounts of bacterial ammonia and nitrogen compounds are utilized by the legumes to synthesize their amino acids, proteins, nucleic acids, and other nitrogen-rich compounds. These accumulate in the legume and can be consumed by animals, completing the cycle. Peas, beans, soybeans, alfalfa, and clover are examples of protein-rich legume plants. They are well known for their nutritious quality, a factor due to the production of nitrogen compounds by bacteria living together with the plant. The value of the nitrogen-fixing ability of the bacteria cannot be overemphasized.

NITROGEN CYCLE

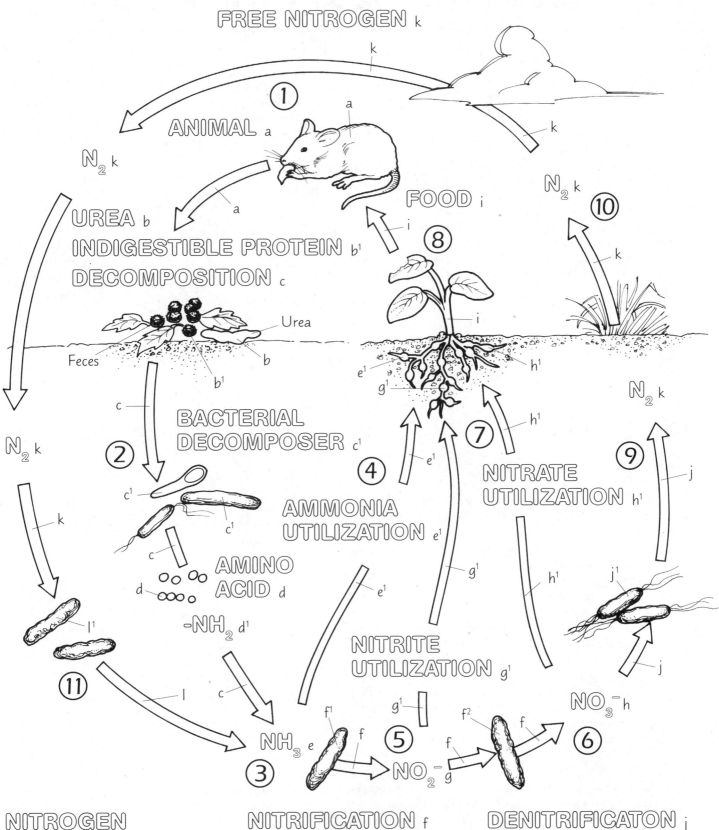

FREE NITROGEN k

k

① ANIMAL a

a

N_2 k

a

FOOD i

i

N_2 k

⑩

k

UREA b

INDIGESTIBLE PROTEIN b^1

DECOMPOSITION c

⑧

Urea

Feces

b

b^1

e^1

h^1

g^1

i

c

BACTERIAL
DECOMPOSER c^1

N_2 k

N_2 k

c^1

c^1

⑦

h^1

② ④

e^1

g^1

⑨

c

AMMONIA
UTILIZATION e^1

NITRATE
UTILIZATION h^1

j

k

AMINO
ACID d

d

e^1

g^1

h^1

j^1

l^1

$-NH_2$ d^1

NITRITE
UTILIZATION g^1

j

⑪

l

c

NO_3^- h

NH$_3$ e

f^1

g^1

f^2

f

f

⑥

③

f

⑤

f

NO_2^-
g

j

NITROGEN
FIXATION l
CLOSTRIDIUM/
CYANOBACTERIA l^1

NITRIFICATION f
NITROSOMONAS f^1
NITROBACTER f^2

DENITRIFICATON j
PSEUDOMONAS j^1

GETTING THE MOST OUT OF COLOR

This book involves coloring. Lots of it. You will be using color to identify a structure and link it to its name (title). Color will be used to differentiate one structure from another, and to show relationships among structures. You will use color to emphasize action, heat, irritation, and it will be used to reflect cool or subdued states. You will give an aesthetic quality to the plates you have colored. What you have colored you will remember for years based partly on the colors you selected. This brief introduction on the use and character of color will give real support to your coloring goals by providing you with a basic understanding of colors and color matching. It will also provide you with the ability to extend a basic collection of twelve hues to thirty six or more colors.

What color will you choose? On what basis will you choose it? How many values of a color do you need and how many do you *have*? How does one pick a color to represent a particular structure or process? How can one use a color to suggest activity? How can you extend your coloring pen/pencil set to make far more colors than you have? Finally, how can you plan the coloring of each plate to get a really pleasing result? Read on.

PRINCIPLES OF COLOR

Sunlight is white light. White light contains all of the colors in the visible spectrum. Visible light represents a very small band in an immensely large band of radiant energy, most of which is not visible to the human eye. If one places a prism in sunlight, an array or spectrum of colors emerge. Light is the essence of color, yet in itself, it is not a color. Without light, there is no color. Night is the absence of light and therefore the absence of color.

Color vision is based on reflectance. White light, as we have mentioned, is composed of all colors. When light strikes an object such as a lemon, most of the spectrum colors in the light are absorbed by the lemon. A small amount is reflected off of the surface of the lemon - the reflected light. This is the color we perceive. It is the color of the object. In the case of the lemon, the reflected color is yellow.

A good example of a spectrum or sequence of color bands can be seen in a rainbow. Rainbows appear when the sun is shining and it is raining. When the white light of the sun passes through raindrops, the light is bent or refracted. When white light is refracted (as by a prism or by raindrops), the colors of the spectrum separate and become visible. Each color of the spectrum has a different wavelength or characteristic. Simply stated, the rainbow spectrum begins with violet and moves to

red, then orange, yellow, green, blue, and back to violet. If we bend the rainbow into a circle and join the violets, we have a color wheel.

To appreciate these color changes, color the rainbow below using the colors indicated. Then color the wheel below the rainbow in the same sequence as the rainbow, starting at the notch with violet.

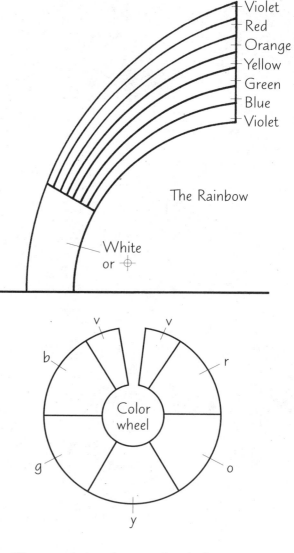

There are three **primary** colors in the spectrum:

red ◯ , **yellow** ◯ , and **blue** ◯

Primary colors cannot be created by mixing other colors. They can be combined (mixed) to make other colors.

By mixing two primary colors you create what is called a **secondary** color:

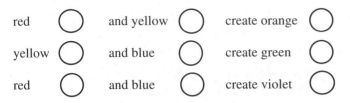

red ◯ and yellow ◯ create orange ◯

yellow ◯ and blue ◯ create green ◯

red ◯ and blue ◯ create violet ◯

This processing can be continued by mixing a primary and a secondary color creating what is known as a **tertiary** color. Tertiary colors have simple names based on the colors combined. Thus mixing red and orange creates the tertiary color red orange. There are six tertiary colors.

Below we have another color wheel made up of three concentric circles. The color wheel is divided into six wedges each marked with a primary or secondary color.

Color each wedge completely with the color indicated. Begin with the primary colors. For the secondary colors, try mixing the primaries instead of using the secondary colors that you may have. This may not turn out well with coloring pens, in which case you may have to use the secondary colors.

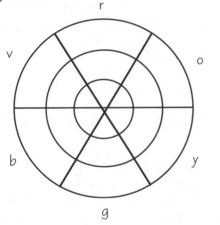

Color is known as **hue**. A pure color means maximum **intensity** or **saturation** of color.

Every pure hue (color) has another set of characteristics known as **value**. Value is the lightness or darkness of a color. Each color has a range of value that extends from very light (near white) to very dark (near black). When we lighten a color, we **tint** it. When we darken a color, we **shade** it. For example, red is a saturated color of maximum intensity. Pink is a tint of red; burgundy is a shade of red.

On your color wheel, color over all of the colors in the outer band or circle with white (or the closest to white that you have). Again, pencils work better than pens in this exercise. Now color all of the colors in the inner band or circle with black, but not enough to obscure the color.

You have now tinted and shaded the primary and secondary colors. There are now three colors for each hue on the wheel. *You can use a color many times by changing its value.* This fact has importance to you as you select various tints and shades of a single color for relating similar structures or related processes in the plates you are working.

Pure or intense colors have different value. Look at the pure colors on your color wheel; notice that blue has a darker value than yellow. Each color has its own value.

Below is a black/white value scale consisting of 11 boxes arranged in a horizontal line (identified by #1). It is called a gray scale. Starting with white (w) at far left, we have added 10% of black to each square progressively until we have pure, 100% black (b) at far right.

Below this scale, numbered #2, there is another 11-box scale that is blank. **Set aside your six primary and secondary colors. One at a time, place the point of one of the pencils/ pens over the gray scale and move it across until you find a gray that has the same value (darkness). Fill the box in under that gray with the matching color.** In the event that more than one color has the same value, color the space under that square.

In the series of boxes identified as #3, #4, and #5, you can make your own value scale from three colors. Leave the boxes at far LEFT uncolored (white, w), and fill in the box at far RIGHT with black (b). Locate one pure (intense) hue from scale #2 and color the same box on scale #3. To the left of the hue, progressively tint the boxes until you reach the white box. To the right, progressively shade the color until you reach black. Repeat the process with two different colors in boxes #4 and #5.

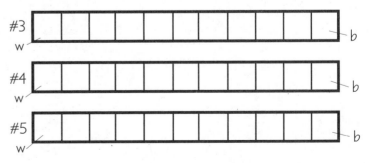

HOW TO USE COLORS

Our next step is to understand how to use color. Colors do many things visually and psychologically. One can create a sense of quiet relaxation, emotional stress, or intellectual excitement. Through color combinations the artist can make one color look like another color, or make a color look brighter than it actually is.

We associate colors with physical phenomena. Colors that are associated with the sun and fire are called "warm colors." Warm colors, such as red, yellow, and orange, visually **advance** or come forward in a scene or painting. "Cool colors" are associated with ice and water; they are blue and green, and they visually **recede**. We cool the far distance in a painting to create atmospheric perspective.

Combinations of colors can have many effects. When we use colors that are next to each other on the color wheel, they create a sense of harmony and are called harmonious or **analogous colors.** Analogous colors have a restful nature. An example of harmonious colors on your color wheel would be red, violet, and blue. Place a patch of these colors on the page next to this paragraph to see the harmony. Pick two more harmonious color schemes and place them on the margin.

Color combinations that use colors located far from each other are contrasting in their nature. Contrasting colors create a greater sense of emotion than harmonious colors do. If we use contrasting colors that are an equal distance apart we have a **triad**. Primary colors are triadic. Red, yellow, and blue will create strong contrast. Secondary colors are also triadic.

Colors directly across the color wheel from each other are **complementary colors**. On the color wheel, red and green are complementary colors. Yellow - violet, and blue - orange are also complementary. When complementary colors are placed next to each other, they intensify the color of each; red is a brighter red and green is a brighter green. This is known as simultaneous contrast.

In the boxes below, color the primary colors on the top bar and their complementary secondary color on the lower bar.

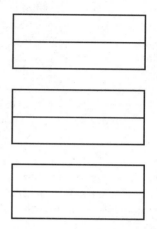

If the colors you used are pure hues, you should be able to observe the effects of simultaneous contrast. Artists like Vincent Van Gogh, Paul Gauguin, and Toulouse-Lautrec were masters of color contrast. Black and white also create simultaneous contrast.

It is interesting to note that while placing complementary colors next to each other, they brighten the color of each. When we mix complementary colors, they dull or neutralize each other.

With the above insights, you can extend your enjoyment and your skill in coloring the plates in this book. Through value, you can show how functions of the same part can change. And you can make six colors do the work of eighteen. Happy coloring!

- Jay and Christine Golik
Napa Valley College, California

REFERENCES

Abbas, AK, Lichtman, AH, and JS Pober, *Cellular and Molecular Immunology,* WB Saunders, Philadelphia, PA, 1991

Alcamo, IE, *AIDS: The Biological Basis,* WC Brown, Dubuque, IA, 1993

Alcamo, IE, *Fundamentals of Microbiology,* 4th ed, Addison-Wesley, Reading, MA, 1994

Brock, TH, Madigan, MT, Marinko, JM, and J Parker, *Biology of Microorganisms,* 7th ed, Prentice Hall, Englewood Cliffs, NJ, 1994

Campbell, MC, and JL Stewart, *The Medical Mycology Handbook*, J Wiley and Sons, New York, 1980

Dorland's Illustrated Medical Dictionary, 28th ed, WB Saunders, Philadelphia, PA, 1994

Elson, LM, *The Zoology Coloring Book,* HarperCollins, New York, 1982

Emmons, CW, *Medical Mycology,* 3rd ed, Henry Kimpton Publishers, London, 1977

Ganong, WF, *Review of Medical Physiology,* 16th ed, Appleton Lange, Norwalk, CT, 1993

Garcia, LS and DA Bruckner, *Diagnostic Medical Parasitology,* 2nd ed, Am. Soc. Microbiology, Washington, DC, 1993

Jawetz, E, Melnick, JW, and EA Adelberg, *Review of Medical Microbiology,* 17th ed, Appleton Lange, Norwalk, CT, 1987

Junqueira, LC, Carneiro, J, and RO Kelley, *Basic Histology,* 7th ed, Appleton Lange, Norwalk, CT, 1992

Kapit, W, and LM Elson, *The Anatomy Coloring Book,* 2nd ed, HarperCollins, New York, 1994

Katz, M, Despommier, DD, and RW Gwadz, *Parasitic Diseases,* 2nd ed, Springer-Verlag, New York, 1989

Kwon-Chung, KJ, and JE Bennett, *Medical Mycology,* Lea and Febiger, Philadelphia, PA, 1992

Merriam-Webster's Collegiate Dictionary, 10th ed, Merriam-Webster, Springfield, MA, 1993

Mims, CA, Playfair, JHL, Roitt, IM, Wakelin, D, and R Williams, *Medical Microbiology,* Mosby, St. Louis, MO, 1993

Netter, F, *Atlas of Human Anatomy,* Ciba-Geigy, Summit, NJ, 1989

Ravel, R, *Clinical Laboratory Medicine,* 5th ed, Mosby, St. Louis, MO, 1989

Stites, DP, and AI Terr, *Basic and Clinical Immunology,* 7th ed, Appleton Lange, Norwalk, CT, 1991

Tortora, GJ, Funke, BR, and CL Case, *Microbiology,* 4th ed, Benjamin-Cummings, Menlo Park, CA, 1992

———, *Life, Death, and The Immune System,* Scientific American (Special Issue), September, 1993

GLOSSARY

This glossary is limited in scope, as many of the texts of the plates explain the definition of structures and processes. See the Index and Table of Contents for guidance to explanation of terms not included here. For expanded definitions and explanations, refer to the texts or articles listed in the section on References. The definitions used here are compatible with those in Dorland's Illustrated Medical Dictionary, 28th edition, WB Saunders, Philadelphia, PA 1994.

A

Adeno-, referring to glands, including lymph nodes.
Adenopathy, enlargement of the lymph nodes.
AIDS, acquired immune deficiency syndrome.
Alga (-e), single-celled, photosynthetic, eukaryotic organism.
Anorexia, loss of appetite; loss of desire to eat food.
Antibiotics, chemical substances produced by microorganisms that are capable of killing other microorganisms. Some antibiotics are synthesized in the laboratory.
Antiseptic, a substance that inhibits the growth and development of microorganisms without necessarily killing them.
Arbovirus, an arthropodborne virus, carried by mosquitoes and ticks in most cases.
Artery, a vessel that conducts blood from the heart to the tissues, directly or indirectly. Except for the pulmonary arteries, they conduct oxygenated blood.
Articular cartilage, the thin cartilage layer at the ends of bones where joints with other bones occur. Articular cartilage does not repair well after injury or disease as it has poor access to oxygenated blood. Cartilage is not vascularized (avascular), and receives its nutrition by diffusion.
Aseptic, without sepsis; sterile; without pathogens or toxins.
Autoantibody, an antibody directed against self-antigens.
Autotroph, an organism capable of synthesizing its own organic nutrients.
Attenuate, lessen the amount; reduce the virulence or vitality.
Atypical, unusual.

B

Bacterium, a prokaryotic cell that typically possesses a cell wall in addition to a cell membrane, and divides by fission.
Biopsy, removal of a piece of tissue for microscopic analysis.
Bursa of Fabricius, a specialized lymphoid organ in the bird near the cloaca. The cloaca is a tubular passageway for the excretion of uric acid crystals and feces, as well as reproductive gametes and eggs. The bursa of Fabricius produces B lymphocytes in birds.

C

Catalyze, to speed up a reaction; enzymes catalyze reactions without becoming a part of the reaction or products.
CD4, cluster of differentiation. Cluster refers to a group of monoclonal antibodies. A receptor on the surface of a lymphocyte, such as a T cell, will bind with one of a cluster of monoclonal antibodies. This attachment permits the identification ("differentiation") of a specific lymphocyte because of the specificity of the lymphocyte-antibody reaction (in effect, an antigen-antibody reaction). The receptor on the lymphocyte that binds with the known antibody is called a CD receptor. CD receptors are numbered (CD1, CD2, and so on). The term "CD" arose in immunology laboratories in conjunction with the need to identify different lymphocytes.
Cellulitis, inflammation of the subcutaneous fibro-fatty tissue (superficial fasciae). Inflammation here spreads rapidly along the tissue plane. Pockets of pus (abscesses) may be associated with cellulitis. Cellulitis is initiated through wounds or burns of the skin. The infective agents in cellulitis are most commonly *Staphylococcus aureus* and group A (beta-hemolytic) streptococci.
Chimera, an organism or organelle, e.g., plasmid, composed of parts from different organisms.
Complementarity, a condition in which the structure of one molecule fits the structure of another, such as the reciprocal shapes of the antigenic determinant of antigens with the antigen binding sites of antibodies; also the reciprocal shapes of adenine and thymine molecules, and cytosine and guanine molecules, in DNA and RNA synthesis. The term is also used with reference to the complementarity between the parent DNA strand and the newly synthesized RNA strand in RNA synthesis.
CO_2, carbon dioxide.
Congenital, existing at birth.
Convertase, an enzyme that converts a substance to its active state.
Copulate, to engage in sexual union between male and female.
Cramps, painful muscle contractions.
Curd, congealed milk.

Curdling, congealing of milk; to form curds.

Cutaneous, referring to skin.

Cyanobacteria, formerly called blue-green algae; prokaryotic, pigment-containing, largely photosynthetic unicellular bacteria.

D

Devitalized, no longer alive; without blood supply or sources of oxygenation.

Disease, a change in normal structure and function, associated with signs and symptoms.

Disinfectant, a substance that removes infective particles from a foreign object; generally kills or inhibits infectious microorganisms.

DNA, deoxyribonucleic acid.

Domain, one section of a globular protein that is distinct and recognizable.

E

Electrolyte, a chemical substance, organic or inorganic (mineral), that is dissociated in the body fluids, including the blood. These atoms or molecules have a charge, hence "electrolytes." Also called ions; e.g., sodium (Na^+), magnesium (Mg^{+2}), nitrate (NO_3^-), and so on.

Etiology, cause.

Eukaryote, an organism whose cells each have a true nucleus enveloped by a nuclear membrane.

Excision, removal by dissection.

F

Facultative, an organism that can grow and reproduce in aerobic as well as anaerobic environments.

Fermentation, a catabolic, energy-yielding, largely anaerobic process in which pyruvic acid is broken down into certain end products, including carbon dioxide, ethanol, and lactic acid.

Fibrosis, fibrotic, referring to the formation of connective tissue fibers. Scarring is fibrosis.

Fungus (-i), eukaryotic organism occurring as a single cell (yeast) or a long, branching filament (mold).

G

Gamete, a sex cell with a single set of chromosomes (haploid).

Generation time, the time it takes for a bacterial population to double.

Genetic engineering, the techniques and mechanisms of manipulating, transferring, and exchanging segments of genetic material in and among organisms.

Genital, relating to the reproductive organs.

Germicide, an agent that kills pathogenic bacteria.

H

Helminth (-es), a parasitic worm.

Hemaglobin, the iron-containing protein of the red blood corpuscle. It attracts and binds oxygen.

Hematogenous, referring to blood.

Hemorrhage, bleeding.

Hepatomegaly, enlargement of the liver.

Hermaphrodite, an organism that has both male and female sex organs.

Heterotroph, an organism requiring consumption or absorption of organic compounds for nutrition and survival.

HIV, human immunodeficiency virus.

Hybrid, an organism produced from two different kinds of parents (of different strains or species).

Hydatid, a cyst.

Hyperbaric, oxygen pressure greater than atmospheric.

Hypersensitivity states, type I (anaphylaxis; immediate reaction when re-exposed to specific antigen), type II (antibody-cell surface antigen reaction; activates complement cascade), type III (immune complex or antibody-antigen reaction with activation of complement cascade), and type IV (T lymphocyte-mediated).

Hypha (-e), long, branching, often intertwining chains of cells, characteristic of fungi.

I

Icosahedral, a geometric figure characterized by twenty faces and twelve corners.

Incubation, an environment in which the temperature and humidity are carefully regulated to develop a population of microorganisms in culture; also the period of time during which microorganisms develop in a host before onset of symptoms and signs.

Infection, the population of body tissues by microorganisms.

Inflammatory cells, cells that are associated with an inflammation, including neutrophils and other phagocytes.

Inoculation, the introduction of disease-producing micro-organisms into living tissue or culture media.

Inoculum, the material to be inoculated (see inoculation).

Intracellular, within a cell.

-itis, a suffix referring to inflammation of the structure represented by the root word, e.g., encephalitis, inflammation of the brain.

J

Jaundice, a condition in which bile pigments enter the circulation, giving a yellowish tint to the skin. Jaundice is perhaps most easily observed in the whites of the eyes (sclerae). Bile is produced in the liver; it is conducted to the gall bladder for storage by numerous small ducts converging to form the hepatic and bile ducts. When these small ducts in the liver are damaged by disease or traumatic processes, the bile flow is blocked, backs up, and damages the cell walls; then the bile easily enters the circulation. Within the liver, the small bile ducts and small arteries and veins are bound together.

K

Kaposi's sarcoma, malignant neoplastic (cancerous) growths of small blood vessels. They grow into the skin and form purple or bluish-red elevations. It is a hemorrhagic disease, and occurs most rapidly in immunodepressed individuals, such as those with AIDS. The disease is also endemic among the people of Eastern Europe and Central Africa.

L

Lactobacillus, Gram-positive, rod-shaped bacteria, wide-spread in nature; they occupy the human mouth and vagina; non-pathogenic. They produce lactic acid by fermentation.

Larva (-e), a developmental stage of certain organisms. Larvae are usually independent and motile; they may be infective.

Latency, a period of time in which there is no activity; a time between stimulus and response.

Ligase, an enzyme that influences the formation of a bond between two molecules.

Lining tissues, those tissues that form a layer or combination of layers between the wall of an organ and the cavity of the organ. All cavities have walls; the surface layer of that wall is a lining tissue (e.g., endothelium, mucosa, serosa, synovium, and so on).

LPS, lipopolysaccharide component of the cell wall of bacteria; forms a toxic component (endotoxin) in Gram-negative bacteria.

Lumen, cavity.

Lymphadenopathy, enlarged or swollen lymph nodes; also called adenopathy.

Lymphocyte, one of the white blood cells or leukocytes. Characterized by a large round nucleus taking up most of the cell volume, the lymphocyte is found in the circulating blood, making up about 30-40% of the white blood cell population. Lymphocytes are motile, and can enter and leave the blood or lymph circulation. They populate most of the connective tissues, and especially the lymphoid tissues of the liver, spleen, lymph nodes, and mucosal associated lymphoid tissues. They may take the form of immature lymphocytes or lymphoblasts, T lymphocytes or T cells, B lymphocytes or B cells, and large (non-T, non-B) lymphocytes. T and B lymphocytes are associated with cell-mediated and humoral immunity, respectively. They are not phagocytic. Lymphocytes of the blood may be designated "small" or "large." See "white blood cell" in this glossary.

Lysis, lytic, chemical destruction or dissolution.

M

Macule, macula, a spot on the skin that is not raised or elevated off the skin. Seen in German measles.

Malaise, fatigue.

Medium, culture, a nutrient substance used to develop and maintain populations of microorganisms.

Mesentery, mesenteric, the double-layered membrane of peritoneum that connects an abdominal organ, such as the intestines, to the body wall in vertebrate animals, including humans. The mesentery gives support to the organ, and provides a means by which the nerves and blood vessels reach and leave the organ. Some of the nerves and vessels supplying the abdominal organs are called mesenteric nerves, veins, or arteries. The mesenteric veins are tributaries of the hepatic portal vein which brings nutrient-laden blood to the liver for metabolic processing.

Miasma, a disease-producing effect arising from the earth or atmosphere; such a notion was held to be the cause of disease until scientific research proved otherwise.

Microbe, a microorganism; especially, disease-causing microorganisms.

Microorganism, protozoans, fungi, bacteria, and viruses. Generally, any organism that can only be seen with a microscope.

Monomer, a single substance composed of non-repeated units. Such a substance is capable of binding with another similar substance, thus forming a dimer (di, two); or with two other similar substances (trimer), and so on (pentamer, polymer).

Mucosa (-e), the lining tissue of organs with cavities; consists of an epithelial or cellular layer supported by fibrous connective tissue, often with glands. It may include smooth muscle.

N

Nigrosin, an aniline (acidic) dye, $C_{36} H_{27} N_3$.

O

Obligate, an organism that requires fixed conditions for survival, and is not capable of adapting; the opposite of facultative; an obligate anaerobe cannot exist in the presence of air; an obligate aerobe requires oxygen.

Operon, a group of genes that function to turn on or turn off the synthesis of a protein, usually an enzyme; includes structural genes that initiate protein synthesis, operator genes that regulate the activity of structural genes, and regulator genes that control the activity of operator genes. There are enzyme induction operons and enzyme repression operons.

Osmosis, osmotic, the movement of fluid from one compartment to another, where there is a semi-permeable membrane between the two compartments, and that membrane prevents the passage of particles (solute) from one compartment to the other. The movement of fluid through the membrane and into the other compartment is induced by the greater amount of particles in that compartment compared with the first.

P

Papule, papular, a small (less than 1 cm in diameter), circular, solid, raised or elevated spot on the skin.

Parasite, an organism that feeds off another living organism.

Pathogen, a microbe that causes disease.

-penia, few.

Pentamer, a structure or substance consisting of five monomers.

Perianal, the area around the anus.

Phagocyte, a cell that takes up foreign material into its cytoplasm by endocytosis. See phagocytosis. Certain phagocytes of the blood and tissues engulf antigenic material and "present" it to lymphocytes (antigen presenter cells or APC). This activates the lymphocytes, and sets in motion the immune response. Phagocytes include neutrophils and monocytes of the white blood cell population, macrophages, histiocytes, scavenger cells, and "inflammatory cells."

Phagocytosis, the uptake of particulate matter, including microorganisms, by phagocytes.

Pleomorphism, pleomorphic, more than one distinct form.

Polymer, a structure of substance consisting of several monomers.

Prokaryote, an organism whose cells lack a true nucleus.

Protease, an enzyme that breaks down protein into smaller units (polypeptides).

Proteolysis, the disintegration of protein. Usually accomplished by proteolytic enzymes.

Pseudopodium (-ia), a transient extension of the cell membrane permitting movement of certain cells, such as amebae.

Pyruvate, the salt of pyruvic acid. Pyruvate is the anionic or negatively charged form of pyruvic acid. The terms "pyruvate" and "pyruvic acid" are generally used interchangeably.

R

Rash, a temporary skin eruption or exanthum; an area of reddish or skin colored spots. They may be elevated off the skin (papule, vesicle) or not (macule).

Recombinant, refers to a cell with a complement of genes that did not come from either parent cell.

Replication, a duplication or repetition.

Reservoir, a storage site for infection; refers to a host of an infectious microorganism that does not experience the signs and symptoms of the disease. Such a host is called a carrier. A carrier, hosting a reservoir of infections microorganisms, can transmit the pathogen to another person or animal.

Restriction, a point along a strand of DNA where enzymes can cleave or cut the strand.

Rhino-, referring to nose.

RNA, ribonucleic acid.

S

Safranine, a basic, amino derivative consisting of a red dye.

Saprobe, an organism that uses dead organic matter for nutrients.

Scar, scar tissue, a particularly dense area of fibrous tissue in an area of old or recent healing; or a particularly dense mass of fibrous tissue, possibly seen through the skin, reflecting a prior disruption of tissue.

Self-limiting, a disease that terminates without specific treatment and leaves no permanent sequelae. The term implies that the immune system of the host overpowers the pathogen, ending the infectious process, and the related signs and symptoms.

Septicemia, poisoning of the blood with toxins or pathogenic microorganisms.

Seroconversion, the appearance of antibodies in the serum following a period in which there were no apparent antibodies in the serum.

Shock, a condition of the body in which the dynamic equilibrium of the cardiovascular and metabolic state is sharply upset; characterized by low blood pressure, poor perfusion pressure of tissues; clinically, shocky patients are pale and weak with clammy cold skin, rapid breathing, and rapid pulse. There are various kinds of shock, including hypovolemic, cardiogenic, neurogenic, and septic.

Splenomegaly, enlargement of the spleen.

Spontaneous generation, the concept that microorganisms arose spontaneously, as opposed to arising from an ancestral cellular lineage.

Strain, a group of organisms within a species.

Sterilization, the absolute elimination of all microorganisms.

Streptococcus, a genus of Gram-positive cocci that are anaerobic, non-motile, and non-sporeformers. Arranged into four groups, one of which includes the beta-hemolytic pathogenic species.

Suspension, a preparation in which a fine substance is introduced into a liquid vehicle and remains "in suspension" (does not precipitate out) within that vehicle.

T

Template, a pattern; a strand of DNA or RNA that provides the sequence of nucleotides for a strand of DNA or RNA being synthesized.

Terrestrial, referring to land or the earth.

Transcription, synthesis of RNA from DNA.

Translation, the synthesis of protein from messenger RNA.

V

VDRL, Venereal Disease Research Laboratory slide flocculation test for rapid screening of syphilis. The test involves both agglutination and precipitation reactions.

Vegetative, concerned with growth and nutrition, but not reproduction or replication.

Vein, a vessel that conducts blood from the tissues to the heart. Some veins conduct blood from organs to other organs, such as the hepatic portal vein. Venous blood is largely de-oxygenated.

Vesicle, vesicular, a small (less than 5 mm in diameter), circular, fluid-filled sac or spot on the skin. See papule.

Viruses, agents of disease so small that they can be visualized only with the electron microscope; structurally characterized by an envelope and a core of nucleic acid; incapable of metabolism and reproduction without a living host cell.

W

White blood cell, a nucleated cell of the blood, as contrasted with the red blood corpuscles which were enucleated in the bone marrow before entering the circulation. White blood cells make up about 1% of the total blood cell/corpuscle population, and consist of granular leukocytes (neutrophils, eosinophils, and basophils) and agranular leukocytes (lymphocytes and monocytes). White blood cells are significantly larger than red blood corpuscles. They are not involved in oxygen transport, but are concerned with phagocytosis and/or response to certain antigens.

Y

Yeast, a form of fungus characterized by cells without hyphae.

Z

Zygote, the cell formed by the fusion of male and female gametes or sex cells.

INDEX

The main reference in a series of references is underlined.